Climate Change and International History

NEW APPROACHES TO INTERNATIONAL HISTORY

Series Editor: Thomas Zeiler, Professor of American Diplomatic History, University of Colorado Boulder, USA

Series Editorial Board:

Anthony Adamthwaite, University of California at Berkeley (USA)
Kathleen Burk, University College London (UK)
Louis Clerc, University of Turku (Finland)
Petra Goedde, Temple University (USA)
Francine McKenzie, University of Western Ontario (Canada)
Lien-Hang Nguyen, University of Kentucky (USA)
Jason Parker, Texas A&M University (USA)
Glenda Sluga, University of Sydney (Australia)

New Approaches to International History covers international history during the modern period and across the globe. The series incorporates new developments in the field, such as the cultural turn and transnationalism, as well as the classical high politics of state-centric policymaking and diplomatic relations. Written with upper-level undergraduate and postgraduate students in mind, texts in the series provide an accessible overview of international diplomatic and transnational issues, events, and actors.

Published:

Decolonization and the Cold War,
edited by Leslie James and Elisabeth Leake (2015)
Cold War Summits, Chris Tudda (2015)
The United Nations in International History, Amy Sayward (2017)
Latin American Nationalism, James F. Siekmeier (2017)
The History of United States Cultural Diplomacy, Michael L. Krenn (2017)
International Cooperation in the Early 20th Century, Daniel Gorman (2017)
Women and Gender in International History, Karen Garner (2018)
International Development, Corinna Unger (2018)
The Environment and International History, Scott Kaufman (2018)
Scandinavia and the Great Powers in the First World War,
Michael Jonas (2019)
Canada and the World since 1867, Asa McKercher (2019)
The First Age of Industrial Globalization,
Maartje Abbenhuis and Gordon Morrell (2019)
Europe's Cold War Relations, Federico Romero,
Kiran Klaus Patel, and Ulrich Krotz (2019)
United States Relations with China and Iran, Osamah F. Khalil (2019)

Public Opinion and Twentieth-Century Diplomacy, Daniel Hucker (2020)
Globalizing the US Presidency, Cyrus Schayegh (2020)
The International LGBT Rights Movement, Laura Belmonte (2021)
Global War, Global Catastrophe, Maartje Abbenhuis and Ismee Tames (2021)
*America's Road to Empire: Foreign Policy from Independence
to World War One*, Piero Gleijeses (2021)
Militarization and the American Century, David Fitzgerald (2022)
American Sport in International History, Daniel M. DuBois (2023)
Rebuilding the Postwar Order, Francine McKenzie (2023)
Soldiers in Peacemaking, Beatrice de Graaf,
Frédéric Dessberg, and Thomas Vaisset (2023)
From World War to Postwar, Andrew N. Buchanan (2023)
The Fear of Chinese Power, Jeffrey Crean (2024)
An International History of US Immigration, Benjamin C. Montoya (2024)

Forthcoming:
China and the United States since 1949, Elizabeth Ingleson
Sparking the Cold War, John McNay
The United States and the Ends of Empire, Sean T. Byrnes

Climate Change and International History

Negotiating Science, Global Change, and Environmental Justice

RUTH A. MORGAN

BLOOMSBURY ACADEMIC
LONDON • NEW YORK • OXFORD • NEW DELHI • SYDNEY

BLOOMSBURY ACADEMIC
Bloomsbury Publishing Plc
50 Bedford Square, London, WC1B 3DP, UK
1385 Broadway, New York, NY 10018, USA
29 Earlsfort Terrace, Dublin 2, Ireland

BLOOMSBURY, BLOOMSBURY ACADEMIC and the Diana logo are trademarks of
Bloomsbury Publishing Plc

First published in Great Britain 2024

Copyright © Ruth A. Morgan, 2024

Ruth A. Morgan has asserted her right under the Copyright, Designs and Patents Act, 1988,
to be identified as Author of this work.

For legal purposes the Acknowledgments on pp. x–xi constitute an extension of this
copyright page.

Series design by Catherine Wood
Cover image © Massive waves caused by Storm Alex on 2nd October 2020 crashing over
Brixham Breakwater in South Devon, UK. Paul Prestidge/Alamy Stock Photo

All rights reserved. No part of this publication may be reproduced or transmitted
in any form or by any means, electronic or mechanical, including photocopying,
recording, or any information storage or retrieval system, without prior
permission in writing from the publishers.

Bloomsbury Publishing Plc does not have any control over, or responsibility for, any
third-party websites referred to or in this book. All internet addresses given in this
book were correct at the time of going to press. The author and publisher regret any
inconvenience caused if addresses have changed or sites have ceased to exist,
but can accept no responsibility for any such changes.

A catalogue record for this book is available from the British Library.

A catalog record for this book is available from the Library of Congress.

ISBN: HB: 978-1-3502-4013-1
PB: 978-1-3502-4012-4
ePDF: 978-1-3502-4015-5
eBook: 978-1-3502-4014-8

Series: New Approaches to International History

Typeset by Newgen KnowledgeWorks Pvt. Ltd., Chennai, India
Printed and bound in Great Britain

To find out more about our authors and books visit www.bloomsbury.com
and sign up for our newsletters.

Contents

Figures

Acknowledgments

This book simply would not have been possible without the support and generosity of many people—I count my lucky stars that I've found my way into their orbit. This book began as a conversation with Maddie Holder at Bloomsbury Academic and grew into a manuscript thanks to the wise and encouraging advice of three anonymous reviewers. Maddie and her colleagues, Abigail Lane and Megan Harris, have been incredibly patient and I am grateful that they have kept the faith in me.

I thank the Traditional Owners of the unceded lands on whose beautiful Country I have written this book, the Ngunnawal and Ngambri Peoples. Moving to this part of the world has been an exciting adventure, and I feel very privileged to live and work here. My sincere thanks to Rae Frances, Bruce Scates, and Frank Bongiorno for making this possible; and to Andrea Gaynor, Katie Holmes, Tom Griffiths, Libby Robin, and Christina Twomey for their guidance. It has been an honor stepping into the role of Director of the Centre for Environmental History, where I have had the opportunity to work with and learn from rising stars Rohan Howitt and Jess Urwin. I am also grateful to my colleagues in the School of History at the Australian National University, particularly Frank Bongiorno, Josh Black, Emily Gallagher, Romney David Smith, Carolyn Strange, and Angela Woollacott, for their encouragement and guidance that have helped develop my ideas into this book.

I am especially grateful for the research community that has supported this project. The Australian Academy of Science and the Moran Award for History of Science Research allowed me to explore some of the themes that grew into this book. Fiona Williamson convened the 2018 conference, "Asian Extremes: Climate, Meteorology and Disaster in History," which has shaped a great deal of my thinking since. Lisa Brady and John McNeill offered much needed food for thought at a session on climate history for the American Historical Association conference at the height of Covid-19. Joanna Dean generously sponsored my contribution to the 2023 Shannon Lectures in History at Carleton University, where I was able to share some of my research with her vibrant colleagues and students. The British Academy and the Australian Academy of the Humanities afforded the wonderful and productive opportunity to work with Lise Butler, Maria Christou, and Or Rosenboim, as well as Will Tullett and Hannah Murray on the history of the future.

As I embarked on this project, I was contributing to the Intergovernmental Panel on Climate Change's (IPCC's) Sixth Assessment Report. I was very fortunate to have the opportunity to work closely with the indefatigable Aditi Mukherji, Martina Angela Caretta, Richard Betts, Tabea Lissner, Yukiko Hirabayashi, Rodrigo Fernandez, Vishnu Pandey, Elena López Gunn, Francesca Spagnuolo, Tero Mustonen, Patrick Nunn, Lisa Schipper, and Mark Howden, among many others. Their efforts and insights were truly inspirational.

This book would not have been possible without the support and assistance of librarians and archivists in Australia, Canada, Germany, the United States, and the UK. I would especially like to thank the staff of the Chifley Library and Document Supply Service at the Australian National University; the Bildarchiv of the Bundesarchiv; Bob Jones of the Canadian Meteorological and Oceanographic Society (CMOS) Archives; the Canberra office of the National Archives of Australia; the UK National Archives; the National Library of Australia; Bethany Antos of the Rockefeller Archive Center; Lynda Corey Claasen, Director of Special Collections and Archives at the University of California, San Diego; and the UN Photo Library. I would also like to thank Robert Naylor for so generously sharing his findings from the Rockefeller Archive Center.

I am extremely appreciative of the advice, mentoring, and moral support I have received during this project and beyond. Special thanks to the generous folks who took the time to read drafts of this manuscript—especially to Jess Urwin and Stephen Macekura for reading the whole draft and helping me to see the bigger picture. Thank you also to dear friends and colleagues who read and commented on chapters—Sandro Antonello, Mark Carey, Christian Downie, Sarah Dry, Andrea Gaynor, Matthias Heymann, Rohan Howitt, Martin Mahony, Emily O'Gorman, Libby Robin, Lisa Schipper, and Sverker Sörlin. I treasure your kindness and I take full responsibility for any errors that remain.

Andrew Connor, Eloise Dowd, Elizabeth Gralton, Ari Heinrich, Susie Protschky, Agnieszka Sobocinska, and Carly Stevens helped me bring this book together—thank you for putting up with me.

Thank you to my parents, Carolyn and Keith—I owe you an infinite debt. And to my darling Tash, you make everything better and brighter in ways that I did not know were possible. To you, Mack and Min, thank you for keeping me afloat.

Abbreviations

AGGG	Advisory Group on Greenhouse Gases
AOSIS	Alliance of Small Island States
COP	Conference of the Parties to the UN Framework Convention on Climate Change
CDM	Clean Development Mechanism
CMOS	Canadian Meteorological and Oceanographic Society
ECOSOC	UN Economic and Social Council
ENMOD	Environmental Modification Convention (convention on the prohibition of military or any other hostile use of environmental modification techniques)
EMEP	European-wide Monitoring and Evaluation Programme
EU	European Union
FAO	Food and Agriculture Organization
G7	Group of Seven
G8	Group of Eight
G20	Group of 20
G77	Group of 77
GARP	Global Atmospheric Research Program
GEF	Global Environment Facility
ICC	Inuit Circumpolar Council
ICSU	International Council of Scientific Unions
IGY	International Geophysical Year
IIASA	International Institute for Applied Systems Analysis
IIPFCC	International Indigenous Peoples' Forum on Climate Change
INC	Intergovernmental Negotiating Committee
IPCC	Intergovernmental Panel on Climate Change
LRTAP	Convention on Long-Range Transboundary Air Pollution
NASA	National Aeronautics and Space Administration
NATO	North Atlantic Treaty Organization
NGO	Nongovernmental Organization
NOAA	National Oceanic and Atmospheric Administration
OECD	Organisation for Economic Co-operation and Development
OPEC	Organisation of the Petroleum Exporting Countries

ppm	parts per million
REDD	Reducing Emissions from Deforestation and Forest Degradation in developing countries
REDD+	Reducing Emissions from Deforestation and Forest Degradation in developing countries, and the role of conservation, sustainable management of forests, and enhancement of forest carbon stocks
SBSTA	Subsidiary Body for Scientific and Technological Advice
SCOPE	Scientific Committee on Problems of the Environment
SCEP	Study of Critical Environmental Problems
SMIC	Study of Man's Impact on Climate
UN	United Nations
UNCED	United Nations Conference on Environment and Development
UNCTAD	United Nations Conference on Trade and Development
UNDP	United Nations Development Programme
UNEP	United Nations Environment Programme
UNESCO	United Nations Educational, Scientific and Cultural Organization
UNFCCC	United Nations Framework Convention on Climate Change
WCED	World Commission on Environment and Development
WMO	World Meteorological Organization
WTO	World Trade Organization
WWF	World Wildlife Fund for Nature

Introduction

Speaking at the opening ceremony of the United Nations (UN) Climate Summit in September 2014, Marshall Islands poet Kathy Jetñil-Kijiner (Figure i.1) addressed the delegates gathered in New York. Describing the view from Oceania, she performed her poem, "Dear Matafele Peinem," dedicated to her daughter: "We've seen waves crashing into our homes and our breadfruit trees wither from the salt and drought. We look at our children and wonder how they will know themselves or their culture should we lose our islands." Urging the world's governments to commit to "a radical change of course," she called on them to listen to the growing international movement for climate justice: "The people who support this movement are Indigenous mothers like me, families like mine, and millions more standing up for the changes needed and working to make them happen. I ask world leaders to take us all along on your ride. We won't slow you down. We'll help you win the most important race of all, the race to save humanity."[1]

Responsible for just a tiny fraction of total carbon emissions, the Marshall Islands is one of the 198 parties to the UN Framework Convention on Climate Change (UNFCCC). The core objective of the convention is to stabilize atmospheric greenhouse gas concentrations to prevent dangerous human interference with the climate system. Among the parties, just five were responsible for producing over half of the world's carbon dioxide emissions in 2021—China, the United States, India, Russia, and Japan; low-income countries such as the Marshall Islands contributed just 0.5 percent. The main change to the top five since 1995, once the convention came into force, has been the rise of India, which helped to nudge Germany out of fifth place. Germany, together with the United States, China, Russia, and the UK, has produced over half of the world's carbon dioxide emissions since 1750.[2] Even if those emissions are arrested now, the climate will continue to warm as a

---

[1] "Statement and Poem by Kathy Jetñil-Kijiner, Climate Summit 2014—Opening Ceremony," United Nations, September 24, 2014, https://v.gd/wibm8q.
[2] "Annual Share of Global CO_2 Emissions," *Our World in Data*, https://v.gd/plyO0t; "Cumulative CO_2 Emissions by World Region," *Our World in Data*, https://v.gd/lzuawe.

FIGURE I.1 *Kathy Jetñil-Kijiner (left) pictured with her family, having performed her poem "Dear Matafele Pienam" at the opening ceremony of the UN Climate Summit, UN Headquarters, New York, September 23, 2014. UN Photo by Cia Pak.*

result of past emissions; the amount and rate of even further heating depend almost entirely on human actions.

As the historical and present distribution of production emissions suggests, the human activity responsible for this heating has been concentrated in some parts of the world, where the scale of fossil fuel consumption has begun to manifest in environmental changes elsewhere that are exacerbating the challenges that more vulnerable peoples already face. Based on the distribution of production emissions alone, whether taken annually or cumulatively, the involvement of some 190 other parties in the climate convention is a curious arrangement. Curtailing the emissions of the largest, say five, polluters would surely go a long way to protecting the climate system, as the convention seeks to do. Just one might even elect to go it alone. Alternatively, the countries most vulnerable to climate change, such as the Marshall Islands, might impose emissions limits on the biggest polluters. These suggestions beg the question, why engage the entire world to limit greenhouse gas emissions? That is, how did anthropogenic climate change become framed and understood as a problem that warranted the involvement of all states in climate diplomacy? And how has this international approach afforded opportunities for states and others to advance their wider interests?

This book is primarily concerned with why and how governments became conditioned to understand climate change as a problem that warrants ongoing international negotiation through the UN. The Intergovernmental Panel on Climate Change (IPCC) was founded in 1988 under the auspices of the World Meteorological Organization (WMO) and the UN Environment Programme (UNEP). Four years later, in response to a UN General Assembly resolution, the UNFCCC was signed at the UN Conference on Environment and Development in Rio de Janeiro, and it came into force in 1994. The reports of the IPCC, which assess the latest peer-reviewed scholarship related to the science of climate change, as well as impacts of and responses to it, help to inform the negotiations between the parties to the climate convention. The formation of these UN bodies was premised on the shared notion that the problem of climate change required the multilateral engagement of all nations, not only the largest polluters, not only the most vulnerable, not only the most populated, and not only the most scientifically well-resourced.

By the late 1980s, climate change was already understood as a global issue—that is, a common concern of humankind, according to the UN. Yet this framing was a relatively recent development. Building on foundations laid in the nineteenth century, most Western geographers and scientists only came to define and measure climate in such global and dynamic terms after the dawn of the Atomic Age.[3] Recognizing that anthropogenic climate change could unfold within the space of a generation or two eventually arose from the intersection of atomic science, ecological thought, and Cold War internationalism. This scientific framing of global climate change would always be remote from direct human experience and require scientific mediation. New ways of thinking about the environment in planetary terms, as earthly, likewise afforded novel ways of thinking about the relationships between humans—beyond race and nation—and between humans and the planet—as either passengers or pilots of Spaceship Earth.[4]

For many governments and observers in the South, the global framing of climate change was historically cause for suspicion. This outlook reflected not only their prioritization of local problems and aspirations but also their capacity to participate in global Western science. Most, if not all, of

---

[3]For example, Matthias Heymann, "Climate as Resource and Challenge: International Cooperation in the UNESCO Arid Zone Programme," *European Review of History* 27 (2020): 294–320. For the nineteenth-century origins and progenitors of this planetary outlook, see Sarah Dry, *Waters of the World: The Story of the Scientists Who Unraveled the Mysteries of Our Oceans, Atmosphere, and Ice Sheets and Made the Planet Whole* (Chicago: University of Chicago Press, 2021).

[4]See, for example, Sabine Höhler, *Spaceship Earth in the Environmental Age, 1960–1990* (London: Routledge, 2015); Paul Warde, Libby Robin, and Sverker Sörlin, *The Environment: The History of an Idea* (Baltimore, MD: Johns Hopkins University Press, 2018); Etienne Benson, *Surroundings: A History of Environments and Environmentalisms* (Chicago: University of Chicago Press, 2020).

the theoretical research on climate change to the early 1990s had been undertaken in the industrialized nations of the North, especially the United States, the Soviet Union, Sweden, West Germany, and Australia.[5] The international scientific networks to monitor environmental change similarly favored those nations, although satellite technology from the 1970s improved the balance somewhat. Facilitating the greater involvement of scientists from the developing world in international science, particularly the IPCC, has gone a long way to improve scholarly representation. It has also had, perhaps even more importantly, significant implications for climate knowledge and thus for climate negotiations more broadly. Since the 1970s, questions from the South as to the assumptions and results of global, ostensibly neutral, scientific analyses have invited closer scrutiny of the knowledge geopolitics of climate science and allied fields.[6] In foregrounding matters of equity and fairness, their critiques offer an important cautionary note to calls to "listen to the science."

The UN—with its attendant agencies—was just one of the postwar international institutions that provided a forum to explore, articulate, and reinforce emerging global outlooks. Although the Cold War loomed large over these organizations, their international nature reflected a renewed faith in the importance of cooperation between nations to ensure peace, prosperity, and progress for all peoples. Nongovernmental organizations, such as the International Council of Scientific Unions, were similarly instrumental, encouraging collaboration and knowledge exchange in the seemingly apolitical and neutral realm of science. According to the proponents of such postwar internationalism, this cooperation could not be left to chance or to individual

---

[5]Paul N. Edwards, *A Vast Machine: Computer Models, Climate Data, and the Politics of Global Warming* (Cambridge, MA: MIT Press, 2010).

[6]For example, Milind Kandlikar and Ambuj Sagar, "Climate Change Research and Analysis in India: An Integrated Assessment of a South-North Divide," *Global Environmental Change* 9 (1999): 119–38; Clark A. Miller, "Challenges in the Application of Science to Global Affairs: Contingency, Trust, and Moral Order," in Clark A. Miller and Paul N. Edwards (eds.), *Changing the Atmosphere: Expert Knowledge and Environmental Governance* (Cambridge, MA: MIT Press, 2001), 247–87; Myanna Lahsen, "Transnational locals: Brazilian Experiences of the Climate Regime," in Sheila Jasanoff and Marybeth Long Martello (eds.), *Earthly Politics: Local and Global in Environmental Governance* (Cambridge, MA: MIT Press, 2004), 151–72; Myanna Lahsen, "Trust through Participation? Problems of Knowledge in Climate Decision Making," in Mary E. Pettenger (ed.), *The Social Construction of Climate Change: Power, Knowledge, Norms, Discourses* (Burlington: Ashgate, 2007), 173–96; Myanna Lahsen, "A Science-Policy Interface in the Global South: The Politics of Carbon Sinks and Science in Brazil," *Climatic Change* 97 (2009): 339–72; Martin Mahony, "The Predictive State: Science, Territory and the Future of the Indian Climate," *Social Studies of Science* 44, no. 1 (2014): 109–33; Jessica O'Reilly, *The Technocratic Antarctic: An Ethnography of Scientific Expertise and Environmental Governance* (Ithaca, NY: Cornell University Press, 2017); Jean Carlos Hochsprung Miguel, Martin Mahony, and Marko Synésio Alves Monteiro, "'Infrastructural Geopolitics' of Climate Knowledge: The Brazilian Earth System Model and the North-South Knowledge Divide," *Sociologias* 21, no. 51 (2019): 44–74.

governments—it required the expert, rational, and objective guidance of the technocracies that would oversee international institutions, such as the UN and the WMO.[7]

The engagement of these international Institutions in emerging concerns about "the environment" from the late 1960s encouraged the pursuit of atmospheric diplomacy through multilateral means. Since the turn of the twentieth century, air pollution had become understood as a local problem requiring local solutions, which postwar publics increasingly expected their governments to address. Aside from the 1941 Trail Smelter case in North America, pursuing pollution abatement across national borders had few international precedents until the 1960s.[8] The 1963 Nuclear Test Ban Treaty, which prohibits nuclear testing in the atmosphere, outer space, and oceans, suggested that other forms of air pollution could also become the subject of such an international regime.[9]

The first international conference on the environment organized under the auspices of the UN General Assembly, the UN Conference on the Human Environment in Stockholm in 1972, arose in part from Swedish concerns about acid rain, which emanated largely from the UK and West Germany.[10] Thanks to the newly articulated principles of the 1972 Stockholm Declaration, environmental concern and scientific expertise could eventually combine to enable European and North American governments to negotiate an international agreement to limit the emission of the pollutants responsible for acid rain. When ozone-depleting substances became recognized as a risk to human health during the 1970s, a larger group of nations pursued a similar approach to their monitoring and restriction, which became something of a template for negotiating the international regime to tackle greenhouse gas emissions.

By the time negotiations for the climate regime began in earnest in the early 1990s, the ozone model had faded from the view of governments and international organizations. Yet for many contemporaries and others since, its promise as a successful atmospheric regime has been a powerful and alluring

---

[7]Perrin Selcer, *The Postwar Origins of the Global Environment: How the United Nations Built Spaceship Earth* (New York: Columbia University Press, 2018); Clark A. Miller, "Scientific Internationalism in American Foreign Policy: The Case of Meteorology, 1947–1958," in Miller and Edwards (eds.), *Changing the Atmosphere*, 167–218.

[8]John R. McNeill, *Something New under the Sun: An Environmental History of the Twentieth-Century World* (New York: W.W. Norton, 2001).

[9]Marvin S. Soroos, *The Endangered Atmosphere: Preserving a Global Commons* (Columbia: University of South Carolina Press, 1997), 4.

[10]Stephen Macekura, *Of Limits and Growth: The Rise of Global Sustainable Development in the Twentieth Century* (New York: Cambridge University Press, 2015); Rachel Emma Rothschild, *Poisonous Skies: Acid Rain and the Globalization of Pollution* (Chicago: University of Chicago Press, 2019).

one.[11] After all, its negotiation was a remarkable feat: after scientists identified ozone-depleting substances in 1974, governments took precautionary action, and not a moment too soon, with the discovery of a "hole" or thinning in the ozone layer. Corporations cooperated with governments, substitute chemicals were produced, and the 1987 Montreal Protocol, which all countries have ratified, has since led to the dramatic reduction of ozone-depleting substances in the stratosphere. The ozone layer is now recovering and risks to human health have been greatly diminished.[12]

Whether sulfur dioxide, ozone-depleting substances, or greenhouse gases, diplomatic efforts to negotiate the reduction of atmospheric emissions have all encountered the same challenge—of reconciling human and environmental benefit (often remote both in geographic and temporal terms) with the fear of economic disadvantage under conditions of scientific uncertainty. Framed as such, this challenge has been among the defining features of climate diplomacy since its inception: the largest polluters (historically, the industrialized nations) are unlikely to bear the worst consequences of their own emissions. Establishing the IPCC as a means to diminish uncertainty and improve the precision of climate prediction has yet to overcome the profound influence on the negotiations of this uneven distribution of climate emissions and impacts. The dual emergence in the late 1980s of climate change as an object of international diplomacy, and liberal environmentalism as the new means of addressing such global environmental challenges, helped cement the market as the means to protect economic growth in the ensuing negotiations.[13] Since the late 1980s, states have continued to pursue fossil-fueled economic growth and national development in the face of mounting scientific evidence of the need to abate total greenhouse gas emissions.

Only those states that can least afford this outlook have seen the situation otherwise and have pressed their case in existential terms. Low-lying island nations have been especially vigilant, as Jetñil-Kijiner's address suggests. Since the late 1980s, their unique vulnerability to rising sea levels and severe weather events has encouraged these nations to intervene in negotiations to strengthen the ambitions of the less vulnerable. In the wake of September 11 and the Bush administration's withdrawal of the United States from the Kyoto Protocol in 2001, global security became a watchword to signal the importance and urgency of sustaining commitments to reduce greenhouse gas emissions. These appeals to multilateralism gave way to declarations of a

[11]James Gustav Speth, *Red Sky at Morning: America and the Crisis of the Global Environment* (New Haven: Yale University Press, 2004).
[12]World Meteorological Organization, *Scientific Assessment of Ozone Depletion: 2022* GAW Report No. 278 (Geneva: WMO, 2022).
[13]Steven Bernstein, *The Compromise of Liberal Environmentalism* (New York: Columbia University Press, 2001).

climate crisis a decade later, as the need arose to negotiate a successor to the Kyoto Protocol. What these declarations described was less a crisis of climate change per se and more a crisis of climate diplomacy—that the prevailing approach would not avert dangerous interference with the climate system.[14] And if that was the case, then what would?

In these negotiations to limit the impacts of present and future climate change, history has weighed heavily on the dynamics of international climate diplomacy. Under the terms of the 1992 UNFCCC, parties have "common but differentiated responsibilities and respective capabilities" to address climate change. Led by the large industrializing nations of China, India, and Brazil, the South has consistently argued that the problem of climate change is the consequence of the industrialization of the North. Having accrued the benefits of this fossil-fueled process, the South has argued, it is incumbent upon those wealthy nations to reduce their emissions first and allow other nations to develop accordingly. That greenhouse gas emissions linger in the atmosphere for decades was all the more reason for the industrialized nations to take the lead.

Grounded in the Cold War politics of decolonization and development, the South's aspirations for independence and development sovereignty have long been nonnegotiable in the climate regime. Delegations from the South held firm to this position throughout the negotiations of the climate convention and the subsequent Kyoto Protocol, such that only industrialized countries bore commitments to reduce their greenhouse gas emissions. The means of implementing such mitigation occupied negotiators during the late 1990s and early 2000s, while fossil fuel interests sought to derail the climate regime entirely. As greenhouse gas emissions continued to rise into the twenty-first century, participants in the climate regime increasingly came to recognize the need to support adaptation measures to limit the impacts of a warming world, particularly on the most vulnerable.

Since the parties to the climate convention signed the Paris Agreement in 2015, the so-called fire wall between the developed and developing nations has become much less of a defining feature of climate diplomacy. Under the terms of the Kyoto Protocol, only thirty-seven industrialized nations were obliged to reduce their emissions by an average of 5 percent relative to 1990 levels, which became just one of the grounds for its criticism. That criticism reached new heights in the aftermath of the 2009 Copenhagen Climate Conference, when observers feared that the past two decades of climate diplomacy would amount to naught. Under the terms of the Paris Agreement, however, all countries now have international legal

---

[14]Mike Hulme, *Why We Disagree about Climate Change: Understanding Controversy, Inaction and Opportunity* (Cambridge: Cambridge University Press, 2009).

obligations to develop, implement, and regularly strengthen their mitigation actions to collectively limit the increase in the global average temperature to well below 2°C above preindustrial levels—ideally, to limit that increase to 1.5°C. The Paris system encourages nations to make ambitious pledges commensurate with their "common but differentiated responsibilities." With the first five-year global stocktake concluding in 2023, the time will soon come to take the temperature on this version of international climate diplomacy.

Negotiating Science, Global Change, and Environmental Justice

Charting the rise of a global climate that humanity could affect, and the negotiation of an international regime to curb that profound planetary influence, *Climate Change and International History* brings to bear the insights of both environmental history and the history of science on the study of international history.[15] Over the past decade, environmental historians have cast their gaze to global scales, both spatially and temporally, examining how environmental problems came to be understood in planetary terms during the Cold War.[16] Precisely how these global frames arose globally (not merely as an American export) and persisted after the collapse of the Soviet Union deserves further inquiry, however. The global scale, as historians of science and scholars in the field of Science, Technology and Society (STS) have shown, is carefully mediated and maintained such that the geopolitical dynamics of its production are obscured.[17] Since the 1990s, historians of science have analyzed the rise of climate science, many scrutinizing the ways that the field and its practitioners have handled not only challenges to their epistemic authority, but also questions of certainty and risk, and their implications for policy action

[15]See Akira Iriye, "Environmental History and International History," *Diplomatic History* 32, no. 4 (2008): 643–6; Erika Marie Bsumek, David Kinkela, and Mark Atwood Lawrence (eds.), *Nation-States and the Global Environment: New Approaches to International Environmental History* (New York: Oxford University Press, 2013).

[16]See also Jacob Darwin Hamblin, *Arming Mother Nature: The Birth of Catastrophic Environmentalism* (New York: Oxford University Press, 2013); Thomas Robertson, *The Malthusian Moment: Global Population Growth and the Birth of American Environmentalism* (New Brunswick: Rutgers University Press, 2012); Alison Bashford, *Global Population: History, Geopolitics, and Life on Earth* (New York: Columbia University Press, 2014); Alessandro Antonello and Mark Carey, "Ice Cores and the Temporalities of the Global Environment," *Environmental Humanities* 9, no. 2 (2017): 181–203.

[17]For example, Sheila Jasanoff, "Image and Imagination: The Formation of Global Environmental Consciousness," in Miller and Edwards (eds.), *Changing the Atmosphere*, 309–37; Sheila Jasanoff and Marybeth Long Martello (eds.), *Earthly Politics: Local and Global in Environmental Governance* (Cambridge, MA: MIT Press, 2004); Stephen Bocking, *Nature's Experts: Science, Politics, and the Environment* (New Brunswick: Rutgers University Press, 2004).

and inaction.[18] These insights, together with those drawn from the burgeoning historical scholarship on development and postwar internationalism, invite further examination of the contested processes that have contributed to, and continue to shape, climate diplomacy.[19]

This book argues that an international approach to climate change has arisen from, and been sustained by, the historical and ongoing negotiation of science, global change, and environmental justice. Negotiation in this sense extends beyond the formal scientific and political diplomacy of the climate order that has unfolded since Malta advocated for the recognition of climate change as a "common concern of humankind" in the UN General Assembly in late 1988. It was this intervention that prompted member states to embark on deliberations for the UNFCCC, and shaped the discussions of the newly formed IPCC. "Negotiation" in this book frames that form of multilateral climate diplomacy much more broadly—as the international response to a longer history of scientific negotiation to comprehend the climate as a planetary system, which resonated first with the architects of postwar and Cold War internationalism, and then with a growing international environmental movement. Eventually, this global and dynamic climate provided a rationale for an international agreement between governments to "prevent dangerous anthropogenic interference with the climate system"—the objective of the Climate Convention.

This more capacious temporal conceptualization of negotiation allows, then, for closer scrutiny of both the contingent and contested ways this scientifically framed global climate was cultivated, and its encounters with the structures and relationships of the international system.[20] In doing so, this book follows historian Sunil Amrith, who, in the context of the internationalization of public health, argues, "Histories of international institutions, internationalist ambitions and international initiatives all need to be embedded in the broader

---

[18]For example, Naomi Oreskes and Erik M. Conway, *Merchants of Doubt: How a Handful of Scientists Obscured the Truth on Issues from Tobacco Smoke to Global Warming* (New York: Bloomsbury Press, 2010); Gabriel Henderson, "Adhering to the 'Flashing Yellow Light': Heuristics of Moderation and Carbon Dioxide Politics during the 1970s," *Historical Studies in the Natural Sciences* 49, no. 4 (2019): 384–419.

[19]See, for example, Odd Arne Westad, *The Global Cold War: Third World Interventions and the Making of Our Times* (Cambridge: Cambridge University Press, 2007); David Ekbladh, *Great American Mission: Modernization and the Construction of an American World Order* (Princeton: Princeton University Press, 2010); Glenda Sluga, *Internationalism in the Age of Nationalism* (Philadelphia: University of Pennsylvania Press, 2013); David C. Engerman, *The Price of Aid: The Economic Cold War in India* (Cambridge, MA: Harvard University Press, 2018); Sara Lorenzini, *Global Development: A Cold War History* (Princeton: Princeton University Press, 2019).

[20]Joseph E. Taylor III, "Negotiating Nature through Science, Sentiment, and Economics," *Diplomatic History* 25, no. 2 (2001): 335–9; Michael Barnett and Martha Finnemore, *Rules for the World: International Organizations in Global Politics* (Ithaca, NY: Cornell University Press, 2004), 70.

debates to which they emerged as a response: debates about the shape of the world, about inequality and about the (differential) value of human lives."[21] As both real and represented, the post-1945 global framing of the climate has enabled governments, scientists, Indigenous Peoples, corporations, environmentalists, and others to strategically invoke and enroll climate change for different, and sometimes competing, ends—in ways similar to the uses of other global commons, such as Antarctica, outer space, and the ocean.[22]

This book shows how the negotiations of the international climate regime have been, since its formation in the UN General Assembly in 1988, entangled in the globalisms, or global political imaginaries, that shape the contested meanings and experiences of a globalizing world.[23] Each of these global imaginaries—market globalism; justice globalism; and global Indigenism— have found expression in the climate regime; foregrounding them shines a light on how negotiations regarding the global climate and anthropogenic climate change have often been (and remain) "a continuation of politics by other means."[24] Recognizing and probing these politics affords a clearer understanding of the challenges and opportunities of negotiating climate change. As historian Dipesh Chakrabarty has put it, "we can never compose our planetary collectivity by ignoring the intensely politicized and necessarily fragmented domain of the global that understandably converts scientists' statements about humans as the cause of climate change into a charged discussion about moral responsibility and culpability."[25]

The material world shapes the climate regime, nevertheless. Unseasonal conditions or extreme weather events, such as droughts, floods, and cyclones, have long provided real and rhetorical anchors for arguments about the nature and rate of global climate change, and thus participate in the negotiations

---

[21]Sunil S. Amrith, "Internationalizing Health in the Twentieth Century," in Glenda Sluga and Patricia Clavin (eds.), *Internationalisms: A Twentieth Century History* (New York: Cambridge University Press, 2016), 245–64.

[22]See, for example, James Spiller, *Frontiers for the American Century: Outer Space, Antarctica, and Cold War Nationalism* (New York: Palgrave Macmillan, 2015); Alessandro Antonello, *The Greening of Antarctica: Assembling an International Environment* (Oxford: Oxford University Press, 2019); Helen Rozwadowski, "Wild Blue: The Post-World War Two Ocean Frontier and Its Legacy for Law of the Sea," *Environment and History* 29, no. 3 (2023): 343–76.

[23]Manfred B. Steger, *Globalisms: The Great Ideological Struggle of the Twenty-First Century*, 3rd ed. (Lanham: Rowman & Littlefield, 2009); Ken Conca, *An Unfinished Foundation: The United Nations and Global Environmental Governance* (New York: Oxford University Press, 2015); Ronald Niezen, *The Origins of Indigenism: Human Rights and the Politics of Identity* (Berkeley: University of California Press, 2003).

[24]Aant Elzinga, "Features of the Current Science Policy Regime: Viewed in Historical Perspective," *Science and Public Policy* 39, no. 4 (2012): 16–28.

[25]Dipesh Chakrabarty, *The Human Condition in the Anthropocene*, The Tanner Lectures in Human Values, Yale University, February 18-19, 2015, https://v.gd/oVqHmk.

of the climate order.[26] For much of the history of the UN climate regime, such climate variability and change has been largely mediated through the lens of Western science, with the critical insights of Indigenous Knowledge and Traditional Knowledge gaining only belated recognition in international climate change circles. Building on an extensive scholarship that has tended to treat the IPCC and Climate Convention separately, this book brings these institutions together in the same analytical frame as a distinct intervention to demonstrate how government delegations and other organizations have negotiated and contested the synoptic perspective of global science.[27]

This book traces the origins of the UN climate regime to the aftermath of the Second World War. Chronologically structured, each chapter notes an increase in the atmospheric concentration of carbon dioxide. Measured in parts per million (ppm), these amounts represent the accumulation of this long-lasting greenhouse gas in the earth's atmosphere. Such gases trap heat near the earth's surface, raising the temperature of the air and the ocean, and thus altering the workings of the climate system. As discussed in Chapter 1, the International Geophysical Year of 1957–58 offered an opportunity for a young chemist, C. D. "Dave" Keeling, to begin measuring these greenhouse gases from stations in Hawai'i and the South Pole. Charting those measurements of the changing composition of the atmosphere provided clear scientific evidence that, over time, human activity was rapidly altering the earth's chemistry (Figure i.2). Keeling's story arose from an episode of scientific internationalism—the International Geophysical Year—that engaged scientists and their governments in the study of the earth's geophysical systems in unprecedented scale and detail. The planet as a whole slowly came into scientific view as a result of such research programs, which governments on both sides of the Iron Curtain hoped would afford them strategic advantage in the Cold War. Observations, meanwhile, that large swathes of the world had been experiencing warmer temperatures prompted closer scrutiny as to their cause, leading scientists to speculate that the newly globalized climate was both more dynamic and manipulable than they had imagined.

Growing concern about the environmental effects of industrialization, meanwhile, instilled new and alarming meaning in Keeling's carbon dioxide

---

[26]See, for example, Julia F. Irwin, "Our Climatic Moment: Hazarding a History of the United States and the World," *Diplomatic History* 45, no. 1 (2021): 421–44; Sheldon Ungar, "The Rise and (Relative) Decline of Global Warming as a Social Problem," *Sociological Quarterly* 33, no. 4 (1992): 483–501; Gretchen Heefner, "The Accidental Environmental Historian," *Diplomatic History* 46, no. 4 (2022): 659–74.

[27]See, for example, Joyeeta Gupta, *The History of Global Climate Governance* (New York: Cambridge University Press, 2014); John J. Kirton and Ella Kokotsis, *The Global Governance of Climate Change: G7, G20, and UN leadership* (Farnham: Ashgate, 2015); Kari De Pryck and Mike Hulme (eds.), *A Critical Assessment of the Intergovernmental Panel on Climate Change* (New York: Cambridge University Press, 2022).

FIGURE I.2 *In a photograph taken at the Scripps Institution of Oceanography in April 1996, chemist Charles David Keeling points to a graph representing growing levels of atmospheric concentrations of carbon dioxide since the late 1950s. SIO Photographic Laboratory Collection. SAC 44. Special Collections & Archives, UC San Diego, bb6656832b.*

measurements. Brokers of postwar scientific and environmental internationalism interpreted their significance through a lens of social and ecological interdependence that connected distant places and peoples. In the United States and Europe, the human and environmental impacts of pollution summoned state regulation and the closer scientific scrutiny of atmospheric change. Air pollution's disregard for territorial borders, as Chapter 2 shows, offered material evidence of the need for cooperation between governments to manage shared environmental problems. By the late 1960s, concerned governments had elevated these concerns to the UN General Assembly, from where they became the subject of negotiations at the 1972 UN Conference on the Human Environment in Stockholm.

Developing countries, meanwhile, were less than convinced that the environmental concerns that animated the North warranted their own interventions. At the 1972 Stockholm Conference, delegations from the South were careful to ensure that Northern environmental policymaking would not curtail their independence or development sovereignty. These divisions only deepened during the 1970s, in the face of demands for a New International Economic Order, and a cascade of energy and food crises. Chapter 3 examines the collision of these crises with anxieties about imminent global climate change. The geopolitical implications of such change summoned its close scientific scrutiny within and outside the WMO, leading to the first World Climate Conference in 1979.

During the following decade, Chapter 4 shows, the vulnerability of the atmosphere and the climate became the subject of sustained scientific and intergovernmental negotiation and debate. As relations soured between the superpowers, the fear of nuclear war combined old concerns of atomic weather modification with newly global climate science. The UNEP meanwhile steered international efforts to regulate ozone-depleting substances, which encouraged internationalist scientists to press their governments to similarly cooperate to curtail the emission of greenhouse gases. Although governments sponsored further scientific study of the problem through the auspices of the newly established IPCC, it was from the UN General Assembly that the nascent international regime emerged to negotiate a multilateral approach to climate change.

The focus of *Climate Change and International History* then shifts to the study of the negotiation of the international climate regime from the late 1980s to the Paris Agreement in 2015. Meeting as the Soviet Union collapsed, governments prepared the foundations of what would become the UNFCCC in 1992. Cold War divisions of East and West gave way to renewed tensions between North and South, which continued to shape the climate regime into the 2010s. Chapter 5 examines the arguments that delegates from the South mounted to ensure that the Climate Convention would hinder neither their sovereignty nor their economic development.

Despite the concerns of the European Union and the developing countries, Chapter 6 shows how the United States joined with Brazil to craft the market mechanisms at the center of the Kyoto Protocol. The Protocol eventually came into force in 2005, but the absence of Washington's commitment and the continued rise of greenhouse gas emissions led European countries and the small island states to seize on the security implications of climate change to raise international ambitions for the agreement that would succeed the Kyoto Protocol. Upholding the multilateralism of the regime, as Chapter 7 demonstrates, required negotiating the growing proportion of emissions from the South, while abiding with the principle of "common but differentiated responsibilities." The book's final chapter examines the negotiation of the successor to Kyoto, from Copenhagen to the Paris Agreement, in which governments, scientists, Indigenous Peoples, and nongovernmental organizations shaped the international climate regime anew.

The rising curve that Keeling's data-points chart has become a potent symbol of climate change—a stark visualization of a phenomenon that can only be understood through the insights of atmospheric science. When the chemist began his study in 1958, the level of carbon dioxide in the atmosphere read 315 ppm. Now, that level has surpassed 400 ppm for the first time in several million years. Reflecting on his career in the late 1990s, the curve's architect reflected on his creation:

> The consumption of fossil fuel has increased globally nearly three-fold since I began measuring CO_2 and almost six-fold over my lifetime. In Southern California it isn't necessary to look at statistics to sense this enormous

increase. ... I have watched the number of lights steadily increase; lights from new homes, lights from commercial enterprises, lights from vehicles after an eight-lane highway was built ... Almost all of these lights, and the activities they support, depend on fossil fuel energy.[28]

As this book shows, those lights—and their many meanings for global change and environmental justice—have been and remain at the heart of international climate diplomacy.

---

[28]C. D. Keeling, "Rewards and Penalties of Monitoring the Earth," *Annual Review of Energy and Environment* 23 (1998): 76.

1

A Climate for Peace

1961: 317.64 ppm

Within months of the US bombings of Hiroshima and Nagasaki, proponents of the atom's power believed that weather control might now be in reach. "It would be within man's power to change the earth's climate," marveled a Pulitzer Prize-winning science journalist. Melting the polar ice cap "would give the entire world a moister, warmer climate."[1] A colleague similarly enthused: "No city will experience a winter traffic jam because of heavy snow. Summer resorts will be able to guarantee the weather and artificial suns will make it as easy to grow corn and potatoes indoors as on the farm."[2] Across the Atlantic, the United Kingdom's *Picture Post* saw in the "potentialities of atomic power" the means to "even challenge the vagaries of weather,"[3] a sentiment that the 1947 film *La vie commence demain* echoed.[4] Such hopes extended to India, where on the eve of independence, the English-language *Illustrated Weekly of India* anticipated that atomic power would "make whatever climate we prefer [by] turning lakes into gigantic radiators to dispel cold weather."[5] Meanwhile in Stalin's Soviet Union, where the development of its own nuclear capabilities was well underway, an engineer marveled that "atomic energy permits us to dream about the alteration of the climate of whole countries" by enabling the creation of a "winter sun."[6]

[1]John O'Neill, *Almighty Atom: The Real Story of Atomic Energy* (New York: I. Washburn, 1945), 60.
[2]David Dietz, *Atomic Energy in the Coming Era* (New York: Dodd, Mead, 1945), 17.
[3]*Picture Post*, March 1946, cited in Christoph Laucht, "'Dawn – or Dusk?': Britain's Picture Post Confronts Nuclear Energy," in Dick van Lente (ed.), *The Nuclear Age in Popular Media: A Transnational History, 1945–1965* (New York: Palgrave Macmillan, 2012), 134.
[4]Brian R. Johnson, "French Cinema vs the Bomb: Atomic Science and a War of Images circa 1950," *Historical Journal of Film, Radio and Television* 42, no. 2 (2022): 207; Gabrielle Hecht, *The Radiance of France: Nuclear Power and National Identity after World War II* (Cambridge, MA: MIT Press, 2000).
[5]*Illustrated Weekly of India*, July 14, 1946, cited in Hans-Joachim Bieber, "Promises of Indian Modernity: Representations of Nuclear Technology in the *Illustrated Weekly of India*," in van Lente (ed.), *The Nuclear Age*, 209.
[6]A. Morozov, "Razrushiteli," *Smena* 9–10 (1946): 11–12, cited in Paul Josephson, "Rockets, Reactors, and Soviet Culture," in Loren R. Graham (ed.), *Science and the Soviet Social Order* (Cambridge, MA: Harvard University Press, 1990), 174.

The seeds of this atomic optimism had been sown in the early twentieth century. In a series of lectures delivered at the University of Glasgow in 1908, an English chemist had shared his vision of the power of the atom.[7] Frederick Soddy, who would later win the 1921 Nobel Prize for his contributions to the study of the chemistry of radioactive substances, believed that unleashing the atom's energy could allow humanity to "transform a desert continent, thaw the frozen poles, and make the whole world one smiling Garden of Eden."[8] Published as *The Interpretation of Radium*, his lectures were translated and shared across Europe and North America, inspiring H. G. Wells to depict a future war waged with what he called "atomic bombs."[9] Such dystopian visions rarely extended to dreams of weather control, yet their proliferation in Hiroshima's wake reflected a world grappling with the political, philosophical, and scientific implications of the atom's Promethean power.

This dawn of the Atomic Age beckoned for "one world or none," as the Federation of American Scientists put it in 1946.[10] Theirs was, as historian Glenda Sluga has described it, a "curiously utopian" moment that arose not in spite of, but because of, the horrors of the Second World War.[11] Spurning unbridled nationalism, intellectual and political elites—Wells, Einstein, and Gandhi among them—drew on an enduring fin de siècle vocabulary of internationalism as they placed their faith in what they termed "world citizenship," "world government," and "one world."[12] Their world-mindedness imbued the intergovernmental organizations that sprouted after the war, seeking in the shared values of prosperity, cooperation, and liberalism the containment of totalitarianism. Through the specialized agencies of the United Nations (UN), their architects believed, experts would deliver an end to the familiar challenges of scarcity and instability that had sparked the recent war and, unless checked, would likely do so again.[13] They would apply their

---

[7]Richard E. Sclove, "From Alchemy to Atomic War: Frederick Soddy's 'Technology Assessment' of Atomic Energy, 1900–1915," *Science, Technology and Human Values* 14, no. 2 (1989): 163–94.

[8]Frederick Soddy, *The Interpretation of Radium* (London: John Murray, 1909), 244.

[9]H. G. Wells, *The World Set Free* (New York: E.P. Dutton, 1914), 109.

[10]Dexter Masters and Katharine Way (eds.), *One World or None* (New York: McGraw Hill Book, 1946).

[11]Glenda Sluga, *Internationalism in the Age of Nationalism* (Philadelphia: University of Pennsylvania Press, 2013), 87.

[12]See, for example, John Partington, "H.G. Wells and the World State: A Liberal Cosmopolitan in a Totalitarian Age," *International Relations* 17, no. 2 (2003): 233–46; Manu Bhagavan, *India and the Quest for One World: The Peacemakers* (New York: Palgrave Macmillan, 2013).

[13]Stephen Macekura, *Of Limits and Growth: The Rise of Global Sustainable Development in the Twentieth Century* (New York: Cambridge University Press, 2015), 24–6; Perrin Selcer, *The Postwar Origins of the Global Environment: How the United Nations Built Spaceship Earth* (New York: Columbia University Press, 2018), 20.

scientific and technical expertise to foster a social, economic, and political climate for peace; as this chapter shows, aspirations to understand the atmosphere and even overcome the vagaries of the weather were key to this internationalist project.

As postwar visions of atomic weather control suggest, climatic concerns of the physical kind occupied the "minds of men." Meteorologist Athelstan Spilhaus, who would oversee the first US military atomic bomb tests, saw in weather control "the difference between famine and plenty"—"a new goal worthy of the greatest effort."[14] British biologist Julian Huxley, who would the following year become the first director general of the UN Educational, Scientific and Cultural Organization (UNESCO), declared in mid-1945 to a crowd at New York City's Madison Garden that "atomic dynamite" could blast off the polar caps and change the earth's climate. For Huxley, such speculation was less of a personal ambition than an illustration of his position on the role of science in the postwar international order. "Atomic power could make an old dream come true. Over and over again it has been asserted that if science and technology were rationally utilized, there would be a reasonable standard of life for every human being on earth," the *New York Times* reported. "To Dr Huxley both are wrapped up in international control of atomic energy and hence in social planning on a world-wide scale."[15]

Unbounded by territorial borders, weather and climate were ideally suited materially to this postwar spirit of world unity. As Swedish meteorologist Carl-Gustav Rossby put it, "the winds blow and the storms move without regard for political boundaries … This indivisibility of the atmosphere and its problems makes meteorology a particularly interesting field in which to experiment with the organization of cooperative research projects on an international basis."[16] In the context of nuclear weapons and the emerging Cold War rivalry, this indivisibility also rendered the atmosphere a space for geopolitical anxiety and competition, as well as cooperation. Understanding the workings of the weather would afford strategic advantage, even the means of deliberate weather modification. Developing the postwar institutions of meteorological diplomacy between states would ensure the means for governments to both pursue scientific internationalism and their domestic interests.

---

[14]Athelstan Spilhaus, 4, in James R. Fleming, "Introduction: Outline of Weather Proposal, October 1945," *History of Meteorology* 4 (2008): 57–8.
[15]Waldemar Kaempffert, "Science in Review: Julian Huxley Pictures the More Spectacular Possibilities That Lie in Atomic Power," *New York Times*, December 9, 1945, 77.
[16]Carl-Gustav Rossby, "Note on Cooperative Research Projects," *Tellus* 3 (1951): 212.

Meteorological Diplomacy: Founding the World Meteorological Organization

When members of the International Meteorological Organization met for the first time in Washington, DC, in September 1947, their host, the US assistant secretary of state, reminded them, "The world needs your help in solving some of its problems and your efforts surely will bring greater comfort and safety and will contribute to a higher standard of living for all peoples."[17] That extreme weather had exacerbated food shortages across much of the world, to the extent that half a billion people faced famine conditions in 1945, only strengthened the appeal of improving meteorological knowledge and cooperation.[18] Although agricultural productivity recovered to prewar levels in both Asia, Western Europe, and the Americas by decade's end, these catastrophic circumstances established uncertain weather conditions as a challenge to lasting peace.[19] In the prospect that Soviet influence might flourish in such impoverished postwar conditions, for instance, lay the foundations for both the Marshall Plan and the Point Four Program. These committed the United States to deliver American expertise and financial aid to ensure the recovery of Western Europe and to kick-start the resource and economic development of what President Harry Truman in 1949 called the "underdeveloped" world of Asia, Africa, and Latin America.[20]

The members of the International Meteorological Organization had gathered in Washington to negotiate the World Meteorological Convention, which would transform what had been a private, professional body into an intergovernmental agency of the newly formed UN. Formed at the Second International Meteorological Congress in 1879, the International Meteorological

---

[17]Garrison Norton, September 1947, cited in Clark Miller, "Scientific Internationalism in American Foreign Policy: The Case of Meteorology, 1947–1958," in Clark Miller and Paul Edwards (eds.), *Changing the Atmosphere: Expert Knowledge and Environmental Governance* (Cambridge, MA: MIT Press, 2001), 186.

[18]Amy L. Bentley, "Uneasy Sacrifice: The Politics of United States Famine Relief, 1945–48," *Agriculture and Human Values* 11 (1994): 4–18; Jenny Leigh Smith, "The Awkward Years: Defining and Managing Famines, 1944–1947," *History and Technology* 31, no. 3 (2015): 206–19.

[19]Amy L. S. Staples, *The Birth of Development: How the World Bank, Food and Agriculture Organization, and the World Health Organization Changed the World, 1945–1965* (Kent, OH: Kent State University Press, 2006); Ruth Jachertz and Alexander Nützenadel, "Coping with Hunger? Visions of a Global Food System, 1930–1960," *Journal of Global History* 6, no. 1 (2011): 99–119; Benjamin R. Siegel, *Hungry Nation: Food, Famine, and the Making of Modern India* (New York: Cambridge University Press, 2018); Gregg Huff, "Causes and Consequences of the Great Vietnam Famine, 1944–5," *Economic History Review* 72, no. 1 (2019): 286–316; Abhijit Sarkar, "Fed by Famine: The Hindu Mahasabha's Politics of Religion, Caste, and Relief in Response to the Great Bengal Famine, 1943–1944," *Modern Asian Studies* 54, no. 6 (2020): 2022–86.

[20]Macekura, *Of Limits and Growth*, 27–9.

Organization had been primarily concerned with coordinating meteorological methods and instruments across nations and empires since first meeting together in Vienna six years earlier. This drive to collectively standardize techniques was part of a wider movement of scientific internationalism that aspired to "a vision of science and technical knowledge as a creed without borders."[21] The formalization of meteorology and meteorological data collection had advanced in lock-step with the spread of telegraphy and rail, and crossing national and imperial borders offered new opportunities to expand meteorological observing networks.[22] During the interwar years, the prospect of forming an intergovernmental organization appealed to many of the participating meteorologists, who saw this development as a means to acquire further resources, authority, and access to a wider array of meteorological data. The measurement, collection, and sharing of this data, they also hoped, would become better standardized and coordinated.[23]

As far as the United States was concerned, the science of meteorology was ideally suited to its postwar vision of international cooperation. By inviting the International Meteorological Organization to meet for the first time outside Europe, the State Department and US Weather Bureau could shape the creation of the new World Meteorological Organization (WMO) on home soil. The worldwide scale of such an international agency appeared ideally suited to the global scale of the atmosphere, which meteorologists could together comprehend through their shared technical language. As STS scholar Clark Miller puts it, the United States envisioned the WMO as the postwar standard-bearer to ensure enduring peace: "Binding standards meant better-quality and higher-frequency data exchange. Better data exchanges meant better weather forecasts and better climatological data. These in turn meant better agricultural production and safer air and sea travel. The result would be greater economic prosperity in the West, more opportunities for cross-cultural exchange, and more social stability in the face of the Cold War."[24]

The intergovernmental nature of this proposed agency raised concerns among the professional meteorologists gathered in Washington, some of whom were wary of its implications for their activities and scientific independence. More difficult was the question of membership to this new agency: as part of the UN, it would be required to admit only those "sovereign states" recognized by international law. This rule would preclude from membership the newly divided Germany, as well as the People's Republic of China, colonial

[21]Mark Mazower, *Governing the World: The History of an Idea* (New York: Penguin, 2012), 192.
[22]Paul Edwards, *A Vast Machine: Computer Models, Climate Data and the Politics of Global Warming* (Cambridge, MA: MIT Press, 2010), 49–60.
[23]Edwards, *A Vast Machine*, 187–8.
[24]Miller, "Scientific Internationalism," 186.

territories, and the Soviet republics, as postwar geopolitics became grafted onto the geography of the atmosphere.[25] This thorny question arose at the Organization's First World Meteorological Congress in Paris in 1951, when China requested membership. Excluding its participation would run contrary to the information globalism, to use historian Paul Edwards's term, that the Organization sought to foster. To overcome this legal hurdle, nonmember countries could send "observers" to the Organization's meetings.[26]

Along with the likes of UNESCO, the Food and Agriculture Organization (FAO), and other international organizations, the WMO also shared in the spirit of postwar development, negotiating with the UN's program of technical assistance to fund both projects and technical staff as well as their supervision. The first recipient was the Government of Libya, where the Organization's largesse would support the development of a national meteorological service to meet its own requirements, while fulfilling its international obligations, particularly to international civil aviation.[27] As independence loomed in 1951, the likelihood grew that the departure of non-Libyan meteorological personnel might disrupt local weather services.[28] Until local staff could be trained, the WMO recruited foreign technicians to record and disseminate weather conditions.[29] Ensuring the unbroken continuity of the instrumental record and developing local meteorological expertise had gained new significance by the early 1950s, which subsequent requests from Yugoslavia and Israel emphasized. Bordering the Mediterranean, where communism appeared to be gaining ground, these nations were also strategically valuable to the United States, the UK, and their allies. By helping to create meteorological networks in these countries, where there had previously been none, the WMO provided political legitimacy to the nation-building projects of such newly independent nations.[30]

Atomic Weather: Nuclear Fallout and the Atmosphere

The postwar optimism of atomic weather control faded as the Cold War set in. Mathematician John von Neumann, who had contributed to the joint Anglo-American-Canadian Manhattan Project, had initially found the prospect "most

---

[25]Paul N. Edwards, "Meteorology as Infrastructural Globalism," *Osiris* 21, no. 1 (2006): 229–50.
[26]World Meteorological Organization, *First Congress of the World Meteorological Organization* (Geneva: World Meteorological Organization, 1951), 7.
[27]"Technical Assistance Programme," *WMO Bulletin* 1, no. 1 (1952): 20.
[28]Edwards, *A Vast Machine*, 199.
[29]Agnese Lockwood, "Libya: Building a Desert Economy," *International Conciliation* 31 (1957): 342.
[30]Gretchen Heefner, "'A Tract That Is Wholly Sand': Engineering Military Environments in Libya," *Endeavour* 40, no. 1 (2016): 38–47; Edwards, "Meteorology as Infrastructural Globalism," 241.

interesting and inspiring" when RCA Laboratory researcher Vladimir Zworykin shared with him in late 1945 a proposal to "channel the world's weather … in such a way as to minimize the damage from catastrophic disturbances, and otherwise to benefit the world to the greatest extent by improved climatic conditions where possible."[31] Having contributed to producing the first computerized weather forecast in the intervening decade, von Neumann had since become more wary of the prospect of deliberately changing the weather. In the US magazine *Fortune*, he warned in 1955 that once "global climate control becomes possible … [those means] will be exploited."[32]

When Zworykin had made his proposal after the war, the United States was the sole nuclear power; since then, the Soviet Union and the UK had developed testing programs of their own. With the outbreak of the Korean War in June 1950 came the rush for the superpowers to add thermonuclear bombs to their growing arsenals: President Truman announced the success of his government's first hydrogen bomb tests two years later. By 1953, the United States had conducted forty-three atmospheric nuclear tests in the deserts of New Mexico and Nevada as well as in the southwest Pacific, the Soviet Union eight in Kazakh SSR, and Britain three off Australia's coast and the continent's red center.[33] Shrouded in secrecy and supposedly remote as these tests were, officials and scientists gave little credence to suggestions that their effects might not be so local and contained as governments had led their civilians to believe.[34]

Although the marginalized and dispossessed peoples of the testing grounds were largely disregarded, others were afforded the means to share their growing unease more widely.[35] Their anxieties were especially concerned with environmental change and their own health. Were unusual or extreme weather conditions, they wondered, a consequence of atomic tests? Nearly a third of the US public thought as much in 1953.[36] More Britons attributed recent bad weather to the hydrogen bomb in both 1955 and 1958, and over half considered that fallout from "Russian H-bombs" might endanger them or

---

[31]Vladimir Zworykin, 1945, 6, in Fleming, "Introduction," 57–8.

[32]John von Neumann, "Can We Survive Technology?," in F. Bródy and T. Vámos (eds.), *The Neumann Compendium* (River Edge, NJ: World Scientific, 1995), 658–73.

[33]Toshihiro Higuchi, *Political Fallout: Nuclear Weapons Testing and the Making of a Global Environmental Crisis* (Stanford: Stanford University Press, 2020), 16.

[34]See, for example, Togzhan Kassenova, "Banning Nuclear Testing: Lessons from the Semipalatinsk Nuclear Testing Site," *Nonproliferation Review* 23 (2016): 329–44; Martha Smith-Norris, *Domination and Resistance: The United States and the Marshal Islands during the Cold War* (Honolulu: University of Hawai'i Press, 2016); Eve Vincent, "Nuclear Colonialism in the South Australian Desert," *Local-Global* 3 (2007): 103–12.

[35]Jacob Darwin Hamblin and Linda M. Richards, "Beyond the *Lucky Dragon*: Japanese Scientists and Fallout Discourse in the 1950s," *Historia Scientiarum* 25, no. 1 (2015): 36–56.

[36]George Gallup, "29% Blame Weather on A-Blasts, Gallup says," *Los Angeles Times*, 19 June 1953, 28.

their children in 1961.[37] As for "odd" weather in the South Pacific, the *Pacific Islands Monthly* reported in mid-1956 that "there are a lot of people still unconvinced that 'them atom bombs' haven't got something to do with it."[38] And what of the effects of the controversial new technique of cloud seeding that had likely inspired Zworykin's proposal for weather control in the first place? Zworykin had been among those assembled in upstate New York in 1942 to watch General Electric scientists Irving Langmuir and Vincent Schaefer put their newly acquired skills in cloud physics to the test.[39]

Concerns about the unintended effects of atomic tests grew only louder, however, in the wake of the fatal contamination of a Japanese fishing boat, the *Daigo Maru* (*Lucky Dragon Five*), from the radioactive fallout of the US Castle Bravo test in the Marshall Islands in March 1954. The US commanders had ignored a weather forecast predicting strong high-altitude winds near the site, despite the established practice of relying on such information to contain the health hazards already known to be associated with nuclear fallout. Bravo's yield was three times more powerful than anticipated, poisoning Marshall Islanders, US service personnel, and the crew of the *Daigo Maru*. Despite the attempts of the US government to cover up the extent of the fallout, Japanese research, contaminated fish stocks, and unusual weather events could not be ignored (Figure 1.1).[40] As Alistair Cooke told listeners in his weekly BBC radio program *Letter from America*, "this is obviously a turning point in history that cannot be shrugged off or pacified with appeals to decent feeling."[41]

French physicist Charles-Noël Martin, with the support of Albert Einstein, cautioned against the escalation of nuclear explosions, which he feared might initiate "a rising disequilibrium of the natural conditions."[42] Under the weight of such public pressure, the WMO commenced its own investigation into the association of "apparently unusual weather conditions" with atomic testing.[43] Under the directorship of the US weather chief, the agency was unwilling, however, to heed Japan's request to condemn such tests. At home, the Joint Committee on Atomic Energy considered the possible weather impacts of the

[37]"1955: Hydrogen Bomb," "1958: Hydrogen Bomb," "1961: Hydrogen Bomb," in George Gallup (ed.), *The Gallup International Public Opinion Polls: Great Britain, 1937–1975, Volume One, 1937–1964* (New York: Random House), 363, 480, 610.

[38]"Droughts Break at Samarai and Lai," *Pacific Islands Monthly*, June 1, 1956, 157.

[39]James R. Fleming, *Fixing the Sky: The Checkered History of Weather and Climate Control* (New York: Columbia University Press, 2010), 139–40.

[40]Matthias Dörries, "Testing the Precautionary Argument after the *Lucky Dragon* Incident," *Disaster Prevention and Management* 30, no. 1 (2021): 64–75.

[41]Alistair Cooke, "Bikini Island Detonation: Letter no. 358, 1 April 1954," in Alistair Cooke and Susan Cooke Kittredge, *Reporting America* (Woodstock, NY: Overlook Press, 2008), 63.

[42]Charles-Noël Martin, 1954, cited in Dörries, "Testing the Precautionary Argument," 64–75.

[43]"Technical Notes," *WMO Bulletin* 3, no. 4 (1954): 121.

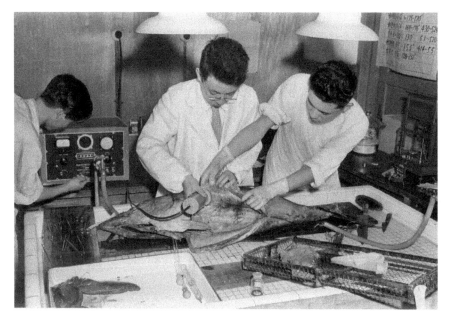

FIGURE 1.1 *Researchers check the radiation levels of a tuna, July 19, 1954, in Tokyo, Japan. Photo 185131422, The* Asahi Shimbun *via Getty Images.*

detonations. The chief of the Scientific Services Division of the US Weather Bureau, Harry Wexler, referred to the sanguine findings of his colleagues and was adamant that "from the standpoint of physics" there were no connections between the explosions and the weather.[44] These findings, combined with the assurances of the US National Academy of Sciences and Soviet reports, encouraged the WMO to close its own investigation in 1956 and turn instead to the study of radiation in the atmosphere.[45]

As the interest of the WMO suggests, the study of nuclear fallout and radiation were of wider scientific value than for weapons tests alone. Tracking fallout through the atmosphere allowed meteorologists to now follow the motion of individual parcels of air, which informed global circulation studies. Fallout also served as a tracer for dust, and monitoring programs sampled stratospheric carbon dioxide and carbon-14 for the first time, providing detailed insight into global atmospheric chemistry and circulation.[46] Such studies

---

[44]"Statement of Dr Harry Wexler, Chief, Scientific Services Division, Accompanied by Dr Lester Machta, Chief Special Projects Section, United States Weather Bureau, Washington, D.C.," in Joint Committee on Atomic Energy, *Health and Safety Problems and Weather Effects Associated with Atomic Explosions* (Washington, DC: Government Printing Office, 1955), 47.
[45]Dörries, "Testing the Precautionary Argument," 64–75.
[46]Edwards, *A Vast Machine*, 209–10.

revealed that fallout spread globally throughout the stratosphere within about two years.[47] These findings, together with *Time* magazine's report of Martin's paper to the French Academy of Sciences, went on to inspire author Nevil Shute to imagine the catastrophic consequences of nuclear fallout in his novel *On the Beach*, the 1959 film version of which starred Gregory Peck and Ava Gardner as the doomed protagonists.[48]

Climate Changes: Warming the World

Although he had been unswayed by concerns about atomic weather, the US Weather Bureau's Wexler was nonetheless well aware of unfolding climatic changes that were widely understood to be more sanguine in nature. The extent and cause of these changes became the subject of postwar scientific inquiry and international collaboration. In June 1947, Sweden's Hans Wilhelmsson Ahlmann of the Stockholms Högskola (now Stockholm University) had presented three seminars to the US Weather Bureau in Washington, which Wexler himself had organized.[49] There, Ahlmann had shared the findings of several decades of field work studying the glaciers of the north Atlantic, where he interpreted their retreat as evidence of "climatic improvement."[50] Sea ice in the European Arctic appeared to have declined, and in the early 1930s, the northeast passage often remained unusually open with steamships circumnavigating Novaya Zemlya and Franz Josef Land without encountering pack ice.[51] As to the cause of what he termed "climatic fluctuations," Ahlmann speculated that a poleward movement of warm tropical air was likely responsible for changes that he understood as the likely result of solar activity, and thus beyond human

---

[47]Paul N. Edwards, "Entangled Histories: Climate Science and Nuclear Weapons Research," *Bulletin of the Atomic Scientists* 68, no. 4 (2012): 31.

[48]"Science: The Unmentionable Subject," *Time*, December 20, 1954, https://v.gd/BR06qn; Gideon Haigh, "Shute the Messenger," *Monthly*, June 2007, https://v.gd/hPi0dj.

[49]Harry Wexler, "Seminars on Meteorological and Related Topics Presented at the US Weather Bureau, Washington, DC. During the Year Beginning July 1, 1946," *Bulletin of the American Meteorological Society* 29, no. 4 (1948): 160–1. While in the capital, Ahlmann also provided secret testimony on polar melting to the Pentagon's Research and Development Board's earth sciences panel, see Ronald E. Doel et al., "Strategic Arctic Science: National Interests in Building Natural Knowledge – Interwar Era through the Cold War," *Journal of Historical Geography* 44 (2014): 60–80.

[50]Hans W. Ahlmann, "Researches on Snow and Ice, 1918–40," *Geographical Journal* 107, no. 1/2 (1946): 11–25; Hans W. Ahlmann, "The Present Climatic Fluctuation," *Geographical Journal* 112, no. 4/6 (1949): 165–93.

[51]Brandon Luedtke, "An Ice-Free Arctic Ocean: History, Science, and Scepticism," *Polar Record* 51, no. 2 (2013): 130–9.

influence or timescales.[52] This work was well received by Huxley, who visited Iceland in the summer of 1949 and associated "desiccation in lower latitudes" to a "world-wide change in climate."[53]

Soviet scientists had long shared Ahlmann's interest in the polar warming trend. They had studied diminishing ice in the Arctic since the early 1920s, which had promising implications for Russian shipping and even agricultural production.[54] The Swedish glaciologist had visited Leningrad in late 1934 in the wake of the Second Polar Year (1932–33), during which unusually warm conditions had allowed both Soviet and Swedish-Norwegian expeditions to navigate into the high Arctic. By now, Stalin had committed to exploring and developing the Soviet Arctic via a new and briefly powerful organization, the Main Administration of the Northern Sea Route, *Glasevmorput*.[55] At its All-Union Arctic Institute in Leningrad, Ahlmann delivered two lectures on glaciology to audiences among whom his work was already familiar, and he connected the changing Arctic conditions to the intensity of the atmospheric circulation.[56] Having afterward cofounded the Society for the Promotion of Cultural and Economic Relations between Sweden and the Soviet Union, Ahlmann maintained his scientific network over the course of the following decade and returned in mid-1945 to celebrate the 220-year jubilee of the Academy of Sciences in Moscow.[57]

Later in 1945, Ahlmann cited the work of Russian scientists among others when he addressed the Royal Geographical Society in London. Presenting his "researches on snow and ice," he described the extent of "climatic improvement" and surmised that its cause was "a general intensification of the atmospheric pressure gradient," which he hoped exploration in Antarctica might further elucidate.[58] In the discussion that followed, the president of the Royal Meteorological Society, Gordon Manley, described this work as "of the

---

[52]Sverker Sörlin and Erik Isberg, "Synchronizing Earthly Timescales: Ice, Pollen, and the Making of Proto-Anthropocene Knowledge in the North Atlantic Region," *Annals of the American Association of Geographers* 111 (2021): 720.

[53]Julian Huxley, *Natural History in Iceland* (Washington, DC: US Government Printing Office, 1950), 336–8.

[54]Sverker Sörlin and Julia Lajus, "An Ice-Free Arctic Sea? The Science of Sea Ice and Its Interests," in Miyase Christensen, Annika Nilsson, and Nina Wormbs (eds.), *Media and the Politics of Arctic Climate Change: When the Ice Breaks* (New York: Palgrave Macmillan, 2013), 70–92; Julia Lajus and Sverker Sörlin, "Melting the Glacial Curtain: The Politics of Scandinavian-Soviet Networks in the Geophysical Field Sciences between Two Polar Years, 1932/33–1957/58," *Journal of Historical Geography* 44 (2014): 44–59.

[55]John McCannon, *Red Arctic: Polar Exploration and the Myth of the North in the Soviet Union* (New York: Oxford University Press, 1998), 33–4.

[56]Lajus and Sörlin, "Melting the Glacial Curtain," 44–59.

[57]Hans W. Ahlmann, "Geography in the Soviet Union," trans. M. Burnett, *Geographical Journal*, 106 (1945): 217–21.

[58]Ahlmann, "Researches on Snow and Ice, 1918–40," 11–25.

greatest climatological significance."[59] Manley had recently dismissed the work of outsider, Guy Stewart Callendar, who had revived the idea of human-caused climate change on the eve of the Second World War.[60] A steam engineer, Callendar had earlier published his findings in the Society's journal, where he explained a relationship between the artificial production of carbon dioxide and warming trends.[61]

The thermodynamic or radiative properties of certain gases, such as carbon dioxide, had been separately identified by American scientist Eunice Foote and Irish physicist John Tyndall in the mid-nineteenth century.[62] Contemporary concerns as to climate changes over geological time, such as ice ages, also later informed the work of Swedish chemist Svante Arrhenius. Drawing on Tyndall's results, Arrhenius had shown that alterations to the concentration of carbon dioxide in the atmosphere could affect ambient temperatures, such that diminishing carbon dioxide would precipitate cooling.[63] His colleague, Nils Ekholm, noted that the "present burning of pit-coal" would "undoubtedly cause a very obvious rise of the mean temperature of the earth," should it continue for "some thousand years"—a notion that Arrhenius popularized in his 1906 *Worlds in the Making*.[64] Translated into English, German, Polish, and Italian, the 1914 Japanese edition likely inspired writer Miyazawa Kenji's 1932 story of a young man who seeks to increase the world's temperature to overcome the effects of cold weather on local harvests.[65]

---

[59]Hans W. Ahlmann, "Researches on Snow and Ice, 1918–40: Discussion," *Geographical Journal* 107 (1946): 27.

[60]Gordon Manley, "Some Recent Contributions to the Study of Climate Change," *Quarterly Journal of the Royal Meteorological Society* 70 (1944): 197–219; Georgina Endfield, Lucy Veale, and Alexander Hall, "Gordon Valentine Manley and His Contribution to the Study of Climate Change: A Review of His Life and Work," *WIREs Climate Change* 6, no. 3 (2015): 287–99.

[61]Guy Stewart Callendar, "The Artificial Production of Carbon Dioxide and Its Influence on Temperature," *Quarterly Journal of the Royal Meteorological Society* 64 (1938): 223–40.

[62]Eunice Foote, "Circumstances Affecting the Heat of the Sun's Rays," *American Journal of Science and Art* 22 (1856): 382–3; John Tyndall, "Note on the Transmission of Heat of Different Qualities through Gases of Different Kinds," *Proceedings of the Royal Institute* 3 (1859): 155–258; Roland Jackson, "Eunice Foote, John Tyndall and a Question of Priority," *Notes and Records: The Royal Society Journal of the History of Science* 74, no. 1 (2020): 105–18.

[63]Svante Arrhenius, "On the Influence of Carbonic Acid in the Air upon the Temperatures of the Ground," *Philosophical Magazine* 31 (1896): 237–76.

[64]Nils Ekholm, "On the Variations of the Climate of the Geological and Historical Past and Their Causes," *Quarterly Journal of the Royal Meteorological Society* 27 (1901): 1–61; Svante Arrhenius, *Worlds in the Making: The Evolution of the Universe*, trans. H. Borns (New York: Harper, 1908), 54–63, first published in Swedish as *Världarnas utveckling* (1906).

[65]Miranda A. Schreurs, "Shifting Priorities and the Internationalization of Environmental Risk Management in Japan," in Social Learning Group (ed.), *Learning to Manage Social Risks: A Comparative History of Social Responses to Climate Change, Ozone Depletion and Acid Rain* (Cambridge, MA: MIT Press, 2001), 191–212; Toshihiro Yamada, "Geoscientists and Buddhist Thought: The Popularization of Science and 'Cosmic Consciousness' in 1920s Japan," *Physis* LVI (2021): 235–56.

The carbon dioxide theory subsequently fell out of favor, as contemporaries in the United States and Sweden cast doubt on the extent that carbon dioxide could influence the climate at a global scale. Callendar too had been unconvinced by the findings when he commenced his own studies. But he demonstrated that the concentration of carbon dioxide in the atmosphere had increased artificially since 1900, to which he attributed part of the prevailing warming trend.[66] According to his model, doubling the atmospheric concentration of carbon dioxide by the combustion of fossil fuels would induce 2°C global warming. This outcome, he deemed, would be "beneficial to mankind," allowing for agriculture in the northern latitudes, greater plant growth, and delaying the advance of "deadly glaciers."[67] For German meteorologist Hermann Flohn, the outbreak of war prevented his further exploration of Callendar's theory, on which he published his own findings in 1941 as part of a wider survey of the effects of human activity on climate.[68]

After the war, research on carbon dioxide and atmospheric chemistry resumed in Scandinavia.[69] In his work, Callendar had used carbon dioxide measurements undertaken by Finnish chemist Kurt Buch, who was receptive to the English engineer's findings and adopted them in his own studies of ocean chemistry. Buch then shared them with his colleague, Swedish meteorologist Carl-Gustav Rossby.[70] On Ahlmann's recommendation, Rossby had returned home from the United States in 1947 to establish and direct the International Institute of Meteorology, where he hosted Buch and encouraged chemist Erik Eriksson to investigate further atmospheric concentrations of carbon dioxide.[71] With Buch's support, Eriksson devised a program of air sampling across Scandinavia as well as Hawai'i, which Rossby speculated

[66]James R. Fleming, *The Callendar Effect: The Life and Work of Guy Stewart Callendar (1898–1964), the Scientist Who Established the Carbon Dioxide Theory of Climate Change* (Boston: American Meteorological Society, 2007), 68–72.

[67]Callendar, "Artificial Production," 236.

[68]Heinz-Dieter Schilling and Andreas Hense, *Hermann Flohn: Meteorologie im Übergang – Erfahrungen und Erinnerungen (1931–1991)* (Bonn: Dümmlers Verlag, 1992). See Hermann Flohn, "Die Tätigkeit des Menschen als Klimafaktor," *Zeitschrift für Erdkunde* 9 (1941): 13–22.

[69]See, for example, British climatologist C. E. P. Brooks's declaration in 1951 that the theory of carbon dioxide had been "abandoned": "Geological and Historical Aspects of Climate Change," in Thomas F. Malone, *Compendium of Meteorology* (Boston: American Meteorological Society, 1951), 1016; Hans W. Ahlmann, *Glacier Variations and Climatic Fluctuations: Bowman Memorial Lectures* (New York: American Geographical Society, 1952).

[70]Maria Bohn, "Concentrating on CO_2: The Scandinavian and Arctic Measurements," *Osiris* 26, no. 1 (2011): 165–79. The likes of John Dalton and others had undertaken measurements of the atmospheric concentration of carbon dioxide since the early nineteenth century, see James R. Fleming, *Historical Perspectives on Climate Change* (New York: Oxford University Press, 1998), 126.

[71]Sverker Sörlin, "Narratives and Counter-Narratives of Climate Change: North Atlantic Glaciology and Meteorology, c.1930–1955," *Journal of Historical Geography* 35 (2009): 237–55. See Erik Eriksson, "Report on an Informal Conference in Atmospheric Chemistry Held at the Meteorological Institute, University of Stockholm, May 24–26, 1954," *Tellus* 6, no. 3 (1954): 302–307.

could be expanded through the upcoming International Geophysical Year (IGY). Rossby further suggested that a recently devised method using radioactive carbon (thanks to nuclear science) might help to overcome local variations in the air measurements.[72]

Changes in the Air: Measuring Carbon Dioxide

In his two decades in the United States, Rossby had transformed the field of meteorology. Working first for the Weather Bureau thanks to an American-Scandinavian Foundation fellowship, and then establishing the nation's first meteorology school at MIT (where he taught Wexler), Rossby brought with him the insights of Norway's Bergen school.[73] Featured on the cover of *Time* magazine in 1956, of Rossby the accompanying article declared, "Most leaders of modern meteorology are friends or past pupils of Dr. Rossby's … The history of modern meteorology is inescapably paralleled by Rossby's career."[74] Highly regarded on both sides of the Atlantic as a "skilled institution-builder and political advisor," his return to Stockholm a decade earlier was part of Ahlmann's wider plan to revive Scandinavian influence in the Arctic to protect Norwegian and Swedish interests from Soviet and US advances after the war. The Swedish glaciologist and diplomat pursued internationally renowned Scandinavian scientists like Rossby to build and lead major research organizations, which he believed would improve environmental knowledge about the Arctic and shape the Cold War contours of polar diplomacy.[75]

Located in ostensibly neutral Sweden, Rossby's Institute at Stockholm University attracted visiting researchers from both East and West.[76] During a 1954 meeting of scientists from Finland, Sweden, the United States, and Hawai'i's Pineapple Research Institute, Buch shared his recent work on carbon dioxide and referred to unpublished "computations" that confirmed the role of the gas in the observed warming trend.[77] Those computations had

---

[72]Carl-Gustav Rossby, "Current Problems in Meteorology," in Bert Bolin (ed.), *The Atmosphere and the Sea in Motion: Scientific Contributions to the Rossby Memorial Volume* (New York: Rockefeller Institute Press, 1959), 14–15.

[73]Norman A. Phillips, "Carl-Gustaf Rossby: His Times, Personality, and Actions," *Bulletin of the American Meteorological Society* 79, no. 6 (1998): 1097–112; Sörlin, "Narratives and Counter-Narratives," 237–55; James R. Fleming, "Carl-Gustaf Rossby: Theorist, Institution Builder, Bon Vivant," *Physics Today* 70 (2017): 51–6.

[74]"Science: Man's Milieu," *Time*, December 17, 1956, https://v.gd/49AIQx.

[75]Doel et al., "Strategic Arctic Science," 75.

[76]Sörlin, "Narratives and Counter-Narratives," 253.

[77]Eriksson, "Report on an Informal Conference," 304.

been undertaken by a US physicist, who combined his research on infrared absorption with new spectroscopic measurements and new computer technology. With funding from the US Office of Naval Research, Gilbert Plass had pursued his calculations first at Johns Hopkins, then at the University of Michigan, and while at the Lockheed Aircraft Corporation, he finalized his results for publication in 1956 in Rossby's *Tellus* and elsewhere.[78] His work had showed that additional carbon dioxide in the atmosphere would absorb infrared radiation, which would raise the average global temperature "at the rate of 1.1°C per century." Reporting Plass's early findings, UNESCO's *Courier* welcomed a future where "vast areas of Canada, Alaska and Siberia may become available for food production."[79] Warning that such warming could become undesirable a few centuries hence, Plass understood this carbon dioxide theory as a means to account for geological climate changes in the deep past and returned to his research on infrared physics shortly afterward.[80]

Rossby died unexpectedly the following year. At the time of his death, he was chair of the Swedish Polar Year Committee for the upcoming IGY and was corresponding with a Swedish scientist based at the Scripps Institution of Oceanography in San Diego. That scientist, Gustaf Arrhenius, grandson of Svante Arrhenius, had earlier informed Rossby of US plans to undertake an inventory of atmospheric carbon dioxide concentrations to provide a baseline for further study. As Wexler, now chief scientist for the US expedition to the Antarctic for the IGY, later reported, "our industrial civilization is pouring great quantities of carbon dioxide into the atmosphere and we do not really know yet what happens to it: how much is dissolved in the oceans, how much is taken up by plant life, and how much remains in the atmosphere."[81] Their exchange prompted close collaboration between Rossby and Arrhenius's colleagues at Scripps to ensure that techniques in the United States and Sweden were coordinated, and the precision of the latter's greatly improved.[82]

The program of air sampling that ensued was merely one of many scientific initiatives of the IGY. Overshadowed by the successful launch of the first artificial earth satellite in October 1957, the Soviet *Sputnik 1*, other programs together undertook a series of global geophysical activities from July 1, 1957, to December 31, 1958—a high point in the eleven-year cycle

---

[78]Spencer R. Weart, "Global Warming, Cold War, and the Evolution of Research Plans," *Historical Studies in the Physical and Biological Sciences* 27, no. 2 (1997): 319–56. See Gilbert Plass, "The Carbon Dioxide Theory of Climatic Change," *Tellus* 8 (1956): 140–54.

[79]Gerald Wendt, "Science Chronicle," *UNESCO Courier* no. 2 (1954): 26–7.

[80]Gilbert Plass,"Effect of Carbon Dioxide Variations on Climate," *American Journal of Physics* 24 (1956): 376–87; Fleming, *Historical Perspectives*, 121–2.

[81]Harry Wexler, "IGY Meteorology Program," in *National Science Foundation: Hearings before the Subcommittee of the Committee on Appropriations; Review of the First Eleven Months of the International Geophysical Year* (Washington, DC: US Government Printing Office, 1958), 39.

[82]Bohn, "Concentrating on CO$_2$," 176.

of sunspot activity over a period that would encompass a full field season at both poles.[83] Coordinated by the International Council of Scientific Unions (ICSU) with UNESCO and the WMO, preparations for the IGY began in 1950 on the suggestion of a Third Polar Year by US physicist and science entrepreneur Lloyd Berkner. That year, he had advised the US State Department to undertake a more international approach to its science policy, such that scientific cooperation between nations would provide the grounds for not only scientific and economic advancement but also peace.[84] The scientific and military desire for global geophysical data also appealed to the Soviet Union in the wake of Stalin's death.[85] Berkner's vision for "world-wide synoptic observation and analysis" would eventually engage sixty-seven nations "tired of war and dissension" in the study of eleven major scientific areas, including meteorology, oceanography, and glaciology.[86]

These areas of study intersected in the nascent agenda to establish a baseline of the atmospheric concentration of carbon dioxide. Such a baseline would provide the scientists the means to study what Scripps director Roger Revelle and geochemist Hans Suess had described on the eve of the IGY as a "large scale geophysical experiment [by humans] of a kind that could not have happened in the past nor be reproduced in the future." They continued, "This experiment, if adequately documented, may yield a far-reaching insight into the processes determining weather and climate."[87] An oceanographer, Revelle was concerned with questions of ocean circulation, ocean chemistry, and the dispersal of atomic radiation, which he continued to study after overseeing a scientific survey of the aftermath of the Operation Crossroads nuclear tests in the Marshall Islands in July 1946.[88] To his mind, postwar scientific and military objectives were intertwined: "It has become apparent that the society which

---

[83]Roger D. Launius, "Towards the Poles: A Historiography of Scientific Exploration during the International Polar Years and the International Geophysical Year," in Roger Launius, James Fleming, and David DeVorkin (eds.), *Globalizing Polar Science: Reconsidering the International Polar and Geophysical Years* (New York: Palgrave Macmillan, 2010), 66–7.

[84]Miller, "Scientific Internationalism," 174.

[85]Elena Aronova, "Geophysical Datascapes of the Cold War: Politics and Practices of the World Data Centers in the 1950s and 1960s," *Osiris* 32 (2017): 311–12.

[86]Launius, "Towards the Poles," 67; Lloyd Berkner, "International Scientific Action," *Science*, April (1954): 570, 575.

[87]Roger Revelle and Hans Suess, "Carbon Dioxide Exchange between Atmosphere and Ocean and the Question of an Increase of Atmospheric CO_2 during the Past Decades," *Tellus* 9 (1957): 18–27.

[88]Jacob Darwin Hamblin, *Arming Mother Nature: The Birth of Catastrophic Environmentalism* (New York: Oxford University Press, 2013), 93. See Ronald Rainger, "Science at the Crossroads: The Navy, Bikini Atoll, and American Oceanography in the 1940s," *Historical Studies in the Physical and Biological Sciences* 30, no. 2 (2000): 340–71; Laura A. Bruno, "The Bequest of the Nuclear Battlefield: Science, Nature, and the Atom during the First Decade of the Cold War," *Historical Studies in the Physical and Biological Sciences* 33, no. 2 (2003): 237–60.

knows the most about its environment and how to turn it to account, is going to be the more likely to win the next war."[89]

The prevailing scientific view that the ocean acted as a vast sponge absorbing carbon dioxide had discouraged Callendar's contemporaries from pursuing his findings further.[90] This position reflected a wider belief in the natural world's tendency toward balance and self-regulation (or in contemporary ecological terms, "climax community"), which human activities were considered too small and localized to disrupt (see Chapter 2).[91] Together, Revelle and Suess demonstrated that as far as the ocean's absorption of carbon dioxide was concerned, such balance could not be guaranteed as the gas remained in the atmosphere far longer than had been previously assumed. At 1950s rates of consumption, the ocean could absorb only about 80 percent of the carbon dioxide produced by fossil fuel combustion, which could account for findings such as Callendar's, that the concentration of the gas in the atmosphere had increased since the turn of the twentieth century.[92] Soviet meteorologist Evgeny Fedorov, meanwhile, saw in rising carbon dioxide levels in 1958 evidence that "human society has already become an involuntary climatological factor."[93]

By the end of the IGY, measurements of the concentration of carbon dioxide in the atmosphere had confirmed the gas's annual increase. With the support of Wexler's Weather Bureau and the US military, newly appointed Scripps chemist C. D. "Dave" Keeling had established air sampling stations at two sites ostensibly under US jurisdiction: the South Pole in Antarctica and on the northern flank of Hawai'i's Mauna Loa volcano. These sites, he deemed, were sufficiently distant from the contaminating influence of industrial areas, urban pollution, and vegetation. Each of these stations provided a glimpse into the atmospheric chemistry of the northern and southern hemispheres, which Revelle had considered would reflect the spatial variability of carbon dioxide concentrations in the air.[94] Two years of data, however, revealed the same overall trend: neither the ocean nor vegetation was absorbing the entirety of industrial emissions; the level of carbon dioxide in the atmosphere had

[89]Roger Revelle, cited in Rainger, "Science at the Crossroads," 369.

[90]Weart, "Global Warming, Cold War," 319–56. See, for example, Stig Fonselius, Folke Koroleff, and Kurt Buch, "Microdetermination of CO_2 in the Air, with Current Data for Scandinavia," *Tellus* 7 (1955): 259–65.

[91]See Frank B. Golley, *A History of the Ecosystem Concept in Ecology: More than the Sum of the Parts* (New Haven: Yale University Press, 1993).

[92]Edwards, "Entangled Histories," 28–40.

[93]E. K. Fedorov, "Vozdeistvie cheloveka na meteorologicheskie protsessy," *Voprosy filosofii* 4 (1958): 144, translated by Jonathan Oldfield, "Climate Change as an Issue of Global Environmental Concern: Emerging Debates Concerning Climate and Climate Change amongst Soviet Geographers, c.1945–1960s," Working paper (2012), https://v.gd/f7SCLe.

[94]C. D. Keeling, "Rewards and Penalties of Monitoring the Earth," *Annual Review of Energy and Environment* 23 (1998): 39.

perceptibly increased.[95] The sophistication of Keeling's equipment and his fastidious approach to air sampling surpassed the Scandinavian measurements, which, having been obtained by more traditional methods and showing a high degree of variability, *Tellus* ceased reporting shortly afterward.[96]

Climate Hypotheses: Sharing Climate Knowledge

Keeling and his carbon dioxide measurements at Mauna Loa would later become synonymous with scientific and lay understandings of anthropogenic causes of climate change. In 1961, however, the role of carbon dioxide in changing the planet's climate was not the hegemonic position among scientists. In October that year, over a hundred scientists from thirty-six countries gathered in Rome at the invitation of the Italian government to discuss "changes of climate with special reference to the arid zones" at the headquarters of the FAO. As the conference organizer, Swedish climatologist Carl-Christian Wallén observed,

> many of those [arid] areas, where we are now trying to extend settlement and agriculture, are considered by many to have been fertile and used for dry-farming or irrigated agriculture in ancient times. A highly interesting aim of future development within the field of climatic fluctuations will, therefore, be an attempt to establish the changes of the margin of the arid and semi-arid lands which have occurred in historical and prehistorical times.[97]

The delegates were meeting under the auspices of UNESCO's Arid Zone Research Project and the WMO, and their nationalities reflected British prime minister Harold Macmillan's "winds of change," as representatives of new nations India, Pakistan, Israel, Ghana, and Tunisia met with colleagues from the Soviet Union, the United States, Argentina, South Africa, Norway, and Australia, among others. Already, the WMO had identified a role for climatologists in the Freedom from Hunger campaign that the FAO had launched earlier that year, whereby their expertise would contribute to the cultivation of dry farming

---

[95]C. D. Keeling, "The Concentration and Isotopic Abundance of Carbon Dioxide in the Atmosphere," *Tellus* 12 (1960): 200–3.

[96]Keeling, "Rewards and Penalties," 38, 41.

[97]C.-C. Wallén, "Aims and Methods in Studies of Climatic Fluctuations," in *Changes of Climate: Proceedings of the Rome Symposium Organized by UNESCO and the World Meteorological Organization* (Paris: UNESCO, 1963), 472.

crops in the arid climes of Iran, Iraq, Jordan, Lebanon, and the United Arab Republic.[98]

Ahlmann's warming pattern in the northern hemisphere was a topic for discussion in Rome.[99] Such conditions had been observed in both hemispheres from the 1920s to the 1940s, and his student, Wallén (and others), had since noted cooler temperatures.[100] One delegate, J. Murray Mitchell of the US Weather Bureau, had recently published his own study of this cooling trend, on which he elaborated in Rome.[101] Having been involved in the study of the meteorological effects of nuclear tests, Mitchell had turned to the study of global temperatures. Unconvinced that the atmospheric carbon dioxide could account for the cooling trend, he considered both the possible effects of volcanic activity and nuclear tests. Volcanic dust, which he estimated had similar atmospheric effects to radioactive debris, could lower temperatures after volcanic eruptions.[102]

The assembled delegates in Rome noted, meanwhile, that over the past decade, droughts had occurred in the Polish lowlands, northeast Brazil, the US Great Plains, and the Canadian Prairie Provinces. The conference proceedings reveal the difficulties of understanding the causes of widespread climate variability at this time. Different disciplines were engaged in its study, ranging from plant ecology and vegetation history to meteorology and oceanography, while different theories abounded as to the scale and cause of variability over time—human, geophysical, or otherwise. Hence, in the words of the Ukrainian delegate, it was "extremely necessary to unite the efforts of research workers in this field under the direction of the World Meteorological Organization."[103]

Participants in Rome discussed a range of theories that might account for the climate fluctuations, including, but not limited to, the differing approaches of Callendar and Ahlmann. Callendar himself was alive to the ongoing speculation about his carbon dioxide theory. In a paper published early that year, he noted, "This matter of atmospheric CO_2 increase is highly controversial at the present

---

[98]"Applied Climatology – a 'Tool in the Freedom-from-Hunger Campaign," *WMO Bulletin* 10, no. 1 (1961): 14–17.

[99]See Gordon Cartwright, "Interview with C.C. (Carl Christian) Wallén, 21 November 1995, 9 December 1995," American Meteorological Society, https://v.gd/4vYwgx. On observations of changes in the "Australian region," see E. L. Deacon, "Climatic Change in Australia since 1880," *Australian Journal of Physics* 6, no. 2 (1953): 209–18.

[100]C.-C. Wallén, cited in J. Murray Mitchell, "Discussion," in *Changes of Climate*, 181.

[101]J. Murray Mitchell, "Recent Secular Changes of Global Temperature," *Annals of the New York Academy of Sciences* 95 (1961): 235–60; J. Murray Mitchell, "On the World-Wide Pattern of Secular Temperature Change," in *Changes of Climate*, 161–81.

[102]Matthias Dörries, "In the Public Eye: Volcanology and Climate Change Studies in the 20th Century," *Historical Studies in the Physical and Biological Sciences* 37, no. 1 (2006): 103.

[103]I. E. Buchinsky, "Climatic Fluctuations in the Arid Zone of the Ukraine," in *Changes of Climate*, 94.

time, and several authors ... have expressed doubt as to the possibility of a CO_2 increase approaching the amount ... added by fossil-fuel combustion."[104] Nor were the emissions of "the smoke-stacks of our factories and the exhaust pipes of our automobiles" necessarily a cause for concern: glaciologist James Dyson described carbon dioxide as a "gift" to the air in his popular account of ice core research in the wake of the IGY.[105] With his director Helmut Landsberg, Mitchell had meanwhile responded to Callendar's findings in the *Quarterly Journal of the Royal Meteorological Study*, noting that the dust hypothesis required closer attention.[106] In Rome, participating Danish meteorologist Leo Lysgaard, who had been with Ahlmann a proponent of "climate amelioration," considered that its causes were most likely the combination of "variation in solar radiation; urbanization (increase of carbon dioxide, water vapour, smoke and dust in the air); and volcanic eruptions."[107]

Although the scientists gathered in Rome were necessarily concerned with the implications of climate change and variability for the world's arid regions and their development, the wider climate fluctuations that informed their discussions remained a source of scientific curiosity and cautious optimism. Ahlmann's research, meanwhile, had been widely welcomed in Britain and Scandinavia for its anticipation of warmer conditions, and Soviet scientists were similarly hopeful.[108] A year earlier, the Soviet delegation at the 19th International Geographical Congress in Stockholm had presented Ahlmann with the Great Gold Medal of the Geographical Society of the USSR at the Soviet embassy.[109] Having attended Ahlmann's London lecture in 1945 and published his own assessment of contemporary climate trends, Fedorov joined with climatologist Mikhail Budyko to convene a meeting in mid-1961 to discuss Soviet meteorological science, including weather modification. In his address on the subject, Fedorov cited Budyko's calculations that the industrial production of thermal energy would eventually affect the planet's heat balance. Anticipating that this accumulation of energy would take "a century or two," Fedorov placed his faith in technological advancement and "assumed that by then this problem should not present any insurmountable

---

[104]G. S. Callendar, "Temperature Fluctuations and Trends over the Earth," *Quarterly Journal of the Royal Meteorological Society* 87, no. 371 (1961): 9.

[105]James L. Dyson, *The World of Ice* (New York: Knopf, 1962), 99, 187.

[106]H. E. Landsberg and J. M. Mitchell, "Correspondence: Temperature Fluctuations and Trends over the Earth," *Quarterly Journal of the Royal Meteorological Society* 87 (1961): 435–46.

[107]L. Lysgaard, "Recent Climatic Fluctuations," *Nature* 161 (1948): 442–3; L. Lysgaard, "On the Climatic Variation," in *Changes of Climate*, 157.

[108]See Sverker Sörlin, "The Anxieties of a Science Diplomat: Field Coproduction of Climate Knowledge and the Rise and Fall of Hans Ahlmann's 'Polar Wwarming'," *Osiris* 26, no. 1 (2011): 83.

[109]"Report of the 19th International Geographical Congress and 10th General Assembly of the International Geographical Union and the Participation of Soviet Geographers," *Soviet Geography* 2 (1961): 40–6.

difficulties."[110] On the matter of polar warming, Budyko speculated that once the ice cover of the Arctic disappeared, the resulting change in climate would not only prevent it from reforming but also provide favorable conditions for "the growth of temperate and subtropical vegetation along the shorelines and on the islands of the Arctic."[111] As far as the conference proceedings reveal, the role of carbon dioxide and its increasing concentration in the atmosphere was not a topic of Soviet discussions.

Speaking at the Rome meeting's conclusion, Wallén noted, "In this symposium we have not been able to treat the controversial question of man's possibilities to foresee the future development of or to influence climate by artificial means." He continued, "So much may be said today that it seems unlikely with our actual knowledge about the general circulation of the atmosphere that large-scale changes of climatic conditions will be possible by artificial means, but I wish to emphasize that we do not know how large can be the consequences of smaller influences induced by man."[112] In these comments, Wallén was gesturing to remarks by US president John F. Kennedy just weeks earlier. As construction of the Berlin Wall commenced, the president had addressed the UN General Assembly in Geneva. Acknowledging the progress of Soviet space exploration, the president advocated for "further cooperative efforts between all nations in weather prediction and eventually in weather control."[113]

Governing the Skies: Atmospheric Rivalries

In the months before President Kennedy delivered his UN address, Soviet scientists and their counterparts in the communist world had gathered in Leningrad to discuss matters of weather modification.[114] While US scientists had developed the means of weather modification by cloud seeding after the war, their Soviet counterparts had successfully undertaken the dispersal of fog and prevented hailstorms.[115] Weeks after the UNESCO conference in

---

[110]Y. K. Fedorov, "Status and Prospects for the Solution of the Problem of the Control of Weather and Climate," in M. I. Budyko (ed.), *Proceedings of the All-Union Scientific Meteorological Conference, vol. 1,* NASA Technical Translation (Washington, DC: NASA, 1964), 70; N. Rusin and L. Flit, *Man versus Climate,* trans. Dorian Rottenberg (Moscow: Peace Publishers, 1960),15.

[111]M. I. Budyko, "The Heat Balance of the Earth," in Budyko (ed.), *Proceedings,* 178–9.

[112]Wallén, "Aims and Methods," in *Changes of Climate,* 473.

[113]John F. Kennedy, "Address before the General Assembly of the United Nations, September 25, 1961," John F. Kennedy Presidential Library and Museum, https://v.gd/nYDFWc.

[114]Joint Publications Research Service, *Soviet-Bloc Research in Geophysics, Astronomy and Space,* no. 83 (Arlington: Joint Publications Research Service, 1964), 7–8.

[115]K. T. Loginov, "Opening Address of the Deputy Director of GUGMS," in Budyko (ed.), *Proceedings,* 40; Rusin and Flit, *Man versus Climate,* 168–9.

Rome, the 22nd Congress of the Communist Party Congress had adopted a new program, in which "the development of climate-control methods was listed among 'the most urgent problems of Soviet science'."[116] As the Soviet Union appeared to be winning the space race, with its launch of *Sputnik* and the failure of the Vanguard launch system, the United States had become increasingly concerned that its rival might also have a more advanced program of weather modification, even with its more limited computing power. Under such circumstances, US meteorologists such as MIT's Henry Houghton feared, "An unfavorable modification of our climate in the guise of a peaceful effort to improve Russia's climate could seriously weaken our economy and our ability to resist."[117]

By the end of 1961, the UN had passed a resolution on the peaceful uses of outer space, which included recommendations to advance the state of atmospheric science and technology in order to provide greater insight into the workings of the climate and to explore the possibility of large-scale weather modification. This regime drew on the Antarctic Treaty, signed in the wake of the IGY in 1959. Antarctica had been overlooked in the earlier polar years (in 1882–83 and 1932–33), but the synoptic vision of geophysics combined with Cold War geopolitics to cast the continent in a newly important light. Already, Ahlmann had instigated the 1949–52 Norwegian-British-Swedish expedition to determine whether the climatic changes he had observed near the Arctic were similarly underway in Antarctica.[118] Contained with the boundaries of Norway's Antarctic sector, the expedition enabled the participating nations to strengthen their claims over the ice and fend off other claimant states.[119]

On the eve of the IGY in February 1956, less than a year after the Bandung Conference of newly independent Asian and African nations, India's delegation attempted to address the "Antarctica Question" in the UN General Assembly.[120] This matter referred to the uncertain and contested governance of Antarctica, which was of utmost concern to the continent's claimant nations. Prime Minister Jawaharlal Nehru elaborated in the Lok Sabha: "We are not challenging anybody's rights there. ... [B]ut as it has become important

[116]Mikhail Budyko, 1962, cited in J. O. Fletcher, *Changing Climate* (Santa Monica: Rand Corporation, 1968), 18.

[117]Henry Houghton, "Other Aspects of Weather Modification: Present Position and Future Possibilities of Weather Control," in Advisory Committee on Weather Control, *Final Report of the Advisory Committee on Weather Control, Volume II* (Washington, DC: Government Printing Office, 1957), 288.

[118]Sverker Sörlin, "Circumpolar Science: Scandinavian Approaches to the Arctic and the North Atlantic, ca. 1920 to 1960," *Science in Context* 27 (2014): 275–305.

[119]Simon Naylor, Katrina Dean, and Martin Siegert, "The IGY and the Ice Sheet: Surveying Antarctica," *Journal of Historical Geography* 34, no. 4 (2008): 574–95.

[120]"Letters Dated 17 February 1956 Addressed to the Secretary-General by the Permanent Representative of India to the United Nations," A/3118, United Nations, https://v.gd/PT5Xjp.

and more especially because of the ... possible experimentation of atomic weapons and the like, we feel that the matter should be considered by the UN and not be left in a slightly chaotic stage with various countries trying to grab it."[121] India's Permanent Representative to the UN also indicated that his government feared that nuclear tests would affect the vitally important monsoon.[122] Although the Latin American bloc convinced India to withdraw its proposal, the episode indicated to the United States and the seven claimant nations that an Antarctic regime had become necessary to cement their claims and to deter other contenders.

New Delhi was not alone in its concern about nuclear weapons tests in Antarctica. As the IGY got underway, Canberra too became fearful that Soviet bases being established in the continent's Australian sector might become launch sites for nuclear weapons.[123] Australia's weather might even "be altered by artificially changing it down south," one influential pundit wondered.[124] By now, the United States had also established three bases, including the prestigious South Pole, where Keeling would undertake his carbon dioxide measurements. Having met in late 1957 with officials from Australia, Aotearoa New Zealand, and the UK, the United States invited the foreign ministers of the claimant nations, as well as Japan, Belgium, West Germany, and the Soviet Union to a conference to discuss a political solution for Antarctica that would allow "a continuation of the international scientific operation which is being carried out so successfully during the IGY."[125] The rhetoric of scientific internationalism became the grounds to negotiate the Antarctic Treaty and to prevent the militarization of the southern continent.

The adoption of the outer space treaty two years later in December 1961 similarly cast the atmosphere as another environment in which to pursue scientific internationalism. The UN resolution encouraged further investigation into both the "basic physical forces" affecting the climate and the possibility of large-scale weather modification. It also identified the need to develop weather forecasting capabilities and encouraged member states to make use of those capabilities. The world's first weather satellite, TIROS-I (Television Infrared Observation Satellite), launched by the United States the previous year, had already demonstrated the value of such technology to weather forecasting and the collection of meteorological data.[126] Gesturing to the Soviet

[121]Jawaharlal Nehru, *14 April 1956, Lok Sabha Debates: Part I – Questions and Answers, Vol. II, 1956, Twelfth Session* (New Delhi: Lok Sabha Secretariat, 1956), 2212.

[122]Adrian Howkins, "Defending the Polar Empire: Opposition to India's Proposal to Raise the 'Antarctica Question' at the United Nations in 1956," *Polar Record* 44 (2008): 35–44.

[123]Simone Turchetti et al., "On Thick Ice: Scientific Internationalism and Antarctic Affairs, 1957–1980," *History and Technology* 24 (2008): 351–76.

[124]Donald Horne, "We Warn the Tsar," *Observer*, February 22, 1958, 1.

[125]Cited in Turchetti et al., "On Thick Ice," 359.

[126]Edwards, *A Vast Machine*, 221; Roger D. Launius, "'We Will Learn More about the Earth by Leaving It than By Remaining on It': NASA and the Forming of an Earth Science Discipline in

Union that the United States had only peaceful interests in outer space, the satellite's launch also reinforced the postwar ethos of modernization through the application of scientific research.

As their nations locked horns over Cuba, Wexler (now the chief of the US Weather Bureau) and Soviet academician Victor Bugaev of the Hydrometeorological Service met in Geneva. Joining them were dignitaries from the WMO, as well as representatives of the ICSU, UNESCO, and the International Union of Geodesy and Geophysics. Together, they embarked on what Paul Edwards calls a "technopolitics of altitude"—the use of technology to achieve political goals in relation to the upper atmosphere and outer space.[127] Their negotiations culminated in 1963 in the WMO's World Weather Watch program, which coordinates the efforts of member states by combining international oceanic and space-based observing systems, telecommunication facilities, and data-processing and forecasting centers to make available meteorological and related environmental information to provide weather services in all countries.[128] As Wallén had earlier predicted in his closing remarks in Rome, statistical analyses of the program's satellite data would soon "play a dominant role in the future development of studies of the causes of climate changes."[129]

Two of the program's World Meteorological Centers were stationed in the capitals of the United States and the Soviet Union (the only countries with the financial resources for expensive satellite programs), while a third in Melbourne, Australia, was established to provide information from the data-sparse southern hemisphere and to offset the influence of the Cold War superpowers.[130] During the IGY, the South African Weather Bureau had been responsible for analyzing data from the Southern Hemisphere, but the rise of apartheid afforded Australia the opportunity to become "the leading meteorological nation" in the region.[131] Bill Gibbs of the Commonwealth Bureau of Meteorology had been elected to the WMO's executive committee in 1963, and he understood Australia's rising influence in geopolitical terms: "Although Australia is clearly aligned with the west she nevertheless has the image of a small, rapidly developing, independent nation which not too recently has been relieved of the 'colonial yoke'. This image appeals to the newly-independent

the 1960s," in Thomas Heinz and Richard Muench (eds.), *Innovation in Science and Organizational Renewal: Historical and Sociological Perspectives* (New York: Palgrave, 2016), 211–42.

[127]Harry Wexler, "Global Meteorology and the United Nations," *Weatherwise* 15 (1962): 141–67; Edwards, *A Vast Machine*, 225.

[128]Fleming, *Fixing the Sky*, 215–16.

[129]Wallén, "Aims and Methods," in *Changes in Climate*, 472.

[130]Edwards, *A Vast Machine*, 232–3.

[131]John Zillman, "A Hundred Years of Science and Service – Australian Meteorology through the Twentieth Century," in D. Trewin (ed.), *Year Book Australia 2001*, no. 83 (Canberra: Australian Bureau of Statistics, 2001), 22–50.

nations of Africa. Furthermore, Australia is not aligned overtly with one of the blocs in W.M.O."[132] Partly on account of his own cordial relations with other meteorological services, Gibbs believed there was "the general recognition of Australia as the leading meteorological service in the southern hemisphere and the spokesman for the other services in this hemisphere."[133]

Conclusion

These discussions across the agencies of the UN during the final months of 1961 reflected a transition underway in the scientific framing and study of climates. Wallén's symposium in Rome reflected a well-established geographical approach of local and regional empirical research, while the 1961 UN resolution on outer space was shaped by an emerging geophysical outlook that was increasingly quantitative, planetary in scope, computer-driven, and future-focused.[134] As this book reveals, the latter's ascent would only continue, as the geopolitical concerns that had primarily prompted the outer space treaty gave way to environmental anxieties about the social, economic, and ecological effects of human-caused climate variability and global change.

The attentions of scientists, policymakers, and commentators would also shift toward the causes of such change and the degree to which humankind could alter the composition of the atmosphere, and, thus, the planet's climate. By decade's end, as the next chapter shows, the old outlook discussed at Wallén's meeting would mostly give way to the new with the development of the computer climate model that combined, or "coupled," the ocean and atmosphere. Humans, who had long been understood to affect changes to climates at local scales through changes to vegetation, were now believed to have the means of planetary influence.

---

[132] "World Meteorological Organization: Sixteenth Session of the Executive Committee – 1964," World Weather Watch and World Centre at Melbourne, A1209, 1970/6162, National Archives of Australia (hereafter, NAA) Canberra.
[133] "Confidential Report on Overseas Visit by Director of Meteorology, June 1964," World Weather Watch and World Centre at Melbourne, A1209, 1970/6162, NAA Canberra.
[134] Matthias Heymann, "The Evolution of Climate Ideas and Knowledge," *WIREs Climate Change* 1 (2010): 581–97.

2

Anxious Air

1972: 327.45 ppm

Within months of the resumption of atmospheric nuclear tests in April 1962, the *New Yorker* magazine began its serialization of Rachel Carson's *Silent Spring*. Her expose of the "unseen and invisible" pollution of synthetic pesticides and their insidious effects on the "vast web of life" prompted concern about the organochlorine DDT .[1] Yet it had been strontium-90, not a pesticide, that Carson first described to explain the way that pesticides entered the soil "in a chain of poisoning and death."[2] After the detonation of the bomb implicated in the *Lucky Dragon*'s misfortune, Japanese scientists had identified the presence of the long-lasting radioactive isotope strontium-90 in the boat's fish catch—already known to accumulate in soil, milk, and human and nonhuman bones, and possibly lead to cancer. Just a few years after the publication of *Silent Spring*, biologists identified pesticides in Antarctic penguins and seals, indicating the extent of the atmospheric and ocean transport of such chemicals.[3] The air was full of anxiety.

Carson's sentiments, and those her work aroused in others, resonated with an emerging discontent in Western societies. J. K. Galbraith's *Affluent Society*, Jean Dorst's *Before Nature Dies*, Rolf Edberg's *On the Shred of a Cloud*, Barry Commoner's *Science and Survival*, Jock Marshall's *The Great Extermination*—all gestured to growing concerns with the environmental and moral consequences of industrialization. Their outlook contrasted sharply with the dawn of the "UN Decade of Development" under the newly appointed Secretary-General U Thant of Burma, the first to represent the "Third World."

---

[1] Rachel Carson, *Silent Spring* (Boston: Houghton Mifflin, 1962), 41, 293.

[2] Carson, *Silent Spring*, 6; Ralph Lutts, "Chemical Fallout: Rachel Carson's *Silent Spring*, Radioactive Fallout, and the Environmental Movement," *Environmental Review* 9 (1985): 210–25.

[3] Rachel Rothschild, *Poisonous Skies: Acid Rain and the Globalization of Pollution* (Chicago: University of Chicago Press, 2019), 23. See, for example, William J. L. Sladen, C. M. Menzie and W. L. Reichel, "DDT Residues in Adelie Penguins and a Crabeater Seal from Antarctica," *Nature* 210 (1966): 670–3; J. L. George and D. E. H. Frear, "Pesticides in the Antarctic," *Journal of Applied Ecology* 3 (1966): 155–67; J. O'G. Tatton and J. H. A. Ruzicka, "Organochlorine Pesticides in Antarctica," *Nature* 215 (1967): 346–8.

This program aimed for the developing countries to achieve at least 5 percent annual growth rates by 1970.[4] Guiding this vision was the 1960 "non-communist manifesto" of Walt Rostow, in which he argued that the "passage of a traditional to a modern growing society" required the recognition of the physical environment "as an ordered world which, if rationally understood, can be manipulated in ways which yield productive change … and progress."[5]

For Rostow and his kin, the modern state deployed science and technology to dominate its environment and maximize its resources. Yet further scientific evidence, measurement, and modeling suggested that some human activities were already impacting the earth and each other. The atom had not only made the atmosphere and other processes newly legible, but it had also revealed the atmosphere to be newly vulnerable to human activities. The revelations of strontium-90 had likewise revealed that human bodies were not immune to the changes in the air. New ways of conceptualizing the earth as a closed, interconnected system invited new ways of understanding the relationships between humans and their wider environment.

A role for international organizations in mediating these relationships quickly emerged. Environmental problems could no longer be understood as locally confined within borders; they became recast as shared and in need of careful international stewardship. As newly independent countries swelled the ranks of the United Nations (UN), however, the seemingly common ground of environmental concern proved less unifying than the internationalists of the developed world had anticipated.

Ecology and the Atmosphere: Articulating Interdependence in a Crowded World

Dave Keeling's findings from the South Pole and Mauna Loa found an engaged audience at a modest New York meeting convened by the Conservation Foundation in March 1963. The subsequent report brought his results into conversation not only with contemporary concerns about world population growth and resource exploitation, but also with the nascent framing (in the Anglophone world at least) of the "biosphere." Among the small coterie of participants in the New York meeting were Keeling, Erik Eriksson, and Gilbert

---

[4]Michael Adas, *Machines as the Measure of Men* (Ithaca, NY: Cornell University Press, 1989), 413–14; Nils Gilman, *Mandarins of the Future: Modernization Theory in Cold War America* (Baltimore, MD: Johns Hopkins University Press, 2003), 5; Stephen Macekura, *Of Limits and Growth: The Rise of Global Sustainable Development in the Twentieth Century* (New York: Cambridge University Press, 2015), 142–3.
[5]Walt Rostow, *The Stages of Economic Growth: A Non-Communist Manifesto* (New York: Cambridge University Press, 1960), 19.

Plass, who was now working for the Ford Motor Company. The subsequent report on the "implications of rising carbon dioxide content of the atmosphere" centered Keeling's findings from the data he had collected from Mauna Loa and the South Pole, and it called for further research funding to sustain his own research program as well as to broaden the study to include other countries.[6] Increasing levels of carbon dioxide would raise the temperature of the earth, warned the report; an increase of 3.5°C, for instance, would melt the polar ice caps, inundating coastal areas, including the cities of New York and London, while devastating terrestrial and marine life near the equator.[7]

Cofounded in 1948 by Henry Fairfield Osborn Jr., the Conservation Foundation pursued the ecological problems that the well-connected Wall Street banker had emphasized in his recent book, *Our Plundered Planet*. Published in the same year, the book warned "all who care about tomorrow" that "man's destructiveness has turned not only upon himself but also upon his own good earth—the wellspring of his life."[8] Osborn Jr. returned to such neo-Malthusian themes in his *Limits of the Earth* (1953) and the 1962 edited collection, *Our Crowded Planet*, which included essays by scientists, historians, writers, and others sympathetic to liberal internationalism on both sides of the Atlantic. Among them were John Boyd Orr, the first director general of the Food and Agriculture Organization; Solly Zuckerman, a British scientific advisor responsible for coining the term "environmental sciences" to describe the study of natural resources; and Julian Huxley, by 1962 the president of the British Eugenics Society and cofounder of the newly established World Wildlife Fund for Nature.[9]

Huxley had earlier invited Osborn Jr. to represent the UN Educational, Scientific and Cultural Organization (UNESCO) at the first meeting of the UN on environmental matters, the 1949 Scientific Conference on the Conservation and Utilization of Resources at its temporary headquarters at Lake Success, New York. That conference had its origins in the Point Four Program of US president Harry Truman, who had envisioned that such a meeting might foster resource conservation as "a major basis for peace" and "a major basis for

---

[6]Conservation Foundation, *Implications of Rising Carbon Dioxide Content of the Atmosphere: A Statement of Trends and Implications of Carbon Dioxide Research Reviewed at a Conference of Scientists* (New York: Conservation Foundation, 1963), 11, 13. Securing funding for the carbon dioxide measurement program was an ongoing struggle for Keeling, see C. D. Keeling, "Rewards and Penalties of Monitoring the Earth," *Annual Reviews of Energy and Environment* 23 (1998): 25–82.
[7]Conservation Foundation, *Implications of Rising Carbon Dioxide Content*, 6.
[8]Henry Osborn Jr., *Our Plundered Planet* (Boston: Little Brown, 1948), 11.
[9]On Zuckerman, see Paul Warde and Sverker Sörlin, "Expertise for the Future: The Emergence of Environmental Prediction, c. 1920–1970," in Jenny Andersson and Eglė Rindzevičiūtė (eds.), *The Struggle for the Long-Term in Transnational Science and Politics: Forging the Future* (London: Routledge, 2015), 38–62.

world prosperity," particularly "for the benefit of under-developed areas."[10] As director general of UNESCO, Huxley was otherwise occupied with his own rival conference in the same town. Having mandated the previous year a new organization, the International Union for the Protection of Nature, which understood its mission as "the preservation of the entire world biotic community," UNESCO sponsored the International Technical Conference on the Protection of Nature.[11] Its purpose was to counter the "wise use" or efficient approach to conservation that was guiding postwar development and modernization, in favor of an ecologically informed respect for "the living communities which form [a population's] environment and from which sustenance is derived."[12] Huxley's UNESCO conference consequently warned scientists that "the idea that all phenomena is actually one phenomenon and that an abrupt change in any of the factors in play can only have profound repercussions on the complex whole even if he has not been able to anticipate the repercussions in his imagination."[13]

In convening the UNESCO conference, Huxley had been assisted by American ornithologist William Vogt, whose *Road to Survival* was published just prior to Osborn Jr.'s *Our Plundered Planet*. Head of the Conservation Section of the Pan-American Union, Vogt drew on his work in Central America to argue for the "rational use of land" and "control of man's demand for fruits of the earth" by limiting population.[14] Later a director of the Planned Parenthood Federation of America and secretary of Osborn Jr.'s Conservation Foundation, Vogt warned against "elementalistic" approaches to the world's many postwar challenges, which he understood as "dynamically interrelated." "We form an earth-company," he argued, "and the lot of the Indiana farmer can no longer be isolated from that of the Bantu ... An eroding hillside in Mexico or Yugoslavia affects the living standard and probability of survival of the American people."[15] For Western Europeans and North Americans,

---

[10]Harry Truman, cited in "Background and Objectives of the Conference," *Proceedings of the United Nations Scientific Conference on the Conservation and Utilization of Resources, vol. 1 Plenary Meetings* (Lake Success: UN Dept of Economic Affairs, 1950), vii.

[11]"Constitution: Preamble," *Conference for the Establishment of the International Union for the Protection of Nature, Fontainebleau, France, 30 September – 7 October 1948*, NS/UIPN/12+Annex 1, UNESCO, https://v.gd/cEqFln. See Paul Warde, Libby Robin, and Sverker Sörlin, *The Environment: A History of the Idea* (Baltimore, MD: Johns Hopkins University Press, 2018), 73.

[12]Jean-Paul Harroy, "Introduction," in *Proceedings and Papers of the International Technical Conference on the Protection of Nature, Lake Success, 22–29 August, 1949* (Paris: 1950), ix; Thomas Jundt, "Dueling Visions for the Postwar World: The UN and UNESCO 1949 Conferences on Resources and Nature, and the Origins of Environmentalism," *Journal of American History* 101, no. 1 (2014): 44–70.

[13]Harroy, "Introduction," ix; Thomas Robertson, "'This is the American Earth': American Empire, the Cold War, and American Environmentalism," *Diplomatic History* 32 (2008), 561–84.

[14]William Vogt, *Road to Survival* (New York: William Sloane, 1948), 107.

[15]Vogt, *Road to Survival*, 53–4, 285.

environmental threats abroad necessarily imperiled their lives at home; their international cooperation and transnational activism stemmed thus from their concerns as to how global changes would impact the well-being of their own nations.

Echoing similar ideas in foreign policy and economics such as liberal internationalism and Keynesianism, Huxley, Osborn Jr., and their ilk stressed the need for the wider recognition of interconnection, or the "interdependence of all living things."[16] These trans-Atlantic intellectuals together drew on a legacy of cosmopolitan interwar thought in the United States, Germany, India, and Britain. In doing so, they brought ecology to bear on the world population problem, while fostering the postwar ideals of world citizenship and scientific humanism on which the universal concept of "Mankind" was founded.[17] In botanist Arthur Tansley's 1935 term "ecosystem," by which humans could be understood as organisms that belonged to integrated systems, these thinkers had found an ecological concept befitting of their atomic age.[18]

What had been a seemingly "soft" science now hardened, as ecologists seeking social relevance turned to quantitative and computational techniques to analyze living systems as akin to machines. By the 1950s, they had begun to apply these techniques to the study of the processes that defined ecosystems: the cycling of nutrients or chemicals within those systems and the flow of energy through the system. Leading the way in this regard was Yale ecologist George Evelyn Hutchinson. A participant in the postwar interdisciplinary Macy Conferences in New York with others, including anthropologist Margaret Mead and physicist John von Neumann, he would become well versed in the new concept of cybernetics.[19] In quantitative terms, Hutchinson followed carbon through a biogeochemical system, applying his calculations in 1948 to "a slight rise in the CO_2 of the air, at least in the northern hemisphere."[20] His demonstration showed how scientists might determine if human activity was affecting the regulatory mechanisms that maintained the stability of the composition of the atmosphere.[21]

---

[16]Osborn Jr., *Our Plundered Planet*, xii.

[17]Alison Bashford, *Global Population: History, Geopolitics, and Life on Earth* (New York: Columbia University Press, 2014), 157–9; Glenda Sluga, "UNESCO and the (One) World of Julian Huxley," *Journal of World History* 21 (2010): 393–418; Jenny Andersson and Sibylle Duhaotois, "Futures of Mankind: The Emergence of the Global Future," in Rens van Munster and Casper Sylvest (eds.), *The Politics of Globality since 1945: Assembling the Planet* (London: Routledge, 2016), 106–25.

[18]Sharon Kingsland, *The Evolution of American Ecology, 1890–2000* (Baltimore, MD: Johns Hopkins University Press, 2005), 184–5.

[19]Peter Taylor, "Technocratic Optimism, H.T. Odum and the Partial Transformation of Ecological Metaphor after World War II," *Journal of the History of Biology* 21, no. 2 (1988): 213–44.

[20]G. Evelyn Hutchinson, "Circular Causal Systems in Ecology," *Annals of the New York Academy of Sciences* 50 (1948): 223.

[21]Kingsland, *The Evolution of American Ecology*, 187.

To Hutchinson, these global, self-regulatory cycles were the mechanisms of the "biosphere," a concept that he had introduced to English-language readers in the final months of the war. The biosphere had been earlier developed by Russian biochemist Vladimir Vernadsky in the 1920s to describe the "domain of life," including "the whole atmospheric troposphere, the oceans, and a thin layer in the continental regions." Borrowing from Austrian geologist Eduard Suess and influenced by Svante Arrhenius, Vernadsky had preempted the systems thinking that would prove so influential in the early decades of the Cold War: "No living organism exists on earth in a state of freedom. All organisms are connected indissolubly and uninterruptedly."[22] Ecological analysis, as historian Sharon Kingsland notes, now "provided a way to link local human actions to large-scale global processes."[23]

The biosphere, then, was a concept ideally suited to the synoptic vision of the participants at the 1955 meeting at Princeton University, *Man's Role in Changing the Face of the Earth*. Among them were the Conservation Foundation's Osborn Jr. and Frank Fraser Darling, who had already invited Hutchinson to serve on the Conservation Foundation's advisory council.[24] Hutchinson was later responsible for organizing the 1963 meeting to which Keeling had shared his findings. The resulting report articulated the relevance of the atmospheric changes Keeling was documenting to the concerns of the Conservation Foundation's membership. In addition to the emission of carbon dioxide from fossil fuels, Hutchinson's report noted that the "increase in land in agriculture which is accompanying the increase in the world's population is almost surely decreasing the total terrestrial biomass." Pointing to the toll of agriculture on tropical forests, the report emphasized the importance of such vegetation for "controlling the rise in CO_2," thanks to photosynthesis.[25]

The participants in the Conservation Foundation meeting understood the rising atmospheric levels of carbon dioxide as representative of broader trends of "man's ability to change the environment." This ability, they believed, was already producing diminishing returns: "Overdevelopment and concomitant overpopulation in many areas of the world have made the problem not one of increasing productivity but one of preventing further decreases." Rejecting the prevailing modernization agenda, the report concluded, "In terms of the health of the planet there are no under-developed countries, only an increasing number of overdeveloped ones."[26] This pessimistic tone arising from the

---

[22]Vladimir Vernadsky, "The Biosphere and the Nöösphere," trans. G. Evelyn Hutchinson, *American Scientist* 33, no. 1 (1945): 1, 4.
[23]Kingsland, *The Evolution of American Ecology*, 187.
[24]Nancy Slack, *G. Evelyn Hutchinson and the Invention of Modern Ecology* (New Haven: Yale University Press, 2010), 297.
[25]Conservation Foundation, *Implications of Rising Carbon Dioxide Content*, 9.
[26]Conservation Foundation, *Implications of Rising Carbon Dioxide Content*, 15.

Conservation Foundation's meeting contrasted with the report of the UN Conference on the Application of Science and Technology for Less Developed Areas held earlier that year in Geneva. Convened for the UN's Development Decade, its proceedings noted that the climatic changes wrought by "Man's interference with the atmosphere ... have to be understood and foreseen," rather than approached fatalistically.[27]

Intervening in the Atmosphere: Air Pollution and Weather Modification

Among the papers shared at Princeton in 1955 were two that considered intentional and unintentional forms of weather modification. The first, by Vincent Schaefer, discussed cloud seeding, while Helmut Landsberg of the US Weather Bureau reflected on the "climate of towns." Their contributions shed light on the possibilities and problems that Cold War science and technology offered in terms of development, industrialization, and urbanization. "Almost all climatic elements are affected by pollution," Landsberg explained, citing Belgium's Meuse Valley (1930), Pennsylvania's Donora (1948), and London (1952) as just several examples of a "continuous, insidious process" of escalating pollution in Western European and North American towns and cities.[28] Just as Carl-Gustav Rossby had earlier advocated the reduction of urban air pollution as "the most practical step cities can now take toward weather control,"[29] Landsberg suggested that urban planning and "adequate community action against air pollution" could "lead to at least a tolerable if not optimal bioclimate for the inhabitants."[30]

In Western Europe, concerns about urban air composition had already prompted the Organisation for European Economic Cooperation (OEEC, precursor to the Organisation for Economic Co-operation and Development (OECD)) to establish in 1957 a working group to measure the different components of air pollution.[31] Formed to distribute Marshall Plan funds and to

---

[27]United Nations, *Science and Technology for Development: Report on the UN Conference on the Application of Science and Technology for the Benefit of the Less Developed Areas, Vol. 1* (New York: United Nations, 1963), 7.

[28]H. E. Landsberg, "The Climate of Towns," in William Thomas (ed.), *Man's Role in Changing the Face of the Earth* (Chicago: University of Chicago Press, 1956), 586.

[29]"Science: Weather Control?," *Time*, July 5, 1943, https://v.gd/0y1ZAl; see Rachel Rothschild, "Burning Rain: The Long-Range Transboundary Air Pollution Project," in James R. Fleming and Ann Johnson (eds.), *Toxic Airs: Body, Place, Planet in Historical Perspective* (Pittsburgh: University of Pittsburgh Press, 2014), 183–4.

[30]Landsberg, "The Climate of Towns," 603.

[31]Rothschild, *Poisonous Skies*, 20.

encourage the economic integration of Western European states, the OEEC was also a means of international scientific exchange across the Atlantic in order to study shared problems.[32] In the wake of the Great London smog of 1952, which had claimed the lives of over four thousand people, air pollution had become increasingly understood in terms of monetary loss—"leaving aside the accompanying human misery," as a World Health Organization report put it.[33] Focused on the visible smoke associated with burning coal, Britain's Clean Air Act of 1956 was an early attempt to ameliorate the human as well as economic and environmental toll of postwar urbanization and industrialization.[34]

In the United States, physicist Edward Teller had noted such concerns about the "smoke and smog" of coal and oil when he presented a lecture to the "Energy and Man" symposium, co-organized by the Columbia Graduate School of Business and the American Petroleum Institute in November 1959. Invited to help mark the centenary of the oil business in the United States, Teller advocated for the use of nuclear energy by drawing the attention of the conference to a pollutant of a different kind, "invisible, transparent, you can't smell it, it is not dangerous to health, so why should one worry about it." That pollutant—carbon dioxide—was accumulating in the atmosphere such that "a temperature rise corresponding to a 10 per cent increase in carbon dioxide will be sufficient to melt the icecap and submerge New York," he warned.[35]

This was not the first time the American Petroleum Institute had engaged with the prospect of rising carbon dioxide levels. The organization had earlier commenced funding a study by geochemist Harrison Brown (another participant of the 1955 Princeton meeting and Manhattan Project veteran), whose research proposal had noted the rising atmospheric concentration of carbon dioxide due to industrialization.[36] This proposal had coincided with his publication in 1954 of *The Challenge of Man's Future*, in which he had speculated that increasing threefold the carbon dioxide content of the atmosphere might double the growth rates of crops and thus double world food production. Wary of the prospect of unchecked population growth,

[32]Alan S. Milward, "The Reconstruction of Western Europe," in Charles Maier (ed.), *The Cold War in Europe: Era of a Divided Continent* (Princeton: Markus Wiener, 1991), 241–70; John Krige, *American Hegemony and the Postwar Reconstruction of Science in Europe* (Cambridge, MA: MIT Press, 2006), 15–39.

[33]Rothschild, *Poisonous Skies*, 20; World Health Organization, *Air Pollution: Fifth Report of the Expert Committee on Environmental Sanitation* (Geneva: World Health Organization, 1958), 13.

[34]Peter Thorsheim, *Inventing Pollution: Coal, Smoke and Culture in Britain since 1800* (Athens: Ohio University Press, 2018).

[35]Edward Teller, "Energy Patterns of the Future," in Columbia University Graduate School of Business and the American Petroleum Institute, *Energy and Man: A Symposium* (New York: Appleton Century Crofts, 1960), 58.

[36]Harrison Brown et al., 1954, cited in Ben Franta, "Correspondence: Early Oil Industry Knowledge of CO_2 and Global Warming," *Nature Climate Change* 8 (2018): 1024–6.

Brown was doubtful that such a scenario could unfold: "To double the amount [of carbon dioxide] in the atmosphere, at least 500 billion tons of coal would have to be burned—an amount six times greater than that which has been consumed during all of human history."[37] Besides, he noted, the oceans would absorb the excess carbon dioxide "at an appreciable rate"; his postdoctoral student at Caltech, Keeling, would soon prove that assumption wrong.[38]

Air pollution and carbon dioxide emissions were just two of the issues considered by the US president's Science Advisory Committee in 1965. Earlier that year, President Lyndon Johnson had told Congress, "Air pollution is no longer confined to isolated places. This generation has altered the composition of the atmosphere on a global scale through radioactive materials and a steady increase in carbon dioxide from the burning of fossil fuels."[39] He subsequently tasked the committee to assess the extent and effects of air, soil, and water pollution in the United States, and to recommend strategies for their amelioration in the context of his vision for the Great Society.[40] The committee's Environmental Pollution Panel comprised five subpanels, including one to address the question of "Atmospheric Carbon Dioxide." Chaired by Roger Revelle (who had become head of the new Harvard Center for Population and Development Studies), the subpanel comprised Keeling, meteorologist Joseph Smagorinsky, geologist Wallace Broecker, and geochemist Harmon Craig. Their short report on the "invisible pollutant" showed not only Keeling's rising atmospheric levels of carbon dioxide, but also their direct relationship to the production of fossil fuels, particularly the increasing output of cheap Middle East petroleum, so important to Western Europe's reconstruction after the Suez crisis.[41]

Noting the need for further mathematical modeling using computer technology, the report concluded tentatively, "The climatic changes that may be produced could be deleterious from the point of view of human beings." Deliberate weather modification could also offer the possibility of

---

[37] Harrison Brown, *The Challenge of Man's Future* (New York: Viking Press, 1954), 142–3.

[38] Wallace S. Broecker and Robert Kunzig, *Fixing Climate: What Past Climate Changes Reveal about the Current Threat – and How to Counter It* (New York: Hill and Wang, 2008), 74–5.

[39] Lyndon Johnson, "Special Message to the Congress on Conservation and Restoration of Natural Beauty," February 8, 1965, *The American Presidency Project*, https://v.gd/Yzz6ea.

[40] See Adam Rome, "'Give Earth a Chance': The Environmental Movement and the Sixties," *Journal of American History* 90, no. 2 (2003): 525–54; Ronald E. Doel and Kristine C. Harper, "Prometheus Unleashed: Science as a Diplomatic Weapon in the Lyndon B. Johnson Administration," *Osiris* 21, no. 1 (2006); Martin V. Melosi, "Environmental Policy," in Mitchell Lerner (ed.), *A Companion to Lyndon B. Johnson* (Chichester: Wiley Blackwell, 2011), 187–209.

[41] Environmental Pollution Panel, *Restoring the Quality of Our Environment: Report of the Environmental Pollution Panel of the President's Science Advisory Committee* (Washington, DC: White House, 1965), 112; Christian Pfister, "The '1950s Syndrome' and the Transition from a Slow-Going to a Rapid Loss of Global Sustainability," in Frank Uekötter (ed.), *The Turning Points of Environmental History* (Pittsburgh: University of Pittsburgh Press, 2010), 90–118.

"countervailing climatic changes" to those wrought by burning fossil fuels.[42] Revelle himself was not overly concerned by these findings, noting afterward, "Our attitude toward the changing content of carbon dioxide in the atmosphere that is being brought about by our own actions should probably contain more curiosity than apprehension."[43] Some quarters were apprehensive about the findings of the president's Science Advisory Committee, however. The president of the American Petroleum Institute, for instance, observed that its report would "unquestionably ... fan emotions, raise fears, and bring demands for action," anticipating the pollution control legislation that followed.[44]

That rising levels of atmospheric carbon dioxide might inadvertently affect the weather was now drawing the attentions of the US National Academy of Sciences and the National Science Foundation. The latter's 1965 report pointed to the inadvertent weather modification that might result from increasing atmospheric carbon dioxide, and it called for the further development of general circulation models to forecast those changes more accurately.[45] Deliberate weather modification, meanwhile, continued to be a preoccupation of US meteorologists both at home and abroad.[46] In South Asia and Indochina, covert weather modification programs sought to turn the tide against famine and communism. When India's newly appointed prime minister Indira Gandhi visited Washington in early 1966, President Johnson called for Congress to support emergency food aid to her country: "A sister democracy will not suffer the terrible strains which famine imposes on free government."[47] The recent memory of the Bengal famine of 1943–44 had loomed large over India. In 1955, the Indian-born plant geneticist Janaki Ammal had told her all-male colleagues at the Princeton meeting on *Man's Role in Changing the Face of the Earth* that "Independent India is making great strides to increase productivity of land already under cultivation; the first five-year plan made India self-sufficient in food supply."[48]

By the early 1960s, however, India's fortunes had turned as industrialization outpaced agricultural production. Confrontation with China and then Pakistan,

---

[42]Environmental Pollution Panel, *Restoring the Quality*, 127.

[43]Roger Revelle, "The Role of the Oceans," *Saturday Review*, May 7, 1966, 41.

[44]Frank Ikard, "Meeting the Challenges of 1966," in *Proceedings of the American Petroleum Institute 1965*, American Petroleum Institute, 1966, 12–15, archived in Climate Files, https://v. gd/3SvTCC; Franta, "Correspondence," 1024–6.

[45]Donald Gilman, "The State of the Science and Possibilities for Research," in Donald Gilman, James Hibbs, and Paul Laskin (eds.), *Weather and Climate Modification: A Report to the Chief, US Weather Bureau* (Washington, DC: Weather Bureau, 1965), 7.

[46]James R. Fleming, *Fixing the Sky: The Checkered History of Weather and Climate Control* (New York: Columbia University Press, 2010), 177.

[47]Lyndon Johnson, "Special Message to the Congress Proposing an Emergency Food Aid Program for India," March 30, 1966, The American Presidency Project, https://v.gd/FtflyQ.

[48]E. K. Janaki Ammal, "Introduction to the Subsistence Economy of India," in Thomas (ed.), *Man's Role in Changing the Face of the Earth*, 332.

rapid population growth, the death of Nehru and then his successor, as well as the failure of the monsoon rains in 1965 endangered what Rostow had only recently seen as the young nation's "take-off."[49] The subcontinent was at the center of the Cold War struggle for Asia: its successful development and modernization would be proof positive of the virtues of democracy. Its looming food crisis alarmed the likes of biologist Paul Ehrlich and food expert Lester Brown, whose environmentalism was profoundly shaped by India's challenges.[50] Having earlier suspended such aid in response to the conflict with Pakistan, the United States began a two-year program of daily food aid as part of Johnson's "International War on Hunger." Meanwhile, the US government canvassed the prospects for weather modification to alleviate the drought and famine conditions in the provinces of Bihar and eastern Uttar Pradesh near Pakistan's border. Alive to their neighbor's sensitivities, wary of the intentions of the United States, and mindful of false hope, Gandhi's government insisted this effort remain secret. The onset of monsoon rains combined with the fertilizer and new crop varieties of the Green Revolution averted disaster.[51]

In Indochina, the Johnson administration was already undertaking a similar mission. Following French attempts to disrupt Vietminh resistance in the mid-1950s with artificial rain, the United States had embarked in the early 1960s on a long mission of cloud seeding over Laos, Cambodia, and Vietnam to prolong the monsoon season and flood North Vietnam's supply routes.[52] After newspapers exposed these activities in Indochina a decade later in 1974, critics likened them to "a meteorological Watergate."[53] Castro's Cuba too was alleged to have been a target of US weather modification: according to Defense Department consultant Lowell Ponte, cloud seeding had been deployed to divert rains from sugar crops to the sea to disrupt the harvest and demoralize the nation.[54] The rainmaking efforts over the Ho Chi Minh trail enflamed Cold War anxieties that geophysical warfare, like nuclear warfare, was not only possible but also imminent. Emboldened by President Nixon's demise, the Soviet Union brought to the UN in 1974 a proposal to prohibit environmental warfare. Despite the political theater that ensued, in which the

---

[49]Rostow, *Stages*, 92; I. J. Catanach, "India as Metaphor: Famine and Disease Before and After 1947," *South Asia* 21 (1998): 243–61; Nick Cullather, "Hunger and Containment: How India Became 'Important' in US Cold War Strategy," *India Review* 6, no. 2 (2007): 59–90.

[50]Robertson, " 'This is the American Earth'," 579.

[51]J. O. Fletcher, *Climatic Change* (Santa Monica: Rand Corporation, 1968).

[52]Doel and Harper, "Prometheus Unleashed," 66–85; Fleming, *Fixing the Sky*, 179–83.

[53]See Tri-State National Weather Association, n.d., cited in *Weather Modification: Programs, Problems, Policy, and Potential* (Washington, DC: US Government Printing Office, 1978), 401.

[54]"News in Brief," *Los Angeles Times*, June 27, 1976, 2. Ponte would also contribute to the popular literature on Bryson's global cooling thesis (Chapter 3), see Lowell Ponte, *The Cooling* (Englewood Cliffs, NJ: Prentice-Hall, 1976).

superpowers amplified threats of global environmental change, over twenty nations signed the UN's 1977 treaty to ban military uses of environmental modification (Chapter 3).[55]

Atmospheric Infrastructure: Articulating an Environment

In addition to Keeling's findings, President Johnson's 1965 subpanel on "Atmospheric Carbon Dioxide" had relied on the work of Smagorinsky's Geophysical Fluid Dynamics Laboratory. With its roots in John von Neumann's application of the electronic computer to matters of weather prediction, the laboratory had set out since its establishment in 1955 to devise a three-dimensional, global general circulation model of the atmosphere—that is, a simulation of the way that the atmosphere moves or circulates around the earth. Within a decade, Smagorinsky's team had completed such a model of the northern hemisphere.[56]

To further expand this model to the global scale, Smagorinsky directed meteorologist Syukuro Manabe, recruited from Japan, to develop a radiation algorithm that would emulate the recently proposed role of cloud and water vapor feedback in heating the atmosphere.[57] To test this algorithm, Manabe and his colleague Richard Wetherald simulated the thermal structure of the atmosphere by running both the likely effects of water vapor emitted into the stratosphere by new supersonic jets, and the doubling of the carbon dioxide content of the atmosphere, just as Arrhenius had attempted at the turn of the century. Doubling carbon dioxide from 300 parts per million (ppm) to 600 ppm, they reported, would increase the global average temperature by 2.4°C.[58] Although the accumulation of carbon dioxide in the atmosphere had not been a driver of climate modelling, such simulations "would become an important resource in the emerging climate change discourse," as historians Matthias Heymann and Nils Hundebøl observe.[59]

---

[55]Jacob Darwin Hamblin, *Arming Mother Nature: The Birth of Catastrophic Environmentalism* (New York: Oxford University Press, 2013), 197–216.

[56]Paul N. Edwards, *A Vast Machine: Computer Models, Climate Data, and the Politics of Global Warming* (Cambridge, MA: MIT Press, 2010), 153–4.

[57]Edwards, *A Vast Machine*, 179–80.

[58]Syukuro Manabe and Richard Wetherald, "Thermal Equilibrium of the Atmosphere with a Given Distribution of Relative Humidity," *Journal of the Atmospheric Sciences* 24 (1967): 241–59.

[59]Matthias Heymann and Nils Hunderbøl, "From Heuristic to Predictive: Making Climate Models into Political Instruments," in Matthias Heymann, Gabriele Gramelsberger, and Martin Mahony (eds.), *Cultures of Prediction in Atmospheric and Climate Science: Epistemic and Cultural Shifts in Computer-Based Modelling and Simulation* (London: Routledge, 2017), 105.

By the mid-1960s, Smagorinsky had been drawn into the organization of what would become the Global Atmospheric Research Program (GARP). Meeting formally for the first time in 1968, the program's purpose was to enhance scientific understanding of the general circulation of the atmosphere so as to improve weather forecasting, and "to explore the extent to which weather and climate may be modified by artificial means."[60] Soviet academician Evgeny Fedorov, director of the Soviet Hydro-Meteorological Institute, had addressed the latter subject at the Fifth World Congress of the World Meteorological Organization (WMO) in 1967, where he reported both on the progress of artificial rain-making, fog dispersal, and hail protection, as well as the "considerable influence" that humans were "already exerting" on the composition of the atmosphere. He concluded his lecture by declaring that "the scientist must intervene in this chaotic process to avoid setting off chain reactions in the atmosphere that might produce undesirable or even destructive climatic changes."[61]

Leading the GARP was Swedish meteorologist Bert Bolin, whose own research concerned atmospheric chemistry and the natural carbon cycle. Rossby's student, Bolin had assumed the directorship of the International Institute of Meteorology after his mentor's death and continued Rossby's practice of facilitating international scientific collaboration for the next fifty years. He had been chair of the subcommittee of the International Council of Scientific Unions (ICSU) that had recommended the GARP as a means to elevate the WMO's World Weather Watch from an "operational program" to a research-oriented "scientific program," as a director of the US Weather Bureau later described it.[62] The World Weather Watch's collection of widespread data from participating countries would allow the GARP to pursue its study of "*the atmosphere as a single physical system.*"[63]

At its 1967 Fifth World Congress, the WMO had given the World Weather Watch the green light to proceed to the implementation of a "world weather system."[64] With the new technology of satellites, rockets, and computers, this system would allow for the scientists to share and collect meteorological data from a growing number of newly independent nations in a divided world. The

[60]Sverre Petterssen, *Research Aspects of the World Weather Watch* (Geneva: World Meteorological Organization, 1966), 4.

[61]United Nations Press Services, "Artificial Modification of Weather Discussed at Meteorological Congress," April 22, 1967, World Meteorological Organization—Miscellaneous Group World Weather Watch, A1838, 871/8/13/Part 1, NAA, Canberra.

[62]Robert White, "Science, Politics, and International Atmospheric and Oceanic Programs," *Bulletin of the American Meteorological Society* 63, no. 8 (1982): 926.

[63]Petterssen, *Research Aspects*, 4. Emphasis in original.

[64]World Meteorological Organization, "Press Release: The Essential Elements of World Weather Watch," October 21, 1966, World Meteorological Organization—Miscellaneous Group World Weather Watch, A1838, 871/8/13/Part 1, NAA, Canberra.

membership of the WMO had expanded to 130—a hundred more than in 1951, the year of its establishment (Chapter 1). Some of these countries required financial and technical assistance to "play their full role in the realization of the goals" of the WMO, which was provided by the UN Development Program's technical assistance programs, as well as bilateral or multilateral arrangements, and the Voluntary Assistance Programme of the WMO. The Fifth World Congress stressed that further assistance was necessary and sought to emphasize the role of meteorology to the economic development of developing countries.[65]

Meanwhile, the third World Weather Centre, Melbourne, had recently purchased a large IBM computer complex that would facilitate its role in transmitting meteorological data to both Moscow and Washington, the other centers of the World Weather Watch.[66] Direct telecommunications links between each of the three centers, the WMO's Director General David Davies believed, was "both scientifically and politically important." "It would be unfortunate if the Russians felt they were being given second best treatment for political reasons," he added.[67] Cold War tensions simmered nevertheless: at the following World Congress in 1971, Cuban delegate Rodriguez Ramirez made a declaration criticizing the Western-centric nature of the World Weather Watch, for which he had the support of Byelorussia, Bulgaria, Czechoslovakia, Hungary, Mongolia, Poland, Romania, Ukraine, and the Soviet Union. Much to the chagrin of his US counterpart, Ramirez stated that the exclusion of socialist countries from the WMO "has greatly hindered the aims of the World Weather Watch which are directed towards a full knowledge of the terrestrial atmosphere and which should be of equal benefit to all countries." Despite the organization's appeals to international cooperation for the development of meteorology, he continued, it was "rejecting United Nations agreements on the peaceful uses of the World Weather Watch." He pointed to the toll of the ongoing conflict in Indochina, which, in addition to the chemical defoliation of the region's forests, had "suffered the destruction of nearly half of its meteorological stations, the loss of the lives of more than 100 scientific and meteorological workers … all of this at the hands of the armed invasion forces of the United States and of its allies."[68]

[65]World Meteorological Organization, *Fifth World Meteorological Congress: Abridged Report with Resolutions* (Geneva: World Meteorological Organization, 1967), 15.
[66]Commonwealth Bureau of Meteorology, "Meteorology Computer Use," *Digital Computer Newsletter* 19, no. 4 (1967): 20–21.
[67]B. C. Hill, "Letter to Secretary, Department of External Affairs, Canberra, 8 December 1967," World Meteorological Organization—Miscellaneous Group World Weather Watch, A1838, 871/8/13/Part 1, NAA, Canberra.
[68]Rodriguez Ramirez, cited in World Meteorological Organization, *Sixth World Meteorological Congress: Proceedings* (Geneva: World Meteorological Organization, 1971), 162–3.

Spaceship Earth: Polluting the Planet

It was to the World Weather Watch that US ambassador Adlai Stevenson pointed in July 1965 when he addressed the Economic and Social Council of the United Nations. Reflecting on the progress of the Decade of Development, he lamented, "We are nowhere near conquering world poverty," and called on the gathered representatives to "find the ways of social, moral, and physical control adequate to stem the rising, drowning flood of people." The World Weather Watch, he believed, was "the best example" of the kind of development project that would be overseen not by an individual nation but by the "UN's own family of world wide organizations." In closing, he reminded the council, "We travel together, passengers on a little space ship, dependent on its vulnerable reserve of air and soil ... We cannot maintain it half fortunate, half miserable, half confident, half despairing ... No craft, no crew can travel safely with such vast contradictions."[69]

Just a few months earlier, US-based economist Kenneth Boulding had used a similar metaphor of the space ship in a meeting convened by the Conservation Foundation in Warrenton, Virginia. In his paper "Economics and Ecology," he observed, "Man is beginning to inhabit a pretty small and over-crowded space ship, destination unknown." Faith in an "infinite reservoir of nature" was no longer reasonable; now, "[man] will have to develop a self-sustaining economy on a small scale ... to maintain his life and, in some degree, his comfort." Stevenson's "little space ship" had been a call for international cooperation to advance the cause of modernization. For Boulding, however, the spaceship metaphor conveyed the prospect of resource exhaustion, that "the earth is in fact a total ecosystem of which man's activities are only a part."[70]

Boulding had been invited to join some forty scholars to commemorate the 1955 Princeton meeting held a decade earlier on *Man's Role in Changing the Face of the Earth*. Among them now were other veterans from that earlier conference—geographer Clarence Glacken, urban critic Lewis Mumford, and the Conservation Foundation's vice president, ecologist Frank Fraser Darling.[71] In addition to their own contributions, participants were reminded of the possibilities of weather modification, the role of the electronic computer in developing models of the atmosphere's circulation, and the rise of atmospheric

[69]Adlai Stevenson, "Strengthening the International Development Institutions," July 9, 1965, Adlai Today, https://v.gd/u3l3yB.
[70]Kenneth Boulding, "Economics and Ecology," in F. Fraser Darling and John P. Milton (eds.), *Future Environments of North America* (New York: Natural History Press, 1966), 233.
[71]F. Fraser Darling, "Introduction," in Darling and Milton (eds.), *Future Environments of North America*, 2.

carbon dioxide.[72] Boulding's spaceship was vulnerable to such changes: "It may be fatally easy," he noted, "for man to change the composition of [the oceans and the atmosphere]—which will destroy the existing equilibrium and move us to a new equilibrium which may be much less desirable for man. As human population increases, the chance of man's activities seriously interfering with the whole balance of the earth becomes greater and greater."[73]

As Boulding prepared to elaborate on these themes in his landmark lecture on "the economics of the coming spaceship earth," the Council of Europe formed a committee of experts to address what Boulding had framed as the "exhaustibility" of the atmosphere.[74] In West Germany's Ruhr Valley, air pollution had become a national political issue, when Willy Brandt called for the region's skies to be blue again on his 1961 campaign to become chancellor.[75] As few European nations had passed legislation to improve their air quality since the London smog, the council's committee was tasked with devising a multilateral approach to prevent or abate air pollution from all sources.[76] They sought to retain a regulatory focus, rather than duplicate the scientific research and exchange underway in other intergovernmental organizations, such as the OECD, which had expanded to include the United States and Canada in 1961.

Mounting measurements of the emanations from Europe's chimneys prompted the council to convene in April 1968 its first scientific conference on the effects of air pollution.[77] As a report to the American Petroleum Institute soon warned, "it is clear that we are unsure as to what long-lived pollutants are doing to our environment; however, there seems to be no doubt that the potential damage to our environment could be severe."[78] A year prior, Swedish soil scientist Svante Odén had shared with shocked readers of the daily newspaper *Dagens Nyheter* that the sulfur dioxide in industrial smoke was causing "Nederbördens försurning" (acidification of rain), which was damaging Sweden's lakes, soils, and forests. His analysis partly relied on air and soil measurements collected since the early 1950s by Rossby's European

---

[72]Paul E. Waggoner, "Weather Modification and the Living Environment," in Darling and Milton (eds.), *Future Environments of North America*, 89.

[73]Boulding, "Economics and Ecology," 233.

[74]Kenneth Boulding, "The Economics of the Coming Spaceship Earth," in H. Jarrett (ed.), *Environmental Quality in a Growing Economy* (Baltimore, MD: Johns Hopkins University Press, 1966), 3–14.

[75]John R. McNeill, *Something New Under the Sun: An Environmental History of the Twentieth-Century World* (New York: Norton, 2000), 88. He would later pursue pollution regulation as chancellor of the Federal Republic of Germany.

[76]Gaetano Adinolfi, "First Steps toward European Cooperation in Reducing Air Pollution – Activities of the Council of Europe," *Law and Contemporary Problems* 33 (1968): 421–6.

[77]Rothschild, *Poisonous Skies*, 28.

[78]Elmer Robinson and R. C. Robbins, *Sources, Abundance, and Fate of Gaseous Atmospheric Pollutants: Prepared for the American Petroleum Institute* (Menlo Park, CA: Stanford Research Institute, 1968), 110, in Center for International Environmental Law, https://v.gd/Pht6Ky.

Air Chemistry Network, which now extended beyond Norway, Sweden, and Finland to include West Germany and Belgium. Rossby himself had noted the increasing concentration of rainfall acidity across Scandinavia, which he correctly attributed to the air pollution carried by prevailing winds from British and German industry, but he had not considered the environmental impacts of these changes.

Although scientific reports of reduced agricultural and forestry yields prompted agreement at the council's 1968 conference that air pollution was responsible for environmental degradation, there was scant evidence that sulfur dioxide could travel long distances across national borders to produce such damage. Nor were Odén's colleagues entirely convinced that his alarm was warranted.[79] Coinciding with its creation of the world's first environmental protection agency in 1968, his findings nevertheless struck a chord in Sweden. There, an environmental turn was underway: just a year earlier, the discovery of high mercury levels in the nation's lakes had led to a widely reported ban on selling fish.[80] This coverage brought a new and concerned Scandinavian and Anglophone readership to Rolf Edberg's 1966 work, *On the Shred of a Cloud*, in which he explored issues of nuclear war and population growth.[81]

The nation's news media had also publicized widely the work of neo-Malthusian politician and chemist Hans Palmstierna, as well as a pessimistic collection aimed at Swedish policymakers, *The Predicament of Man*.[82] Among the contributors to that collection was Georg Borgström, whose preoccupation with questions of population, pollution, and food production anticipated the 1968 publications of Paul Ehrlich's *The Population Bomb* and Garrett Hardin's "The Tragedy of the Commons."[83] In *The Hungry Planet*, Borgström returned to Vernadsky's concept to describe the vast ecological footprint of humanity: "Man's total biosphere consumes five times more than man himself."[84] He had spoken on this topic at a 1963 symposium on microbiology in Stockholm, convened by UNESCO, where the concept of the biosphere was also becoming familiar, thanks in no small part to the

[79]Rothschild, *Poisonous Skies*, 28.

[80]David Heidenblad, *The Environmental Turn in Postwar Sweden: A New History of Knowledge* (Lund: Lund University Press, 2021), 27.

[81]Heidenblad, *The Environmental Turn*, 65; Rolf Edberg, *On the Shred of a Cloud*, trans. Sven Åhman (Tuscaloosa: University of Alabama Press, 1969).

[82]David Heidenblad, "Mapping a New History of the Ecological Turn: The Circulation of Environmental Knowledge in Sweden 1967," *Environment and History* 24 (2018): 265–84.

[83]Sunniva Engh, "Georg Borgström and the Population-Food Dilemma: Reception and Consequences in Norwegian Public Debate in the 1950s and 1960s," in Johan Östling, David Heidenblad, and Niklas Olsen (eds.), *Histories of Knowledge in Postwar Scandinavia: Actors, Arenas and Aspirations* (London: Routledge, 2020), 39–58.

[84]Georg Borgström, *The Hungry Planet: The Modern World at the Edge of Famine* (New York: Macmillan, 1965), 66.

director of its Natural Sciences Department, the Soviet soil scientist Victor Kovda.[85] The international proliferation of Vernadsky's concept culminated in September 1968, when UNESCO convened in Paris the "intergovernmental conference of experts on the scientific basis for rational use and conservation of the resources of the biosphere," which foreshadowed the environment's ascent on the international political agenda.[86] Kovda's background paper on the scientific concepts that underpinned the "biosphere" fleshed out Vernadsky's framework of interrelated living matter, emphasizing the importance of improving and applying scientific understandings of the earth's natural systems.[87]

With the support of the UN and other institutions, the Paris meeting had arisen from the International Biological Programme. Inspired by the worldwide data-sharing activities of the International Geophysical Year (IGY), the Programme formally began in 1964 to study the "biological basis of productivity and human welfare."[88] For its US architects, at least, this Programme aimed to collate enormous quantities of raw biological data to transform ecology into a hardened "Big Science," akin to geophysics, that would also be worthy of public funding.[89] While its purpose and aims may have been unclear to many participants, the leader of the US contribution, the oceanographer Revelle, saw the Programme as an extension of the IGY. The insights of the International Biological Programme would now add the earth's living components, its biosphere, where he noted in 1967 that "man is becoming a geological and biological agent."[90]

At the Paris Biosphere meeting, to which the Conservation Foundation's Frank Fraser Darling contributed a key working paper, delegates agreed that pollution had become a major international issue. "A new awareness of the loss of environmental quality is occurring throughout the world," such that

[85]Borgström, *The Hungry Planet*, 73; Jonathan Oldfield and Denis Shaw, "V.I. Vernadskii and the Development of Biogeochemical Understandings of the Biosphere, c.1880s–1968," *British Journal for the History of Science* 46, no. 2 (2013): 303.

[86]UNESCO, *Use and Conservation of the Biosphere* (Paris: UNESCO, 1970).

[87]Oldfield and Shaw, "V.I. Vernadskii," 308; V. Kovda, "Contemporary Scientific Concepts Relating to the Biosphere," in UNESCO, *Use and Conservation of the Biosphere*, 13–29.

[88]Jan-Henrik Meyer, "From Nature to Environment: International Organizations and Environmental Protection before Stockholm," in Wolfram Kaiser and Jan-Henrik Meyer (eds.), *International Organizations and Environmental Protection: Conservation and Globalization in the Twentieth Century* (New York: Bergahn Books, 2019), 31–73.

[89]Elena Aronova, Karen Baker, and Naomi Oreskes," Big Science and Big Data in Biology: From the International Geophysical Year through the International Biological Program to the Long-Term Ecological Research (LTER) Network, 1957–Present," *Historical Studies in the Natural Sciences* 40 (2010): 183–224.

[90]Roger Revelle, 1967, cited in Subcommittee on Science, Research and Development, *The International Biological Program: Its Meaning and Needs* (Washington, DC: US Government Printing Office, 1968), 140.

there was a "popular demand for correction" of air, water, and soil pollution in industrialized nations.[91] The delegates noted,

> The technological developments of man as shown by his achievements in industry, transport, communications and urbanization, all of which are essential aspects of human welfare, have nevertheless resulted in major problems of pollution: the carbon dioxide balance in the atmosphere is being altered and a variety of pollutants, including radioactive materials and a wide range of toxic chemicals, is being added to the biosphere.[92]

Convened just a month after the Soviet invasion of Czechoslovakia, the meeting's assessment echoed the recent publication of an essay by Soviet physicist, Andrei Sakharov. Calling for an end to the nuclear standoff between the Cold War superpowers, the essay also demanded civil rights for African American people and highlighted the worsening "problem of geohygiene" in the Soviet Union and the United States, including the effect of burning coal on the atmosphere.[93]

Keeling too had been reflecting on the biosphere. Speaking in the months after the Earthrise photograph was beamed back from Apollo 8 in late December 1968, he spoke on the effects of rising carbon dioxide levels in the atmosphere. Citing his earlier work with Revelle and Manabe for President Johnson's Scientific Advisory Committee, he noted with some alarm that the emission of carbon dioxide from fossil fuels had doubled every fifteen to twenty years since 1850, save for the Great Depression and the two world wars.[94] Drawing on ecologist Hutchinson's studies of the carbon cycle, Keeling concluded that land clearing and forest fires had also released carbon dioxide at a rate that was likely too great for plants to absorb. Uncertain as to the exact implications of these changes for the climate, he nevertheless stressed that "carbon dioxide is just one index of man's rising activity today." In contrast to those scientists who saw in such changes "unprecedented opportunity" for science and technology, he was concerned as to the longer-term effects of population growth and air pollution. Reflecting on the pessimism of ecologist Frank Egler, who had influenced Rachel Carson, Keeling concluded, "If the human race survives into the 21st century with the vast population increase that now seems

---

[91] UNESCO, *Use and Conservation of the Biosphere*, 196.
[92] UNESCO, *Use and Conservation of the Biosphere*, 210.
[93] Andrei Sakharov, "Text of Essay by Russian Nuclear Physicist Urging Soviet-American Cooperation: Thoughts on Progress, Peaceful Coexistence and Intellectual Freedom," *New York Times*, July 22, 1968, 14–16.
[94] C. D. Keeling, "Is Carbon Dioxide from Fossil Fuel Changing Man's Environment?," *Proceedings of the American Philosophical Society* 114, no. 1 (1970): 10–17.

inevitable, the people living then, in addition to their other troubles, may face the threat of climatic change brought about by an uncontrolled increase in atmospheric carbon dioxide from fossil fuels."[95]

Environmental Internationalism: Development, Decolonization and Diplomacy

Prior to the Paris meeting on the biosphere in September 1968, neutral Sweden had already flagged to the UN Stockholm's interest in hosting a conference on "man and the environment."[96] Support at home for such an event seemed certain, given the wide interest in environmental concerns. Elsewhere, grim news fueled a growing environmental movement: the recent oil spill of the *Torrey Canyon* supertanker in the English Channel had followed closely after the collapse of a coal tip in the Welsh village of Aberfan killed 116 children and 28 adults.[97] Such environmental concerns were becoming widespread across the industrialized world, along with a rising disenchantment with militarism, capitalism, racism, and conservatism that spilled over into protests and demonstrations throughout 1968.[98] In an address drafted by politician Palmstierna, Sweden's permanent representative to the UN Sverker Åström seized his moment that December to remind the General Assembly that addressing shared environmental problems together was "a question of collective self-preservation."[99] The member nations passed Resolution 2398 and agreed to meet in Stockholm for the UN Conference on the Human Environment in June 1972.

The Stockholm Conference proposal in December 1968 marked the first time that human-induced climate change was raised in the UN General Assembly. Sweden's representative Åström had described the increasing atmospheric concentration of carbon dioxide as a "well-known example of pollution," going on to observe that "man has already rendered the temperature equilibrium of the globe more unstable."[100] In sharing this view, Åström foreshadowed the

---

[95]Keeling, "Is Carbon Dioxide," 16. See Linda Lear, *Rachel Carson: Witness for Nature* (New York: Henry Holt, 1997).

[96]Eric Paglia, "The Swedish Initiative and the 1972 Stockholm Conference: The Decisive Role of Science Diplomacy in the Emergence of Global Environmental Governance," *Nature: Humanities and Social Sciences Communications* 8 (2021): 10.1057/s41599-020-00681-x.

[97]John Sheail, "*Torrey Canyon*: The Political Dimension," *Journal of Contemporary History* 42 (2007): 485–504.

[98]Macekura, *Of Limits and Growth*, 106.

[99]Sverker Åström, "Agenda Item 91: Problems of the Human Environment," UN General Assembly, Twenty-Third session, 1732nd Plenary Meeting, December 3, 1968, 6.

[100]Åström, "Agenda Item 91," 2.

elevation of climate change to the raft of global environmental problems that would be tabled in the lead up to the 1972 conference.

MIT's Carroll Wilson, who had served as an assistant of Vannevar Bush during the Second World War, was among the well-connected actors at the center of these activities, along with the likes of Canadian oil executive Maurice Strong and English-born economist Barbara Ward.[101] Assembling a large gathering of scientists in July 1970, Wilson envisioned his month-long "Study of Critical Environmental Problems" (SCEP) would fill a research gap on "global problems such as changes in climate and in ocean and terrestrial ecosystems."[102] He then convened another more focused meeting in Stockholm in 1971 to study "man's impact on climate" in preparation for the UN conference the following year.[103] Whereas the participants of the first conference had been Americans, the second climate-focused conference drew participants from fourteen countries, including India, Israel, Japan, the Soviet Union, the United States, and Western Europe. In light of the findings of these reports, the ICSU outlined a plan for a global environmental monitoring system that would allow for the assessment of changes to the chemistry of the atmosphere arising from human activity, including climate trends.[104]

In the meantime, Wilson had introduced an MIT computer engineer to his colleagues in the Club of Rome, which would lead to the publication of *Limits to Growth* on the eve of the Stockholm Conference.[105] As Wilson's studies (known as SCEP and "Study of Man's Impact on Climate" (SMIC)) had indicated, the use of computer modeling was becoming the foremost means to understand and simulate the complexities of global environmental change, albeit a costly one that was beyond the means of all but a few institutions. Keeling's accumulating atmospheric carbon dioxide data had meanwhile figured across all three studies, providing a clear signal of human-induced

---

[101]On Strong and Ward's mobilization of their business and political networks for the Stockholm conference and afterwards, see Glenda Sluga, "Capitalists and Climate," *Humanity Journal* November 6, 2017, https://v.gd/Xk7tNO; and Ben Huf, Glenda Sluga, and Sabine Selchow, "Business and the Planetary History of International Environmental Governance in the 1970s," *Contemporary European History* 31 (2022): 553–69.

[102]Study of Critical Environmental Problems, *Man's Impact on the Global Environment: Assessment and Recommendation for Action* (Cambridge, MA: MIT Press, 1970), xi.

[103]William Matthews, William Kellogg, and G. D. Robinson (eds.), *Man's Impact on the Climate* (Cambridge, MA: MIT Press, 1971).

[104]SCOPE, *Global Environmental Monitoring: A Report Submitted to the UN Conference on the Human Environment, Stockholm 1972* (Stockholm: Scientific Committee on Problems of the Environment, 1971), 7, 12, 22, 51.

[105]Donella Meadows et al., *The Limits to Growth: A Report for the Club of Rome's Project on the Predicament of Mankind* (New York: Universe Books, 1972). On Wilson's initial encounter with Peccei and the role of the OECD Secretariat in the foundation of the Club of Rome, see Matthias Schmelzer, " 'Born in the Corridors of the OECD': The Forgotten Origins of the Club of Rome, Transnational Networks, and the 1970s in Global History," *Journal of Global History* 12, no. 1 (2017): 26–48.

environmental change and evidence of the value of global atmospheric monitoring. The delegates at the 1972 Stockholm Conference recognized the need for such an international network and recommended its establishment under the auspices of the WMO and the ICSU.[106]

This close interest in the study of global environmental problems had done little to assuage the concerns of developing countries, however. Wilson himself had anticipated as much on the eve of the publication of *The Limits to Growth*. In a letter to the Club of Rome's founder, Italian business executive Aurelio Peccei, he warned in early 1972,

> The implications of *The Limits to Growth* for the less developed countries are very disturbing and (will) generate great hostility. Those who are perceptive can see through the arithmetic and understand just what … [it] means in terms of their aspirations for material standards of living comparable with those of the highly developed countries. Nothing we can say in our rhetoric is going to make any impact on these people.[107]

Wilson urged Peccei to distance the group from the *Limits to Growth* report, or risk reinforcing "the growing and rather wide-spread suspicion that the 'Club of Rome' is a world-wide conspiracy of an elite group working in the shadowy background to change the world in ways which will have the most profoundly disadvantageous consequences for half or two-thirds of the world's people."[108] Regardless of Wilson's concerns, Peccei would be one of the luminaries invited to participate in a "distinguished lecture series" at Stockholm, along with the likes of Thor Heyerdahl, Gunnar Myrdal, Solly Zuckerman, and Barbara Ward.[109]

Despite the efforts of the Stockholm Conference's secretary-general, Maurice Strong, developing nations largely viewed the meeting's agenda of global environmental protection as an attempt to curb their economic growth. Gesturing to the Club of Rome's report *Limits to Growth*, João Augusto de Araújo Castro, Brazil's ambassador to the United States, was skeptical that developing countries would benefit from global policies to restore an "equilibrium of 'spaceship earth'." The environmental ambitions of the developed countries, he observed, would "better succeed if the relative positions of the [first and second] classes were maintained, for the emergence of one single class would presuppose a considerable change in the living

[106]Edwards, *A Vast Machine*, 364–6.
[107]Carroll Wilson to Aurelio Peccei, February 2, 1972, cited in "One of a Kind: Carroll L. Wilson, a Biography by Milton Lomask," in *Carroll L. Wilson, 1910–1983: A Report of the Carroll L. Wilson Awards Committee*, 160. Reproduced from copy T171.M4218.W553.1987 in the Institute Archives and Special Collections, MIT Libraries.
[108]Carroll Wilson to Aurelio Peccei, January 9, 1972, cited in "One of a Kind," 159–60.
[109]See Barbara Ward (ed.), *Who Speaks for Earth?* (New York: W.W. Norton, 1973), 153–68.

standards of the first class, something that may not be attained in the light of present global socioeconomic realities."[110] Brazil's ambassador to the United Nations, Miguel Ozorio de Almeida, argued that the South American nations already contributed to ameliorating the pollution of the North. "The areas of pastures and jungle in Latin America do a great deal towards restoring the atmosphere, by absorbing carbon dioxide and releasing oxygen through photosynthesis," he told a seminar in Mexico City in September 1971.[111]

During the preparations for the Stockholm Conference, Brazil led the developing countries in its strident criticism of environmental protection. These countries had earlier formed a loose coalition known as the Group of 77 (G77) at the first session of the UN Conference on Trade and Development in 1964, understanding that their shared economic interests might be better served by negotiating together as a bloc.[112] That the 1972 Stockholm Conference was being steered by a group of Western industrialized nations, particularly Canada, the Netherlands, Sweden, and the United States, only reinforced the developing countries' skepticism toward the rising environmental agenda. In mid-1970, for instance, Brazil had presented a working paper that insisted that policies for the protection of the environment must "be planned as a means to promote development and not as an obstacle and a barrier to the rising expectations of the underdeveloped world. They must be defined also clearly in terms of economic activities and of channeling to the developing world the additional resources it will need."[113]

Brazil's position enjoyed the support of the Chilean delegation. In late 1970, Chile called for "additional resources" for the South to cover the costs of environmental protection and regulations, which became known as "additionality." Calls for "compensation" soon followed, should the environmental measures adopted by industrialized trading partners affect the economic welfare of the developing countries.[114] With India (the only nation whose head of state attended the Stockholm Conference), Brazil's delegation repeated these calls for additionality and compensation during the conference, which

---

[110]João Augusto de Araújo Castro, "Environment and Development: The Case of the Developing Countries," in David Kay and Eugene Skolnikoff (eds.), *World Eco-Crisis: International Organizations in Response* (Madison: University of Wisconsin Press, 1972), 241.

[111]"Statement by H.E. Ambassador Miguel A. Ozorio de Almeida, Head of the Brazilian Delegation," *Latin American Regional Seminar on Problems of the Human Environment and Development, Mexico City, September 6-11, 1971* (Mexico City: Economic Commission for Latin America, 1971), 6. See also Jacob Darwin Hamblin, "Environmentalism for the Atlantic Alliance: NATO's Experiment with the 'Challenges of Modern Society'," *Environmental History* 15, no. 1 (2010): 54–75.

[112]Macekura, *Of Limits and Growth*, 113.

[113]Cited in Lars-Göran Engfeldt, *From Stockholm to Johannesburg and Beyond: The Evolution of the International System for Sustainable Development Governance and Its Implications* (Stockholm: Edita Västra Aros, 2009), 45.

[114]Macekura, *Of Limits and Growth*, 114–15.

FIGURE 2.1 *Indian prime minister Indira Gandhi and conference secretary-general Maurice Strong meet at the Fokets Huts building to attend the UN Conference on the Human Environment, Stockholm, June 5, 1972. UN Photo by Yutaka Nagata.*

were later adopted as Principles 11 and 12 in the Stockholm Declaration.[115] "On the one hand the rich look askance at our continuing poverty—on the other, they warn us against their own methods. We do not wish to impoverish the environment any further and yet we cannot for a moment forget the grim poverty of large numbers of people," explained Indian prime minister Indira Gandhi (Figure 2.1) in her speech before the conference. She challenged her fellow delegates, "Are not poverty and need the greatest polluters?"[116]

In the absence of the Soviet bloc, which had boycotted the conference in protest of East Germany's exclusion, the developing countries seized the opportunity to steer the conference agenda.[117] Attending its first major UN conference since becoming a member, China's delegation soon asserted itself as a significant leader of the developing countries. Unlike Brazil and the developing countries that attended the Founex seminar in June 1971, Beijing had not been

[115]Macekura, *Of Limits and Growth*, 123.
[116]"Address to the Plenary Session of the United Nations Conference on Human Environment at Stockholm, June 14, 1972," in Indira Gandhi, *Indira Gandhi: Speeches and Writings* (New York: Harper & Row, 1975), 191–9.
[117]Adil Najam, "The South in International Environmental Negotiations," *International Studies* 31, no. 4 (1994): 427–64.

involved in the preparations for the conference.[118] Nevertheless, on the opening day of the Stockholm Conference, Beijing announced its intent to challenge the draft declaration and put forth its own version to the conference newsletter *Eco* that outlined a position it would maintain into the twenty-first century.[119] Led by Tang Ke, vice minister of the Fuel and Chemical Industries Ministry, China's delegation successfully campaigned for participating nations to be permitted to amend the declaration.[120] One such revision changed Principle 9 from stating that development was the best remedy for the environmental deficiencies arising from underdevelopment, to "Environmental deficiencies generated by the conditions of underdevelopment and natural disasters pose grave problems and can be remedied by accelerated development through the transfer of financial and technological assistance as a supplement to the domestic effort of the development countries and such timely assistance as may be required."[121]

During the Stockholm meeting, representatives of China's delegation had met with one of the advisers to their Canadian counterparts, George Manuel, with whom they discussed the prospect of a tour of the People's Republic of China by the executive council of the National Indian Brotherhood.[122] A Secwépemc man from the Neskonlith Indian Band, a government-registered "status" Indian, Manuel had welcomed a late invitation to join the delegation as he believed that the growing environmental crisis represented "a threat to the very existence of the culture and way of life of thousands of native Canadians."[123] At the conference, he met with Aslak Nils Sara, a Sámi chemist and director of the Nordic Sami Institute who was now working in the newly established Norwegian Department of the Environment.[124] Outside the official proceedings of the UN conference, he encountered Navajo and Hopi members of the Black Mesa Defense Fund, who were among the two dozen

[118]On the Founex seminar in Switzerland, see Michael W. Manulak, "Developing World Environmental Cooperation: The Founex Seminar and the Stockholm Conference," in Kaiser and Meyer (eds.), *International Organizations and Environmental Protection*, 103–27.

[119]Wade Rowland, *The Plot to Save the World: The Life and Times of the Stockholm Conference on the Human Environment* (Toronto: Clarke, Irwin, 1973), 92–3.

[120]See Xiaoxuan Wang, "The 1972 Stockholm Conference and China's Diplomatic Response," *Cultures of Science* 6, no. 2 (2023): 146–52.

[121]Cited in Faramelli, "Toying with the Environment and the Poor," 474.

[122]Jonathan Crossen, "Another Wave of Anti-colonialism: The Origins of Indigenous Internationalism," *Canadian Journal of History* 52, no. 3 (2017): 533–59. Manuel also met with the representatives of the Brazilian, West German, Australian, and Tanzanian delegations.

[123]Cited in Peter McFarlane, *Brotherhood to Nationhood: George Manuel and the Making of the Modern Indian Movement* (Ontario: Between the Lines, 1993), 166.

[124]Henry Minde, "The Challenge of Indigenism: The Struggle for Sami Land Rights and Self-Government in Norway, 1960–1990," in Svein Jentoft, Henry Minde, and Ragnar Nilsen (eds.), *Indigenous Peoples: Resource Management and Global Rights* (Delft: Eburon, 2003), 75–106; Jonathan Crossen, "Decolonization, Indigenous Internationalism, and the World Council of Indigenous Peoples," PhD dissertation (Ontario: University of Waterloo, 2014), 79–80.

Native Americans who had traveled to Stockholm to argue for Indigenous self-determination not only in the interests of their own peoples but also for the benefit of the whole world.[125]

Thanks to the connections of his assistant Marie Smallface Marule and her husband Jacob Marule with the Copenhagen-based International Working Group for Indigenous Affairs and Survival International, Manuel also met with Sámi people in the town of Rensjön in Sápmi, northern Sweden. His meeting with Johan Kuhmunen featured on the front page of the *Dagens Nhyeter*, where their shared message was clear: "We have the same problems and should address them internationally."[126] Having also noted similar challenges facing Indigenous peoples in Australia and Aotearoa New Zealand on a tour the previous year, they discussed his idea for an international Indigenous peoples' conference, to which Kuhmunen told Manuel "that the idea had long been a concern of his as well, and that he would be very pleased to participate in such a conference."[127] He returned from Stockholm via Geneva, where he met with representatives of the World Council of Churches, which had recently endorsed the 1971 Declaration of Barbados for the Liberation of Indians and now welcomed Manuel's proposal for a World Conference of Indigenous Peoples. They also discussed the possibility of the National Indian Brotherhood seeking nongovernmental organization status at the United Nations Economic and Social Council (ECOSOC), which had recently commenced a "Study of the Problem of Discrimination against Indigenous Populations." Intending to transfer this status to an Indigenous international organization, the Brotherhood commenced the preparation of a proposal to ECOSOC on Manuel's return to Ottawa.[128]

Although the conclusion of the Stockholm Conference did not dispel the tensions between the South and the North, a new specialized agency that would become the UN Environment Programme (UNEP) was among the outcomes of the proceedings. The United States and other developed nations had hoped it might be headquartered in Vienna, but it was Kenya's capital Nairobi that became home to the first and only major UN agency to reside in a developing country.[129] Delegations representing the South had rallied behind

---

[125]Paul C. Rosier, " 'Modern America Desperately Needs to Listen': The Emerging Indian in an Age of Environmental Crisis," *Journal of American History* 100, no. 3 (2013): 711–35.

[126]Minde, "The Challenge of Indigenism," 75–106. The International Working Group for Indigenous Peoples (1968) and Survival International (1969) had been recently formed by social anthropologists alarmed by the toll on Indigenous peoples of resource extraction in Brazil, Paraguay, Peru, and Colombia. See Henry Minde, "The Destination and the Journey: Indigenous Peoples and the United Nations from the 1960s through 1985," in Henry Minde with Svein Jentoft, Harald Gaski, and Georges Midré (eds.), *Indigenous Peoples: Self-Determination, Knowledge, Indigeneity* (Delft: Eburon, 2008), 49–86.

[127]Cited in McFarlane, *Brotherhood to Nationhood*, 168.

[128]Crossen, "Another Wave of Anti-colonialism," 533–59.

[129]Macekura, *Of Limits and Growth*, 128–9.

Nairobi in an act of Third World solidarity and in an attempt to invest their concerns for "development" in the foundations of this new UN organization.[130]

Conclusion

From the 1950s, the rising influence of ecological thought and the introduction of the "biosphere" concept to the Angloworld invited scientists and a nascent environmental movement to understand the interdependent relationships between humans and their environment. The articulation of these connections prompted some to reappraise the wider ecological impacts of human activity, as well as their implications for human health. This sense of vulnerability extended to the atmosphere, where governments and scientists were taking close interest in the changes that human activity could cause. Measuring and monitoring those changes was an international project for scientists, seeking to map the atmosphere as a single, coherent system.

Yet the planetary gaze that the Apollo missions afforded elided the important national circumstances that preoccupied governments, particularly from the South, as well as the concerns of Indigenous peoples. Their aspirations for self-determination, development and decolonization challenged the North's vision of a global environment at the Stockholm Conference, and, in doing so, established firm foundations for the character of the environmental diplomacy that followed. The subsistence and resource crises of the decade ahead would challenge not only the new UNEP but also the Stockholm vision of "one Earth."[131]

---

[130]Adil Najam, "Developing Countries and Global Environmental Governance: From Contestation to Participation to Engagement," *International Environmental Agreements* 5 (2005): 309.
[131]Barbara Ward and René Dubos, *Only One Earth: The Care and Maintenance of a Small Planet* (New York: Norton, 1972).

3

Endangered Atmosphere

1981: 339.95 ppm

In early 1974, United Nations (UN) secretary-general Kurt Waldheim addressed the Inter-State Committee on Drought Control in the Sahel, in Ougadougou, Upper Volta (what is now Burkina Faso). "In less than 50 years' time," he warned, "the advancing desert threatens to wipe three or four countries of Africa completely off the map."[1] Another year of drought loomed over the region's newly independent republics, as the world began to reel again from unusual weather conditions. Over the past two years, grain production had fallen dramatically thanks to drought and frost in Argentina, Australia, India, and the Soviet Union, which led to the 1973 "wheat deal" between the superpowers that caused soaring international grain prices. In the Philippines, meanwhile, rice and corn production had fallen, frost had hit the coffee plantations of southern Brazil, and the Peruvian anchovy fishery had nearly collapsed, endangering a major source of protein for animal feed in the United States, Japan, and Europe. Now, in 1974, grain production in North America was set to tumble, while skyrocketing fertilizer prices threatened the Green Revolution. According to the Overseas Development Council, global food reserves had plummeted from ninety-five days' worth of world consumption in 1961 to just twenty-six days in 1974.[2]

The implications of this emerging crisis were on the agenda at what became the Sixth Special Session of the UN General Assembly, where the Programme for the Establishment of a New International Economic Order was adopted. Decolonization had transformed the UN, swelling the ranks of the General Assembly with newly independent nations, mostly from the developing world. Recognizing their collective strength in the General Assembly on account of its one-state-one-vote principle, these nations had formed a coalition as the Group of 77 (G77). Some of these nations had already leant their support to

---

[1]Kurt Waldheim, cited in Howard Brabyn, "The Drama of 6,000 Kilometres of Africa's Sahel," *UNESCO Courier* April (1975): 5.
[2]Lester Brown and Erik Eckholm, *By Bread Alone* (New York: Praeger, 1974), 3.

George Manuel's proposal for UN Economic and Social Council (ECOSOC) recognition of the National Indian Brotherhood, which would soon be replaced by the newly formed World Council of Indigenous Peoples. Inspired by the "Third World" movement for decolonization and international anti-colonial collaboration, an emerging "Fourth World" represented what Manuel hoped would be a growing solidarity between Indigenous peoples. Now, just days after the first International Conference of Indigenous People in Georgetown, Guyana, they demanded the reform of the global economy to improve the relative position of the developing nations.[3]

Addressing this Sixth Special Session, US secretary of state Henry Kissinger identified six problem areas that needed to be resolved to "spur both the world economy and world development." Having already proposed an international food conference (which would soon be held in Rome), he called on the International Council of Scientific Unions (ICSU) and the World Meteorological Organization (WMO) to address the "possibility of climatic changes in the monsoon belt and perhaps throughout the world. The implications for global food and population policies are ominous."[4] Agriculturally favorable climate conditions since the mid-1950s no longer appeared certain to continue, as grim news from the Sahel met Garrett Hardin's "lifeboat ethics."[5] These were not the benign changes anticipated by Hans Wilhelmsson Ahlmann and those others who had welcomed a warming world (Chapter 1).

The rising tide of environmentalism had by then established "the environment" as a common ground in what was elsewhere becoming an age of fracture.[6] In a prelude to the 1972 Stockholm meeting and since, national governments from France to India hastened to implement environmental protection legislation and to establish environment departments. Japan's "long environmental sixties," meanwhile, were culminating in the government's capitulation to public pressure to regulate industry and pollution.[7] The first "green" political parties rode this wave of political influence, with activists in Australia, Aotearoa New Zealand (1972), and West Germany (1973) winning

[3]Nils Gilman, "The New International Economic Order: A Reintroduction," *Humanity* Spring (2015): 1–16; Jonathan Crossen, "Another Wave of Anti-colonialism: The Origins of Indigenous Internationalism," *Canadian Journal of History* 52, no. 3 (2017): 533–59.
[4]Henry Kissinger, "Address to the Sixth Special Session of the United Nations General Assembly," *International Organization* 78, no. 3 (1974), 580–1. See United Nations, *Report of the World Food Conference* (New York: United Nations, 1975), 21.
[5]See, for example, Stephen Schneider, in Committee on Foreign Relations, *Foreign Policy Choices for the Seventies and Eighties, Ninety-Fourth Congress, vol. 1* (Washington, DC: US Government Printing Office, 1976), 342–56; Garrett Hardin, "Lifeboat Ethics: The Case against Helping the Poor," *Psychology Today* 8, no. 4 (1974): 800–12.
[6]Daniel Rodgers, *The Age of Fracture* (Cambridge, MA: Harvard University Press, 2011).
[7]Simon Avenell, "Japan's Long Environmental Sixties and the Birth of a Green Leviathan," *Japanese Studies* 32 (2012): 423–44.

votes, while environmentally oriented organizations emerged in the South, with the Environmental Liaison Center (1972) and the Green Belt Movement in Kenya (1977); International, Environment and Development Action in the Third World in Senegal (1976); and Sahabat Alam Malaysia (1977).[8]

International associations focused on the environment had doubled over the decade between 1963 and 1973, from five to ten, and they would more than double again by the early 1980s.[9] A flurry of international environmental agreements had also followed in Stockholm's wake—joining wetlands protection (1971) were conventions on marine pollution (1972), endangered species (1973), and migratory species (1979), as well as protections for the Baltic (1973) and Mediterranean (1976).[10] This environmental wave also swept the WMO: following the devastating cyclone Bhola in the Bay of Bengal, its president in 1971 declared, "Our science has assumed a greatly extended significance in world affairs, since the atmosphere is an essential, if not the prime element in the human environment."[11]

This environmental esprit de corps belied, however, a wider disenchantment with the international order. The successive collapse of Bretton Woods, the first oil embargo in 1973, and the New International Economic Order revived for the North the specter of the Club of Rome's recent predictions (Chapter 2). In the wake of the 1972 Stockholm conference, environmental, economic, and political upheaval combined to demonstrate the risks of the growing interdependence between nations, and between humans and their environment. Transcending national borders, the scale and nature of industrial pollutants invited scientists, governments, and legal experts to draw upon the principles established at Stockholm to better understand, monitor, and regulate the changing atmosphere.

Common Air: Legislating the Atmospheric Commons

"The exploration of outer space has given us a new understanding of the atmosphere which surrounds and protects life on earth, and has given us a

---

[8]Pratap Chatterjee and Matthias Finger, *The Earth Brokers: Power, Politics and World Development* (London: Routledge, 1994), 74.

[9]Margaret E. Keck and Kathryn Sikkink, *Activists beyond Borders: Advocacy Networks in International Politics* (Ithaca: Cornell University Press, 1998), 11.

[10]John R. McNeill, "The Environment, Environmentalism, and International Society in the Long 1970s," in Niall Ferguson et al. (eds.), *The Shock of the Global: The 1970s in Perspective* (Cambridge, MA: Harvard University Press, 2010), 263–80.

[11]Alf Nyberg, in World Meteorological Organization, *Sixth World Meteorological Congress, Geneva, 5–30 April 1971 Proceedings* (Geneva: World Meteorological Organization, 1972), 32–3.

new region to share, in which there are no frontiers, no boundaries, no ways of barricading one part off from another," the US anthropologist Margaret Mead declared before an audience at Teen Murti House in New Delhi in June 1973. Having presented the Joint NGO statement to the UN Conference on the Human Environment in Stockholm the previous year, Mead took the opportunity of delivering the annual Jawaharlal Nehru Memorial Lecture to expand on her "prospects for world harmony"—for a "planetary community." [12] On April 22, 1970, she had joined the likes of Kenneth Boulding, René Dubos, Eugene Odum, Hardin, and Adlai Stevenson among the speakers at the first Earth Day. In her address to those gathered in New York City's Bryant Park, she had reflected on the value of "seeing the earth as small and lonely and blue," for science and technology now afforded the means "to look at the whole earth ... to begin to deal with the whole problem." [13] Reflecting a shift toward a planetary consciousness inspired by space exploration, she now saw in the atmosphere a boundless sphere of common humanity that could inspire a new form of social organization that reflected what she described as "today's interconnected planetary system." [14]

For Mead, the atmosphere was not simply a useful metaphor for a new politics. As president of the American Association for the Advancement of Science, she joined with atmospheric scientist William Kellogg of the US National Center for Atmospheric Research in October 1975 to convene a conference, "The Atmosphere: Endangered and Endangering." The attention of the conference participants, who included Wallace Broecker, Harrison Brown, James Lovelock, J. Murray Mitchell, Edith Brown Weiss, and Stephen Schneider, was focused on the atmospheric effects of human activity and their implications for an increasingly interdependent world. For Mead at least, the atmosphere was "the ultimate international commons," "a domain that must either be shared and responsibly protected by all people or all people will suffer." [15] Citing the ongoing challenges of negotiating the Law of the Sea, Mead and Kellogg called on the conference participants to prepare the scientific case for a "Law of the Air" that might "protect the peoples of the world from dangerous and preventable interference with the atmosphere upon which all life depends." [16]

---

[12] Margaret Mead, "Prospects for World Harmony (1973)," in Robert Textor (ed.), *The World Ahead: An Anthropologist Anticipates the Future* (New York: Berghahn Books, 2005), 288–9.

[13] Margaret Mead, "Earth People," in Environmental Action, *Earth Day: The Beginning – a Guide for Survival* (New York: New York Times, 1970), 245.

[14] Margaret Mead, "A Note on Contributions of Anthropology to the Science of the Future (1971)," in Textor (ed.), *The World Ahead*, 272.

[15] Margaret Mead, "Preface," in William Kellogg and Margaret Mead (eds.), *The Atmosphere: Endangered and Endangering* (Washington, DC: US Department of Health, Education and Welfare, 1977), xxii.

[16] Mead, "Preface," xix.

Such a "Law of the Air" was already taking shape. Conference participant Weiss, a scholar of international law, was in the middle of a five-year research program at the Brookings Institution in Washington, DC, where she was developing a legal means to manage international "commons." Like Antarctica, outer space, and the ocean, the earth's weather and climate remained outside the jurisdiction of any one country, and their very nature ensured their regulation would be elusive. For Weiss and her collaborators, the international regimes for these commons that had been developed in the 1950s and 1960s now appeared "poorly suited to the dramatic recent expansion of human capabilities for using and altering the resources of these realms." Overhauling these regimes was relevant to US foreign policy, their primary research interest, as well as "for the benefit of future generations"—a theme that remained an enduring interest for Weiss and a pillar of the emerging concept of sustainable development.[17]

For Weiss, who was drafting the sections on weather and climate, the negotiations at the 1972 UN Conference on the Human and Environment in Stockholm had augured well. According to the Stockholm Declaration's Principle 21, states "have the responsibility to ensure that activities within their jurisdiction or control do not cause damage to the environment of other States or of areas beyond the limits of national jurisdiction."[18] Its precedent can be traced to the 1941 Trail Smelter Convention, the result of arbitration to limit heavy pollution drifting south into the United States from Canada's industries.[19] Only recently, the governments of Australia, Fiji, and Aotearoa New Zealand had cited the principle in their challenges to the legality of French atmospheric nuclear weapons testing in the Pacific in the International Court of Justice.[20] The claimants argued that the spread of fallout from the further tests would violate their territorial sovereignty and, in Australia's case, Principle 21 of the Stockholm Declaration.[21]

Despite the Test Ban Treaty of 1963, France announced its intent to move its nuclear testing grounds from Algeria to French Polynesia. Three years later, its first Pacific atmospheric tests commenced at Mururoa Atoll and Fangataufa

---

[17]Seyom Brown et al., *Regimes for the Ocean, Outer Space and Weather* (Washington, DC: Brookings Institution, 1977), 3. See, for example, Edith Brown Weiss, "Climate Change, Intergenerational Equity and International Law: An Introductory Note," *Climatic Change* 15 (1989): 327–35.
[18]Cited in Brown et al., *Regimes for the Ocean*, 213.
[19]Arne Kaijser, "Combatting 'Acid Rain': Protecting the Common European Sky," in Anna-Katharina Wöbse and Patrick Kupper (eds.), *Greening Europe: Environmental Protection in the Long Twentieth Century* (Boston: de Gruyter, 2022), 368.
[20]On May 9, 1973, the governments of Australia and New Zealand instituted separate, albeit similar, proceedings against France in the International Court of Justice, in which they each claimed that atmospheric nuclear tests constituted violations of international law. On May 16, 1973, the government of Fiji sought (unsuccessfully) permission to intervene in the Australian case against France and on May 18, 1973, the New Zealand case. See Tim Stephens, *International Courts and Environmental Protection* (New York: Cambridge University Press, 2009), 137–45.
[21]Marvin S. Soroos, *The Endangered Atmosphere: Preserving a Global Commons* (Columbia: University of South Carolina Press, 1997), 104.

Atoll in 1966. In the wake of the Cuban Missile Crisis, Washington, Whitehall, and Moscow had brokered the Treaty Banning Nuclear Weapons in the Atmosphere, in Outer Space and Under Water in mid-1963. Political scientist Marvin Soroos describes the aboveground or atmospheric tests as "the first major air pollution problem to provoke widespread concern in the world community and become the subject of an international regime."[22] Although the International Court of Justice was in favor of Australia and Aotearoa New Zealand, France was unmoved and continued its atmospheric testing program into 1974, after which its nuclear testing regime proceeded underground as per the Test Ban Treaty.[23]

The Stockholm conference had also recommended governments be mindful of activities in which there might be an appreciable risk of climatic effects. In Recommendation 70, states were asked to evaluate carefully the likelihood and magnitude of effects on the climate from a contemplated action, and to consult widely with other states about any potential risks from their activities.[24] Although the Soviet Union had abstained from the Stockholm conference, Moscow later drew on this recommendation in its 1974 draft UN resolution dealing with what would eventually become the convention against environmental modification for the purposes of warfare (Chapter 2). Despite the ambiguities of the subsequent 1977 Environmental Modification Convention (ENMOD) treaty, Weiss interpreted these developments approvingly, noting, "We may be on the verge of a new era in negotiating arrangements to govern man's use of weather and climate."[25]

Shortly after Weiss and her colleagues published their findings, the Organisation for Economic Co-operation and Development (OECD) shared the results in 1977 of its study of the long-range transport of sulfur compounds. This survey of some seventy-six ground stations in eleven member countries confirmed Swedish and Norwegian concerns (Chapter 2), documenting since 1972 the extent and nature of transboundary air pollution across Western Europe. Its report observed that "sulphur compounds do travel long distances (several hundred kilometers or more) in the atmosphere and ... the air quality in any one European country is measurably affected by emissions from other European countries." This pollution spread unevenly, however, affecting

---

[22]Soroos, The Endangered Atmosphere, 4.

[23]Soroos, The Endangered Atmosphere, 107.

[24]Barbara West et al., "Managing the Atmospheric Resource: Will Mankind Behave Rationally?," in Kellogg and Mead (eds.), The Atmosphere, 82–3.

[25]Edith Brown Weiss, "Weather Control: An Instrument of War?," Survival 17, no. 2 (1975): 68; Lawrence Juda, "Negotiating a Treaty on Environmental Modification Warfare: The Convention on Environmental Warfare and Its Impact upon Arms Control Negotiations," International Organization 32 (1978): 975–91.

southern Scandinavia and Switzerland more greatly than neighboring areas owing to differences in precipitation and soil type.[26]

Suspecting that further sources of sulfur dioxide emanated from behind the Iron Curtain, Norwegian scientists pressed their government and the OECD to extend the monitoring network to include sites in Eastern Europe. Under the prevailing détente, Norway advocated for the study of air pollution as a vehicle for East–West environmental cooperation during preparations for the 1975 Helsinki Conference on Security and Cooperation in Europe.[27] Despite the wariness of the Western countries toward engaging with the Soviet Union on the project, Norway successfully obtained the consent of the UN Economic Commission for Europe to begin consultations with Moscow to develop a monitoring network to be overseen by the UN. In early 1975, Austria, Canada, Denmark, Finland, Hungary, the Netherlands, Norway, Poland, Portugal, Sweden, the Soviet Union, and West Germany committed to joining what became the European-wide Monitoring and Evaluation Programme (EMEP), with France, Belgium, and Switzerland joining in 1976, and Britain a year later.[28]

Having meanwhile identified the effects of acid rain within the Soviet Union, Moscow saw in the problem of transboundary air pollution common ground with the Scandinavian countries and invited Norway's environment minister to visit in early 1978.[29] Gro Harlem Brundtland had earlier met with officials in East Germany, Czechoslovakia, and the Soviet Union to discuss the development of the monitoring network, where she had sought their interest in formulating an international agreement to limit the pollutants responsible for acid rain.[30] Establishing an alliance with the Soviet Union to organize such a convention afforded Oslo the means to impose further pressure on the large emitters—Britain, France, and West Germany—which were proving reluctant partners in the network and staunchly opposed to pollution controls.[31] Preliminary results from EMEP afforded Oslo scientific evidence to persuade the Western European countries to assent to an agreement.[32]

In late 1979, thirty-four European and North American countries agreed to the Convention on Long-Range Transboundary Air Pollution (LRTAP) in Geneva. This compromise referred to Stockholm's Principle 21, calling on states in Article 2 to reduce transboundary air pollution as much as economically

---

[26]OECD, *The OECD Program on Long-Range Transport of Air Pollutants* (Paris: OECD, 1977), 9, 10.
[27]Rachel Rothschild, "Détente from the Air: Monitoring Air Pollution during the Cold War," *Technology and Culture* 57, no. 4 (2016): 831–65.
[28]Rothschild, "Détente from the Air," 847.
[29]Valentin Sokolovsky, "Fruits of a Cold War," in Johan Sliggers and Willem Kakebeeke (eds.), *Clearing the Air: 25 Years of the Convention on Long-range Transboundary Air Pollution* (Geneva: United Nations, 2004), 7–17.
[30]Rothschild, "Détente from the Air," 859.
[31]Kaijser, "Combatting 'Acid Rain,'" 372.
[32]Sokolovsky, "Fruits of a Cold War," 7–17.

feasible and to report on their efforts.[33] After its ratification by twenty-four of its signatories, this nonbinding convention came into force in March 1983. "Most importantly," argues historian Arne Kaijser, "the convention established an *international regime* for further negotiation: it formulated a set of rules and created an organizational structure with the Executive Body, comprising government officials from the signatory countries, that would meet once a year, and a permanent secretariat." This regime further provided the firm foundation for the development of transnational research programs and expertise, as well as the continuation of the monitoring program.[34]

Ozone Protection: Human Activity and the Stratosphere

When the Soviet permanent representative to the UN Yakov Malik presented his draft resolution on environmental warfare in October 1974, he had expanded on some of its possible consequences. In addition to producing artificial hurricanes and tsunami, maddening infra-acoustic sound fields, and melting the ice caps, he proffered the creation of "windows" in the earth's ozone layer leading to the increased penetration of ultraviolet radiation: "As a result, all form[s] of life in those areas would be destroyed, and all that territory would be converted into a barren desert."[35] The welfare of the ozone layer had been one of the topics of discussion at Stockholm in 1972, where the stratospheric effects of the exhaust from the new supersonic aircraft, such as the Anglo-French Concorde and Soviet Tupolev-144, were debated and attracted the attentions of the international press.[36] Following the recommendations of the "Study of Man's Impact on Climate" (SMIC), the Stockholm delegates agreed to establish an atmospheric monitoring network of over a hundred stations, which the WMO would coordinate.

On the eve of the Stockholm meeting, the US Department of Transportation had anticipated that by 1985, at least five hundred supersonic planes would be crossing the skies, particularly above the North Atlantic.[37] An American

---

[33]Don Munton et al., "Acid Rain in Europe and North America," in Oran Young (ed.), *The Effectiveness of International Environmental Regimes: Causal Connections and Behavioral Mechanisms* (Cambridge, MA: MIT Press, 1999), 166–7.

[34]Kaijser, "Combatting 'Acid Rain'," 373. Emphasis in original.

[35]Yakov Malik, "Statement by the Soviet Representative (Malik) to the First Committee of the General Assembly, October 21, 1974," in US Arms Control and Disarmament Agency, *Documents on Disarmament 1974* (Washington, DC: US Government Printing Office, 1976), 546–56.

[36]Paul N. Edwards, *A Vast Machine: Computer Models, Climate Data, and the Politics of Global Warming* (Cambridge, MA: MIT Press, 2010), 361–2.

[37]Robert H. Cannon Jr., "Planning a Program for Assessing the Possibility That SST Aircraft Might Modify Climate," *Bulletin of the American Meteorological Society* 52 (9) (1971): 836–42.

fleet would not be among them, however. In March 1971, Congress voted to refuse funds for the development of prototypes after sustained lobbying from environmental groups including the Friends of the Earth, the Environmental Defense Fund, and the Citizens League against the Sonic Boom.[38] Like-minded organizations in the UK (Anti-Concorde Project, the Conservation Society, and the UK Federation against Aircraft Nuisance), France (Association Nationale contre les Bangs Supersoniques), Switzerland (Eidgenössisches Aktionskomitee gegen den Überschallknall Ziviler Luftfahrzeuge), Australia (Anti-Concorde Project), and West Germany (Europäische Vereinigung gegen die schädlichen Auswirkungen des Luftverkehrs) similarly campaigned against the proliferation of such aircraft, while the governments of Denmark, Norway, Iceland, Sweden, Switzerland, Ireland and Japan, were preparing legislation to ban flights above the speed of sound over their territory.[39]

Although the United States had canceled its own supersonic transport program, the Concorde's approaching take-off prompted the Department of Transportation to embark on a study of stratospheric pollution, the Climate Impacts Assessment Program. Over three years, more than one thousand scientists from the United States as well as the UK, France, Canada, Australia, Japan, Belgium, and the Soviet Union contributed to the research that would support the study of the possible environmental effects of supersonic flight. Although its findings were misinterpreted at the time, the study found that a fleet of five hundred aircraft would cause a 16 percent loss of ozone in the northern hemisphere and an 8 percent loss of ozone in the southern hemisphere.[40] With the viability of their supersonic transport program up in the air, France and the UK also undertook their own studies, not least to determine whether there were legitimate grounds for the United States to limit the entry of the Concorde.[41] To prove its safety, the British Meteorological

---

[38]Douglas Ross, "The Concorde Compromise: The Politics of Decision-Making," *Bulletin of the Atomic Scientists* 34, no. 3 (1978): 46–53.

[39]Paul Crutzen, "SST's – a Threat to the Earth's Ozone Shield," *Ambio* 1, no. 2 (1972): 41–51.

[40]Edward Parson, *Protecting the Ozone Layer: Science and Strategy* (New York: Oxford University Press, 2003), 28; A. J. Grobecker, S. C. Coroniti, and R. H. Cannon, *The Effects of Stratospheric Pollution by Aircraft: Final Report* (Washington, DC: Climatic Impact Assessment Program, 1974). In Australia, where the prospect of shorter flights to Europe and North America appealed greatly, the government also commissioned the Australian Academy of Science to undertake its own investigations in 1972. The Academy found little cause for concern. See Australian Academy of Science, *Atmospheric Effects of Supersonic Aircraft* (Canberra: Australian Academy of Science, 1972); Australian Advisory Committee on the Environment, *The Environmental Effects of Supersonic Transport Aircraft* (Canberra: Australian Government Publishing Service, 1974).

[41]Committee on Meteorological Effects of Stratospheric Aircraft (COMESA), *The Report of the Committee on Meteorological Effects of Stratospheric Aircraft (COMESA), vol. 1 and 2* (Bracknell: Meteorological Office, 1976); Comité d'Études sur les Consequences des Vols Stratosphériques (COVA), *Activites 1972–1976: rapport des synthese* (Boulogne: Société Météorologique de France, 1976). See Adrian F. Tuck, "Perspective on Aircraft in

Office developed a new climate model to study the effects of the aircraft on global mean surface temperature. The results of this model alone were not convincing; however, its findings sufficiently aligned with other results and helped to convince Congress to permit the Concorde to fly across the Atlantic.[42]

Although these studies had apparently absolved supersonic transport of atmospheric risk, the ozone layer remained a cause for concern. In the months before the Soviet Union had proposed its environmental warfare convention, an article in the scientific journal *Nature* alerted readers to the risks posed to ozone by chlorofluorocarbons (CFCs). In contrast to the findings of Lovelock, who had recently deemed them benign, Mario Molina and Sherwood Rowland suggested that their steady accumulation would eventually lead to the destruction of atmospheric ozone.[43] These molecules, invented in the 1930s as nontoxic, nonflammable refrigerants, were now also commonly found as propellants in aerosol or spray cans, and in polyurethane foam and solvents. Their production had soared from 54 metric tons in 1934 to 812,522 metric tons in 1974, growing by almost a quarter each year.[44] These findings were especially concerning, as the loss of ozone had become publicly linked to the increased incidence of skin cancer in lower latitudes.

By 1976, the United States, Canada, Sweden, and Norway had announced plans to restrict the production of CFCs and to regulate the use of aerosols.[45] But it was clear that domestic action alone would not suffice—although the United States was responsible for half of the global production of CFCs in 1974, West Germany and then Japan were not far behind.[46] The latter, along with most of Europe, remained unconvinced that intervention was necessary, even as major manufacturing nations, such as the Netherlands and West Germany, began to exercise caution. At the behest of the WMO, the UN

the Stratosphere: 50 Years from COMESA through the Ozone Hole to Climate," *Quarterly Journals of the Royal Meteorological Society* 147, no. 735 (2021): 713–27.

[42]Janet Martin-Nielsen, "Computing the Climate: When Models Become Political," *Historical Studies in the Natural Sciences* 48, no. 2 (2018): 223–45.

[43]James Lovelock, "Atmospheric Fluorine Compounds as Indicators of Air Movements," *Nature* 230 (1971): 379; Mario Molina and Sherwood Rowland, "Stratospheric Sink for Chlorofluoromethanes: Chlorine-Catalysed Destruction of Ozone," *Nature* 249 (1974): 810–12. On Lovelock's research funding and its implications, see Leah Aronowsky, "Gas Guzzling Gaia, or: A Prehistory of Climate Change Denialism," *Critical Inquiry* 47 (2021): 306–27.

[44]Maximilian Auffhammer, Bernard J. Morzuch, and John K. Stranlund, "Production of Chlorofluorocarbons in Anticipation of the Montreal Protocol," *Environmental and Resource Economics* 30 (2005): 378.

[45]Parson, *Protecting the Ozone Layer*, 39–44.

[46]Miranda Schreurs, *Environmental Politics in Japan, Germany and the United States* (Cambridge: Cambridge University Press, 2004), 116–44; Committee on Impacts of Stratospheric Change, *Halocarbons: Environmental Effects of Chlorofluoromethane Release* (Washington, DC: National Academy of Sciences, 1976).

Environment Programme (UNEP) gathered representatives of governments and international organizations in Washington in March 1977 to share resources and coordinate a response. Participants from both developing and developed nations agreed to a World Plan of Action to undertake further monitoring and research, without recommending any changes to production.[47] By year's end, the Programme had formed a Coordinating Committee on the Ozone Layer (Chapter 4), which would become in 1982 the technical and scientific advisory body of its Working Group of Legal and Technical Experts for the Preparation of a Global Framework Convention for the Protection of the Ozone Layer.[48]

Climate Systems: Modeling the Global Climate

As preparations for the World Food Conference in Rome got underway, MIT's Jule Charney presented the 1974 Symons Memorial Lecture of the Royal Meteorological Society. There, he shared his recent work on the "dynamics of desert and drought in the Sahel."[49] Long interested in deserts, he had taken the opportunity of a recent sabbatical in Israel to develop a theory of desertification that might shed light on the drought afflicting the Sahel. "It occurred to me," he recalled, "that the overgrazing which I ascertained occurred over an enormous area ... could influence regional climate."[50] Studying satellite photographs from the new Nimbus 3 weather satellite, Charney hypothesized that overgrazing and deforestation had diminished vegetation, which had increased the reflectivity or albedo of the land's surface, thus exacerbating the region's dryness, "ie: it feeds back upon itself!"[51]

A member of John von Neumann's postwar circle and an architect of the Global Atmospheric Research Program (GARP), Charney's findings found a wide audience. Popularized through works such as Lester Brown and Erik Eckholm's *Spreading Deserts: The Hand of Man*, their resonance with well-established narratives of African land mismanagement carried them throughout

---

[47]UNEP, "World Plan of Action: The Natural Ozone Layer and Its Modification by Man's Activities," in Asit Biswas (ed.), *The Ozone Layer* (New York: Pergamon Press, 1979), 377–81.

[48] Soroos, *The Endangered Atmosphere*, 152–7.

[49]Jule Charney, "Dynamics of Deserts and Drought in the Sahel," *Quarterly Journal of the Royal Meteorological Society* 101, no. 428 (1975): 193–202.

[50]George W. Platzman, *Conversations with Jule Charney* (Cambridge, MA: MIT Libraries, 1987), 145–6.

[51]Charney, "Dynamics of Deserts," 195; Nick Brooks, *Drought in the African Sahel: Long Term Perspectives and Future Prospects*, Working Paper 61 (Norwich: Tyndall Centre for Climate Change Research, 2004), 10.

the UN and development organizations.[52] Charney presented his study at a GARP conference in Wijk, near Stockholm, convened by Bert Bolin, where his approach aligned with the program's objective "to understand the mechanisms that are responsible for the climate fluctuations and to determine the nature of the change in climate caused by given external or internal stimuli (man-made or natural) to the atmosphere-ocean-earth system." As such, "GARP climatic studies should concentrate on those aspects which lend themselves to physical-mathematical (numerical and analytical) model studies."[53]

For Canadian geographer Ken Hare, this meeting represented a turning point for the WMO. The discussions converted the WMO from a meteorological body to a climatic one by framing the "climatic system … as an entire whole, involving interactions between atmosphere, ocean, biota, soils, rocks, ice and human society."[54] Like Charney's approach, which employed simulations of general or global circulation models, GARP's endeavors reinforced climatology's shift to the global scale, away from the localized study that had typified the papers represented at the UN Educational, Scientific and Cultural Organization (UNESCO) Arid Zone Conference in 1961 (Chapter 1).[55] Such mathematical models were proliferating as modelers who had been concerned with numerical forecasting became interested in applying this technique to climatology—as historian Paul Edwards puts it, they were beginning to "simulat[e] the entire Earth system, replicating the world in a machine."[56]

The model that Charney used had been developed at the National Aeronautics and Space Administration's (NASA's) Goddard Institute for Space Studies in New York, where in the age of "small is beautiful" its scientists had begun to turn their mission back to earth.[57] This model allowed him to test his albedo hypothesis in other dryland areas, such as the US Great Plains and the Thar desert.[58] At Wijk, Syukuro Manabe unveiled a new global model

[52]Erik Eckholm and Lester Brown, *Spreading Deserts: The Hand of Man* (Washington, DC: Worldwatch Institute, 1977); Sharon Nicholson, *Dryland Climatology* (New York: Cambridge University Press, 2011); Diana Davis, *The Arid Lands: History, Power, Knowledge* (Cambridge, MA: MIT Press, 2016).

[53]Bert Bolin, *A History of the Science and Politics of Climate Change: The Role of the Intergovernmental Panel on Climate Change* (Cambridge: Cambridge University Press, 2007), 29. See Bert Bolin (ed.), *The Physical Basis of Climate and Climate Modelling* (Geneva: World Meteorological Organization, 1975).

[54]F. Kenneth Hare, "Changing Climate and Human Response: The Impact of Recent Events on Climatology," *Geoforum* 15, no. 3 (1984): 383.

[55]See Matthias Heymann, "The Evolution of Climate Ideas and Knowledge," *WIREs Climate Change* 1 (2010): 581–97.

[56]Edwards, *A Vast Machine*, 139.

[57]Erik Conway, *Atmospheric Sciences at NASA: A History* (Baltimore, MD: Johns Hopkins University Press), 5.

[58]F. Kenneth Hare, "Climate and Desertification," in UN Conference on Desertification (ed.), *Desertification: Its Causes and Consequences* (New York: Pergamon Press, 1977), 100.

of the atmosphere that he had developed with his old collaborator, Richard Wetherald. Theirs was the first three-dimensional global circulation model: a simplified planet that was half land, half "swamp ocean." Just as they had a decade earlier, the pair simulated the impact of doubled carbon dioxide—Dave Keeling's measurements showed that the atmospheric accumulation of carbon dioxide continued to increase and that this trend was a global phenomenon.[59] Manabe and Wetherald's model produced an average warming of around 3°C across the planet. This higher value relative to their 1967 effort was the result of the greater complexity of their new model, which now better emulated the distribution of land and sea as well as the consequences of diminished snow cover in a warmer world. This warming, according to their new model, would be greater at the poles and more modest in the tropics, while cooling the stratosphere.[60] Although Manabe and Wetherald warned readers not "to take too seriously the quantitative aspect" of their results, they were hopeful that their simulation had shed light on some of the mechanisms that direct the responses of the climatic to changes in carbon dioxide.[61] By decade's end, at least nine different general circulation models had been developed, with four American, one British, one Australian, one Canadian, and two Soviet models.[62]

Conceptualizing the earth's atmosphere as a planetary system also found expression beyond the world of such complex models. At Mead and Kellogg's 1975 meeting, freelance English chemist Lovelock had presented his "Gaia hypothesis," that the atmosphere was a homeostatic system regulated by the biosphere.[63] Having earlier identified the increasing accumulation of CFCs in the atmosphere and contributed to NASA's lunar research, Lovelock had embarked with US microbiologist Lynn Margulis on describing the planet as a

---

[59]Charles D. Keeling et al., "Atmospheric Carbon Dioxide Variations at Mauna Loa Observatory, Hawaii," *Tellus* 28, no. 6 (1976): 538–51; Charles D. Keeling et al., "Atmospheric Carbon Dioxide Variations at the South Pole," *Tellus*, 28, no. 6 (176): 552–64.

[60]Syukuro Manabe, "Appendix 2.4: The Use of Comprehensive General Circulation Modelling for Studies of the Climate and Climate Variation," in Bert Bolin (ed.), *The Physical Basis of Climate and Climate Modelling* (Geneva: World Meteorological Organization, 1975), 148–62; Syukuro Manabe and Richard Wetherald, "The Effects of Doubling the CO_2 Concentration on the Climate of a General Circulation Model," *Journal of the Atmospheric Sciences* 32, no. 1 (1975): 3–15; Bolin, *A History*, 30; Mike Hulme, "Classics in Physical Geography Revisited," *Progress in Physical Geography* 25, no. 3 (2001): 385–7.

[61]Manabe and Wetherald, "The Effects of Doubling," 13.

[62]W. Lawrence Gates (ed.), *Report of the JOC Study Conference on Climate Models: Performance, Intercomparison and Sensitivity Studies* (Geneva: World Meteorological Organization, 1979); Matthias Heymann and Nils Randlev Hundebøl, "From Heuristic to Predictive: Making Climate Models into Political Instruments," in Matthias Heymann, Gabriele Grameslberger, and Martin Mahony (eds.), *Cultures of Prediction: Epistemic and Cultural Shifts in Computer-Based Modelling and Simulation* (New York: Routledge, 2017), 100–19.

[63]James Lovelock with J. Dana Thompson, "Appendix II: The Interaction of the Atmosphere and the Biosphere," in Kellogg and Mead (eds.), *The Atmosphere*, 115–23.

self-regulating, living organism.[64] Their concept, while speculative at that point, offered what historian Erik Conway describes as "a view of Earth that could be grasped by systems engineers, a profession that specialized in (nonliving) feedback control systems."[65] Although participants at Mead and Kellogg's meeting were uncertain as to its wider implications, the Gaia hypothesis drew attention to the relationships between biology, atmospheric chemistry, and climate that were yet to be incorporated into models of the general circulation of the atmosphere.

Bolin, who had encouraged Lovelock to publish his first coauthored paper with Margulis in the journal *Tellus*, recognized at Wijk that the GARP would have to "face up to all aspects of the climate problem" and "embrace a range of non-atmospheric disciplines as well" to comprehend the complex chemical and biological processes affecting the climate.[66] How humankind was affecting these processes was the subject of a meeting in Dahlem, West Germany, in late 1976, where convener Swiss chemist Werner Stumm reminded participants of the importance of a holistic framework that combined both sociological and ecological systems. He cited ecologist Crawford "Buzz" Holling, who was working at the International Institute for Applied Systems Analysis (IIASA) in Austria: "It is this interaction that makes it so dangerous, if easier, to take a fragmental view; to study an isolated piece of any ecological or social system. In concentrating on one fragment and trying to optimise it, we find to our surprise, that the rest of the system can respond in unsuspected ways."[67]

Over the course of the 1970s, these conceptualizations of an interconnected global environment that could be scientifically modeled demanded the collection of a wide array of environmental and climatic data. "Almost any new data about the earth are grist to the cyclists' models," Stumm's colleagues noted.[68] Accumulating this environmental data relied on the surveillance technologies of the Cold War and the coordination of the international institutions for which technocratic knowledge continued to serve as the lubricant of intergovernmental cooperation. This infrastructural globalism, to use Edwards's term, found an expression in the Global

[64]James Lovelock and Lynn Margulis, "Atmospheric Homeostasis by and for the Biosphere: The Gaia Hypothesis," *Tellus* 26 (1974): 2–10.

[65]Conway, *Atmospheric Science at NASA*, 119.

[66]James Lovelock, *Homage to Gaia: The Life of an Independent Scientist* (Oxford: Oxford University Press, 2000), x; James Lovelock, *Gaia: A New Look at Life on Earth* (Oxford: Oxford University Press, 2000), xix. Lovelock had earlier published "Gaia as Seen through the Atmosphere," *Atmospheric Environment* 6 (1972): 579–80. See "Basic Principles of Climate Modelling," in Bolin (ed.), *The Physical Basis of Climate*, 13.

[67]Crawford Holling, cited in Werner Stumm, "Introduction," in Werner Stumm (ed.), *Global Chemical Cycles and Their Alterations by Man* (Berlin: Dahlem Konferenzen, 1977), 19–20.

[68]R. M. Garrels and A. Lerman, "The Exogenic Cycle: Reservoirs, Fluxes and Problems," in Stumm (ed.), *Global Chemical Cycles*, 30.

Environmental Monitoring Service that arose from the 1972 Stockholm conference (Chapter 2).[69] Based in the Nairobi headquarters of the new UNEP under the Earthwatch program, this service had its institutional genesis in the ICSU, headed by Roger Revelle and Tom Malone. In the collaborative tradition of the International Geophysical Year (Chapter 1), the service was intended as a centralizing mechanism for existing data collection platforms, such as those already established under the WMO for the World Weather Watch and background air pollution. By the end of the decade, Malone believed that the Earthwatch network might allow for scientists to quantitatively understand "the functioning of the earth's life-support system."[70]

Climate Alarm: Scientific Debate and Controversy

On his return to Canberra from preparatory meetings for the 1974 World Food Conference in Rome, the Australian prime minister's advisor encouraged his government to investigate an issue that had emerged during the proceedings. In Rome he had learned that the polar ice cap was extending over a larger area, which was affecting global weather patterns "to a significant degree," and that the drought conditions in Africa and South Asia were "evidence of this change." "It is, I believe, important that the Government should be informed on the likely effect of a 'Little Ice Age' on the Australian climate if it is in fact, occurring," he wrote. "Such knowledge, even if it could be expressed only in terms of probabilities, could have a profound effect on the pattern of development and investment."[71] In the United States, meanwhile, the CIA was undertaking its own investigations and warned, "The stability of most nations is based upon a dependable source of food, but this stability will not be possible under the new climatic era." Its 1974 report also noted that governments abroad had enhanced their climate research programs in recognition of these concerns.[72] Popular and scientific reports of devastating climate-related disasters from around

---

[69]Edwards, *A Vast Machine*, 23.

[70]Tom Malone, "Introductory Remarks from SCOPE," in Jill Williams (ed.), *Carbon Dioxide, Climate and Society* (Oxford: Pergamon Press, 1978), 18.

[71]"Letter from H.C. Coombs to E.G. Whitlam, November 14, 1974," in Changes in the World's Weather, A463, 1974/4555, NAA Canberra. The Australian Academy of Science undertook the subsequent inquiry, see Committee on Climatic Change, *Report of a Committee on Climatic Change* (Canberra: Australian Academy of Science, 1976).

[72]CIA, *A Study of Climatological Research as It Pertains to Intelligence Problems* (Washington, DC: CIA, 1974); James P. Sterba, "Problems from Climate Changes Foreseen in a 1974 CIA Report," *New York Times*, February 2, 1977, 10.

the world heightened public, scientific, and government concern for climatological issues and their prediction.[73] Was the world's climate indeed changing? If so, how, and to what effect?

In late 1972, US climatologist J. Murray Mitchell reported that global temperatures had continued to decline since the 1940s—a trend that followed a period of worldwide warming from the 1880s that he had identified a decade earlier. Mitchell had shared his initial results at the UNESCO Arid Zone meeting in Rome (Chapter 1), having concluded that the "extant quantitative theories of climatic change, including solar variability, the secular increase of carbon dioxide, and volcanic activity ... appear to be insufficient to account for the recent cooling."[74] Now, even though satellite mapping showed a larger than usual accumulation of snow and ice in the northern hemisphere, and having taken into account the cooling effects of dust particles or aerosols, he concluded that prevailing temperature trend would be only temporary: "It is more likely than not that the net impact of human activities on the climate of future centuries will be in the direction of warming."[75]

Although the 1963 Test Ban Treaty had helped to dissipate political and public interest in the United States in the possible climatic effects of nuclear tests, the rising wave of environmentalism aroused wide concern about the climatic effects of air pollution.[76] At a 1968 meeting on global air pollution in Dallas, where Mitchell had also presented his findings, another team of US scientists, led by Reid Bryson, determined that the climatic effects of a "human volcano" of industrial and agricultural dust outweighed the effects of rising carbon dioxide levels "resulting in a rapid downward trend of temperature."[77] Bryson's findings soon found support from climate modeling undertaken by NASA's S. Ichtiaque Rasool and Schneider, the results of which they published during the SMIC meeting (Chapter 2). If the rate of atmospheric pollution continued to grow, they concluded, it would cause a large decrease in global temperature that would be "sufficient to trigger an ice age."[78] "When is the ice

---

[73]For example, Tom Alexander, "Ominous Changes to the World's Weather," *Fortune* February (1974): 90–5, 142–52; further examples cited in Thomas C. Peterson, William M. Connolley, and John Fleck, "The Myth of the 1970s Global Cooling Scientific Consensus," *Bulletin of the American Meteorological Society* 89, no. 9 (2008): 1325–38.

[74]J. Murray Mitchell, "Recent Secular Changes of Global Temperature," *Annals of the New York Academy of Sciences* 95 (1961): 249.

[75]J. Murray Mitchell, "The Natural Breakdown of the Present Interglacial and Its Possible Intervention by Human Activities," *Quaternary Research* 2, no. 3 (1972): 436–45.

[76]Matthias Dörries, "In the Public Eye: Volcanology and Climate Change in the 20th Century," *Historical Studies in the Physical and Biological Sciences* 37, no. 1 (2006): 87–125.

[77]Reid A. Bryson and Wayne M. Wendland, "Climatic Effects of Atmospheric Pollution," in S. Fred Singer (ed.), *Global Effects of Environmental Pollution* (Dordrecht: D. Reidel, 1970), 130–8.

[78]S. Ichtiaque Rasool and Stephen Schneider, "Atmospheric Carbon Dioxide and Aerosols: Effects of Large Increases on Global Climate," *Science* 173 (1971): 138–41.

CLIMATE CHANGE THREAT TO WORLD?   Long term weather changes, notably a southward migration of the monsoons, combined with a sharp reduction in fertilizer production, threaten food supplies for a large part of the world's population, according to experts who met recently at the Rockefeller Foundation. The multiple problems associated with food production, increasing populations, and what looked to some participants as ominous shifts in rainfall for India and sub-Saharan Africa, led the group to recommend internationally co-ordinated efforts to predict climate at least one year ahead, and the establishment of food reservoirs for inevitable emergencies.

FIGURE 3.1 *Reports of future climate change and food shortages prompted interdisciplinary gatherings, such as the meeting pictured above at the Rockefeller Foundation in 1974. This collage includes photographs of prominent scientists and analysts who appear in this book: (from left to right) Lester Brown, Stephen Schneider, Gilbert F. White, Edith Brown Weiss, Reid Bryson, and F. Kenneth Hare. RF Illustrated, August 1974. Courtesy of Rockefeller Archive Center.*

age coming?," Schneider later recalled journalists from Europe and the United States asking him during a press conference in Stockholm.[79]

Although Rasool and Schneider subsequently revised their findings, Bryson remained steadfastly committed to his cooling position and shared widely his pessimism about its implications for global agricultural production (Figure 3.1). He had become "perhaps the most outspoken and oft-quoted climatological doomsayer," as the *New York Times* put it in 1974.[80] Historian Robert Naylor describes Bryson as a "crisis climatologist," who both shaped, and was shaped by, the catastrophism of the period.[81] Having testified before a US Senate subcommittee on food and foreign policy in late 1973, Bryson joined with his German and British counterparts Hermann Flohn and Hubert Lamb to issue a warning on the eve of the World Food Conference:

If national and international policies do not take these near-certain failures into account, they will result in mass deaths by starvation and probably

---

[79]Stephen Schneider, *Science as a Contact Sport: Inside the Battle to Save Earth's Climate* (Washington, DC: National Geographic, 2009), 62.
[80]Alan Anderson Jr., "Forecasting: Cloudy," *New York Times*, December 29, 1974, 156.
[81]Robert Luke Naylor, "Reid Bryson: The Crisis Climatologist," *WIREs Climate Change* 13, no. 1 (2021): e744.

in anarchy and violence that could exact a still more terrible toll. ... We are aware of differences among experts as to the cause-and-effect relationships of observed climatic facts and, consequently, as to the most likely prognosis. Professionally, the differences are important, but they do not—and should not be allowed to—obscure the larger consensus that the observed changes are neither trivial nor ephemeral.[82]

According to the *New York Times*, "the Bonn statement caused an uproar." Furthermore, "a number of climatologists and meteorologists took issue with the 'consensus' it described." The newspaper cited Manabe, Charney, and Joseph Smagorinsky as each refuting the statement: "One man's trend is simply another man's periodicity," observed the latter.[83]

Soviet climate scientist Mikhail Budyko was likewise unmoved, remaining adamant that human activities were warming, not cooling, the climate.[84] In the wake of the extremely hot and dry summer of 1972 in central Russia, he had written in *Pravda*, "With today's speed of technological development, it will start influencing economic activity not later than in 20–30 years, and in 50–80 years will dramatically change it in many countries."[85] Although he was cautious about these changes, Soviet studies tended to suggest that a warming climate would benefit the Soviet Union, particularly in terms of agricultural productivity, while impeding the United States.[86] The prospect of a global cooling, however, represented a grave threat to Moscow. Another CIA report, to which Bryson's research had also contributed, predicted as much: "A shorter growing season would restrict production in the high latitude areas, like Canada and the USSR."[87] In light of these findings, Soviet scientists and the press criticized the CIA's report on the grounds of insufficient scientific evidence.[88]

Lamb was meanwhile convening a WMO symposium on long-term climatic fluctuations, which Kellogg later described as a "milestone" for largely settling

---

[82]Anderson Jr., "Forecasting: Cloudy," 156.

[83]According to Howe, their response to the Bonn statement was likely a result of divisions within the climate science community based on different methodologies, disciplines, and resources. See Joshua P. Howe, *Behind the Curve: Science and the Politics of Global Warming* (Seattle: University of Washington Press, 2014), 98–9.

[84]Anderson Jr., "Forecasting: Cloudy," 156.

[85]Mikhail Budyko, *Pravda*, June 25, 1972, cited in Anna Mazanik, "Environmental Change and Soviet Media Before 1986: Dissident and Officially Sanctioned Voices," in Marianna Poberezhskaya and Teresa Ashe (eds.), *Climate Change Discourse in Russia: Past and Present* (London: Routledge, 2019), 38.

[86]Katja Doose, "A Global Problem in a Divided World: Climate Change Research during the late Cold War, 1972–91," *Cold War History* 21, no. 4 (2021): 469–89.

[87]CIA, *Potential Implications of Trends in World Population, Food Production, and Climate* (Washington, DC: CIA, 1974), 30.

[88]Mazanik, "Environmental Change and Soviet Media," 39.

the dispute over global cooling. The aerosols from industrial and agricultural activities did not cause cooling of the atmosphere after all, which left "the greenhouse warming to dominate the stage," Kellogg recalled.[89] Affording the 1975 symposium further significance was the publication of findings derived from the ice cores recently drilled from the Greenland ice sheet. Geochemist Wallace Broecker argued that the cooler temperatures of the 1960s were the result of a solar cycle, which would soon give way to a carbon dioxide-induced "pronounced global warming" that would be "beyond the limits experienced during the last 1000 years."[90] The ice core evidence had confirmed what modelers in the Soviet Union and the United States had begun to speculate in the 1960s—the earth was not as geologically stable as they had once supposed.[91]

The CIA's report that had drawn from Bryson's research had reached Whitehall by May 1976.[92] The prospect of an imminent ice age had earlier prompted the House of Commons to seek clarification from the Meteorological Office, after the BBC aired a documentary in 1974 for which Bryson and Lamb had served as consultants.[93] The Cabinet Office now turned again to the Meteorological Office, whose research director John Sawyer described it as a "completely misleading report which undue prominence to the views of a few climatologists with especially alarmist views ... The evidence that a permanent climatic change of significant magnitude is in train is at best exceedingly sketchy."[94] Although Sawyer had earlier published on the "greenhouse effect," he had seen "no immediate cause for alarm about the consequences of carbon dioxide increase in the atmosphere."[95]

His sober assessment persisted in the reluctance of his successors at the Meteorological Office to engage with what they described as "alarmist"

---

[89]William W. Kellogg, "Mankind's Impact on Climate: The Evolution of an Awareness," *Climatic Change* 10 (1987): 113–36; WMO, *Proceedings of the WMO/IAMAP Symposium on Long-Term Climatic Fluctuations* (Geneva: World Meteorological Organization, 1975).

[90]Wallace S. Broecker, "Climatic Change: Are We on the Brink of a Pronounced Warming?," *Science* 189 (1975): 460–3.

[91]Erik Conway, "Planetary Science and the 'Discovery' of Global Warming," in Roger Launius (ed.), *Exploring the Solar System: The History and Science of Planetary Exploration* (New York: Palgrave Macmillan, 2013), 183–202.

[92]Jon Agar, "'Future Forecast – Changeable and Probably Getting Worse': The UK Government's Early Response to Anthropogenic Climate Change," *Twentieth Century British History* 26, no. 4 (2015): 602–28.

[93]Nigel Calder, *The Weather Machine* (Harmondsworth: Penguin Books, 1974); Meteorological Office (Ice Age Predictions), House of Commons Hansard, vol. 882, Column 441, December 3, 1974.

[94]John S. Sawyer, June 18, 1976, cited in Agar, "'Future Forecast'," 612.

[95]John S. Sawyer, "Man-Made Carbon Dioxide and the 'Greenhouse' Effect," *Nature* 239 (1972): 23–6.

views from across the Atlantic.[96] As concerns grew in Whitehall, however, the Cabinet Office directed the Meteorological Office in 1978 to apply the modeling capabilities developed for the Concorde question to the problem of carbon dioxide.[97] The reluctance of the Meteorological Office to engage in the study or prediction of climate change had reputational implications. The Cabinet Office had become aware that the UK had not kept pace with research in the United States and that its scientific leadership was coming under fire.[98] The European Communities Commission had begun to develop its own coordinated climatological research program, focused on the interaction of human activity and the climate.[99] After all, "neither the weather nor the climate respects geographic or political frontiers and that this is an indication of the need for and the desirability of a common research effort at a European level." As historian Janet Martin-Nielsen observes, "to maintain the Meteorological Office's status as a leading European weather service, the United Kingdom now needed to play a central role in climate debates and research."[100] Although divergent views on the direction of climate trends persisted, national governments had become keenly interested in the geopolitical implications of a changing climate.

Collective Concern: The World Climate Conference

In February 1977, Kellogg presented a report to the WMO on the effects of human activities on global climate, "with consideration of the implications of a possibly warmer Earth."[101] By his own admission, he had relied heavily on the SMIC report he had helped prepare in the lead up to the 1972 Stockholm conference (Chapter 2), with the addition of "better estimates of the greenhouse warming for a given increase in carbon dioxide and other

---

[96]Agar, " 'Future Forecast'," 602–28; Martin-Nielsen, "Computing the Climate," 223–45. According to Hamblin, Sawyer was similarly nonplussed about US and Soviet concerns about the risks of weather modification for military purposes, see Jacob Darwin Hamblin, *Arming Mother Nature: The Birth of Catastrophic Environmentalism* (New York: Oxford University Press, 2013), 205–6.

[97]Martin-Nielsen, "Computing the Climate," 223–45.

[98]Martin-Nielsen, "Computing the Climate," 223–45.

[99]E. Holst, *Report Drawn Up on Behalf of the Committee on Energy and Research on the Proposal from the Commission of the European Communities to the Council (Doc. 350/78) for a Decision Adopting a Multiannual Research Programme for the European Economic Community in the Field of Climatology*, EP Working Documents, Document 478/78, December 7, 1978, 5, Archive of European Integration, https://v.gd/I5JHNY.

[100]Martin-Nielsen, "Computing the Climate," 237.

[101]William W. Kellogg, *Effects of Human Activities on Global Climate* (Geneva: World Meteorological Organization, 1977).

related factors."[102] Less than a year earlier, the WMO's executive committee had approved a Statement on Climatic Change prepared by its own expert committee. The WMO had set up this panel in 1974 in response to the urging of the UN General Assembly that it study climatic changes. Chairing the committee was Australia's Bill Gibbs, who had served as the WMO's first vice president between 1967 and 1975; he was joined by Bolin, Flohn, and Mitchell, as well as representatives of UNESCO, the UNEP, and the ICSU. Taking a more cautious approach than Kellogg later would, this committee dismissed the likelihood of a global cooling and acknowledged that human activities, such as burning oil and coal, could alter the climate, as might the release of other chemicals and thermal emissions. They recommended that given the possible consequences of these activities, improved prediction of their impacts on the global climate was necessary.[103]

As Kellogg prepared his report, more extreme weather hit the headlines—this time, in Europe and North America. Drought in northwestern Europe and California caused widespread economic dislocation in 1976, and in the following winter it brought particularly cold conditions to eastern and central North America, which rising fuel costs exacerbated.[104] By making specific predictions, Kellogg had gone further than Gibbs's expert panel, which had stated that "a large carbon dioxide increase would result in a very significant warming of global climate, by several degrees celsius."[105] In his work, Kellogg had combined the effect of increasing carbon dioxide on global temperatures with a range of estimates of future fossil fuel consumption and drew a "temperature versus time scenario" curve out to the year 2050. He later recalled, "Even though this exercise was accompanied by many caveats and warnings not to take this prediction too seriously, it seems to have captured the imagination of a great many people both inside and outside the climatological community."[106] The WMO responded shortly afterward, deciding to convene a scientific and technical World Climate Conference in early 1979 in Geneva.[107]

The WMO had support from other members of the UN family as well as the ICSU, not least as the question of climate had been central to the decade's

[102]Kellogg, "Mankind's Impact on Climate," 123.

[103]"WMO Statement on Climatic Change," *WMO Bulletin* 25, no. 3 (1976): 211; "Technical Report by the WMO Executive Committee Panel of Experts on Climatic Change," *WMO Bulletin* 26, no. 1 (1977): 50–4.

[104]F. Kenneth Hare and W. R. Derrick Sewell, "Awareness of Climate," in Robert Kates and Ian Burton (eds.), *Geography, Resources and Environment: Themes from the Work of Gilbert F. White*, vol. 2 (Chicago: University of Chicago Press, 1986), 210, 219.

[105]*World Meteorological Organization Annex X* (Geneva: World Meteorological Organization, 1976), 185.

[106]Kellogg, "Mankind's Impact on Climate," 123.

[107]*Resolution 15 EC-XXIC World Climate Conference* (Geneva: World Meteorological Organization, 1977), 140–1.

food (1974), water (1976), and desertification (1977) conferences.[108] The organization of the World Climate Conference fell to Canadian geographer Ken Hare, who had been involved in the preparations for the desertification meeting in Nairobi as well as chairing the meetings that resulted in the 1976 report, *Climate Change, Food Production and Interstate Conflict*.[109] The findings of those meetings, funded by the Rockefeller Foundation with its interest in the future prospects of the Green Revolution, reached the administrator of the National Oceanic and Atmospheric Administration (NOAA) in Washington, DC, Robert M. White, who then brought them to the attention of Kissinger and the Ford White House.[110] In Hare's telling, the White House was already "highly sensitized to the issue" of forecasting grain yields, which White leveraged to attain executive support for a climate conference akin to those on water and food. The White House, in turn, convinced the UN secretary-general that "yet another conference" would be valuable, leading to the matter's delegation to the WMO.[111]

White would go on to chair the 1979 conference, where specialists would be invited to discuss climate and associated economic and social questions. As Hare later put it, his colleagues agreed that, in contrast to the intergovernmental nature of the 1972 Stockholm conference with its attendant difficulties, "any question of a political conference should await reasonable expert consensus that the problems were real and that feasible solutions existed."[112] Meeting in Geneva were invited experts from sixteen countries, including the Soviet Union and China, with participants from across the developed and developing nations of Europe, Africa, Asia (including North Korea), Australasia, the Middle East, and South America. At the conclusion of the 1979 conference, the participants called for greater investment in climate research and "to foresee and to prevent potential man-made changes in climate that might be adverse to the well-being of humanity." Given the vulnerability of all countries to climate variability, they advocated for a "common global strategy" and for "global co-operation to explore the future course of global climate and to take

---

[108]Robert M. White, "Climate at the Millennium," in F. Kenneth Hare et al. (eds.), *Proceedings of the World Climate Conference: A Conference of Experts on Climate and Mankind* (Geneva: WMO, 1979), 7.

[109]Hare, "Changing Climate," 387. Those conferences were held in January 1974 in New York and in June 1975 at Bellagio, Italy, see *Climate Change, Food Production and Interstate Conflict* (New York: Rockefeller Foundation, 1976).

[110]F. Kenneth Hare, "Climate: The Neglected Factor," *International Journal* 36 (1981): 378. On the Rockefeller Foundation's sponsorship of climate science, see Edouard Morena, "The Climate Brokers: Philanthropy and the Shaping of a 'US-Compatible' International Climate Regime," *International Politics* 58 (2021): 541–62.

[111]Hare, "Climate," 378–9.

[112]Hare, "Climate," 379.

this new understanding into account in planning for the future development of human society."[113]

Keen to ensure the coherence of the conference proceedings, Hare had brought together the speakers almost a year earlier at the IIASA at Laxenberg, near Vienna.[114] This thinktank had been formed in 1972 to facilitate scientific collaboration across the Iron Curtain on questions of the global future, and it deployed predictive computer models that could be applied to both East and West.[115] The institute had established a formal relationship with the WMO in 1977, which led to a cosponsored meeting on the subject of "carbon dioxide, climate and society" in early 1978.[116] How the current understanding of the impacts of carbon dioxide on the climate might inform energy policy had been the focus of this meeting. There were familiar faces from Europe, North America, and Australia among the one hundred scientists that attended, including speakers Bolin, Schneider, and Flohn, as well as Keeling, Broecker, and Kellogg. Having assembled into three groups to discuss different dimensions of the climate problem, they agreed on the need for simulations of climatic change at the regional level, more sophisticated general circulation models of the climatic system, and the reduction of energy demand and diversification of the sources of energy supply—it was too early to advocate policies requiring reduction in the use of coal and fossil fuels.[117]

From Hare's perspective, this issue of uncertainty demanded closer attention. Reflecting on the World Climate Conference a year later, he blamed "the climatologist's inability to predict" for the "failure" of atmospheric scientists to persuade governments about the importance of anthropogenic climate change. Gloomily, he concluded, "There is no sign that he is about to acquire the skill."[118] As for his assessment of the World Climate Conference, he welcomed the appreciation that climate variability was relevant to many sectors beyond agriculture and the agreement as to the significance of the problem. He was disappointed, nevertheless. The conference, in his view, had yet to make an impact beyond the atmospheric sciences, attracting little interest from economists to the study of climate impacts. Only one economist had contributed to the conference, attempting to assess the impact of

[113]"The Declaration of the World Climate Conference," in Hare et al. (eds.), World Climate Conference, 713–14.
[114]Climate Research Board, National Research Council, International Perspectives on the Study of Climate and Society: Report of the International Workshop on Climate Issues (Washington, DC: National Academies Press, 1978).
[115]Isabell Schrickel, "Control versus Complexity: Approaches to the Carbon Dioxide Problem at IIASA," Berichte zur Wissenschaftsgeschichte 40 (2017): 140–59.
[116]See Jill Williams (ed.), Carbon Dioxide, Climate and Society (Oxford: Pergamon Press, 1978).
[117]Jill Williams and Renee Calderon, "Carbon Dioxide, Climate and Society," Environment International 1, no. 5 (1978): 277–8.
[118]Hare, "Climate," 386.

stratospheric pollution in the United States.[119] Generating more interest would be difficult, in Hare's view, as the WMO had opted to publish the proceedings "in house," rather than seeking a commercial publisher and advertising their "crucial material."[120]

There was cause for hope, however. He welcomed the Laxenberg Institute's ongoing research on the economic impacts of climate change, as well as the recent establishment of the WMO's World Climate Programme.[121] This Programme would oversee and expand the organization's monitoring and data collection activities, while fostering further research on climate change and variability, the uses of climate information for planning purposes, and, finally, the impacts of climatic variability on human and natural systems, which was delegated to the UNEP.[122] Through integrated systems studies, the WMO hoped to engage further with economists, sociologists, and environmental scientists, and to ensure that this work involved decision makers: "We realise that the effects of weather and climate cannot be considered in isolation from other effects—economic, political and social," Gibbs noted.[123]

Fuel for the Future: Energy and the Atmosphere

In a commemorative essay for the twentieth anniversary of Hawai'i's Mauna Loa Observatory in 1978, Keeling noted, "The increasing amount of CO_2 in the atmosphere from the burning of fossil fuels has become a serious environmental concern. ... [A]mounts more than 10% over amounts recorded before the Industrial Revolution, and a rise of 6% in the last 19 years alone." Keeling attributed this rise to the industrial world's ongoing reliance on burning fossil fuels, which met 97 percent of energy demand.[124] Such rising fossil fuel emissions could lead to the depletion of the West Antarctic Ice Sheet and the inundation of coastal cities within just fifty years, a glaciologist had recently

---

[119]Ralph C. d'Arge, "Climate and Economic Activity," in Hare et al. (eds.), *World Climate Conference*, 652–81.

[120]Hare, "Climate," 381.

[121]Hare, "Climate," 382–3.

[122]World Meteorological Organization, *Outline Plan and Basis for the World Climate Programme, 1980–1983* (Geneva: World Meteorological Organization, 1980).

[123]Statement by W.J. Gibbs on behalf of the Secretary-General, World Meteorological Organization, to the Interagency Consultative Meeting on Population, Resources, Environment and Development, Geneva, 28–29 July 1980. Helen Freeman, 15 September 1980, W.M.O – Miscellaneous Group. World Weather Watch, A1838, 871/8/13 Part 4, NAA, Canberra.

[124]C. D. Keeling, "The Influence of Mauna Loa Observatory on the Development of Atmospheric CO_2 Research," in John Miller (ed.), *Mauna Loa Observatory: A 20th Anniversary Report* (Boulder: Environmental Research Laboratories, 1978), 36.

warned.[125] Although oil prices had stabilized in the second half of the 1970s, the oil shock of 1979 renewed anxieties about the availability of the world's energy resources, which the Three Mile Island accident only worsened.

In terms of energy policy, governments were quickly learning to take account of "interdependence," what political scientists Joseph Nye and Robert Keohane described in 1977 as the "reciprocal effects among countries."[126] The decade's volatile weather conditions, meanwhile, had prompted the Ford and Carter administrations in the United States to review the role of climate in their nation's policymaking.[127] "Our present awareness of current climate fluctuations and our ability to predict their occurrences falls far short of fulfilling the needs of the planners and policymakers who must face problems associated with these fluctuations," cautioned the NOAA in 1977.[128] Already, the National Academy of Sciences had invited Revelle to lead a subcommittee to evaluate the state of knowledge on the relationship between fossil fuels and atmospheric change. The resulting 1977 report, *Energy and Climate*, did not sound alarm bells. In characteristically reserved fashion, Revelle's report concluded, "The climatic effects of carbon dioxide release may be the primary reason for limiting energy production from fossil fuels over the next few centuries." To address the "profound uncertainties" of this issue, the report called for an "extraordinarily interdisciplinary" and well-coordinated research effort for which there was no existing institutional mechanism.[129]

Carter's Council of Environmental Quality embarked on its own investigation into the carbon dioxide problem and its "implications for policy in the management of energy and other resources." Chaired by ecologist George Woodwell and involving both Revelle and Keeling, the 1979 report warned,

If we wait to prove that the climate is warming before we take steps to alleviate the CO_2 build-up, the effects will be well underway and still more

---

[125]John Mercer, "West Antarctic Ice Sheet and CO2 Greenhouse Effect: A Threat of Disaster," *Nature* 271 (1978): 321–5. On the history of scientific concern about the vulnerability of the West Antarctic ice sheet, see Jessica O'Reilly, Naomi Oreskes, and Michael Oppenheimer, "The Rapid Disintegration of Projections: The West Antarctic Ice Sheet and the Intergovernmental Panel on Climate Change," *Social Studies of Science* 42, no. 5 (2012): 709–31.

[126]Cited in Daniel Sargent, "The United States and Globalization in the 1970s," in Niall Ferguson et al. (eds.), *The Shock of the Global: The 1970s in Perspective* (Cambridge, MA: Belknap Press, 2010), 51.

[127]Gabriel Henderson, "Adhering to the 'Flashing Yellow Light': Heuristics of Moderation and Carbon Dioxide Politics during the 1970s," *Historical Studies in the Natural Sciences* 49, no. 4 (2019): 384–419.

[128]National Oceanic and Atmospheric Administration (NOAA), *NOAA Climate Program* (Rockville: US Department of Commerce, 1977), 18.

[129]Geophysics Study Committee, *Energy and Climate: Studies in Geophysics* (Washington, DC: National Academy of Science, 1977). See also Roger Revelle, "The Scientist and the Politician," *Science* 187, no. 4181 (1975): 1100–5.

difficult to control. The earth will be committed to appreciable changes in climate with unpredictable consequences. The potential disruptions are sufficiently great to warrant the incorporation of the CO_2 problem into all considerations of policy in the development of energy.[130]

The climatic consequences of this problem were the focus of another report commissioned by the National Research Council. Chaired by Charney, and with Bolin the only contributor from outside the United States, the Ad Hoc Study Group on Carbon Dioxide and Climate delivered in 1979 findings that would be "disturbing to policymakers": "The ocean, the great and ponderous flywheel of the global climate system, may be expected to slow the course of observable climate change. A wait-and-see policy may mean waiting until it is too late."[131]

Elsewhere, governments and intergovernmental organizations took seriously these concerns about carbon dioxide. In 1979, the OECD's Council on Coal and the Environment had recommended that "member countries, in the light of appropriate research results, seek to define acceptable fuel qualities, emission levels or ambient media qualities, as appropriate for carbon dioxide."[132] West Germany's Chancellor Helmut Schmidt had meanwhile shared in Washington that "in his judgement, CO_2 accumulation in the atmosphere represented a major threat to the future of mankind."[133] Similarly, the British Cabinet Office had identified the carbon dioxide problem as "the most serious potential man-made threat to the global climate."[134] With energy at the top of the agenda at the Group of Seven (G7) summit in Tokyo in June 1979, the assembled heads of government declared, "We need to expand alternative sources of energy, especially those which will help to prevent further pollution, particularly increases of carbon dioxide and sulphur oxides in the atmosphere."[135]

For Weiss too, these developments warranted renewed consideration of managing carbon dioxide during what she described as "the century

[130]George Woodwell et al., *The Carbon Dioxide Problem: Implications for Policy in the Management of Energy and Other Resources: A Report to the Council on Environmental Quality* (Washington, DC: US Department of Energy, 1979), 11.
[131]Verner Suomi, "Foreword," in National Research Council, *Carbon Dioxide and Climate: A Scientific Assessment* (Washington, DC: National Academy of Sciences, 1979), viii.
[132]OECD, *Recommendation of the Council on Coal and the Environment*, OECD/Legal/0173, adopted May 8, 1979.
[133]Committee on Governmental Affairs, "CO_2 Symposium: Findings and Recommendations," in *Carbon Dioxide Accumulation in the Atmosphere, Synthetic Fuels and Energy Policy: A Symposium* (Washington, DC: Government Printing Office, 1979), iv.
[134]Cited in Martin-Nielsen, "Computing the Climate," 223–45.
[135]G7 Summit, "Declaration: Tokyo, June 29, 1979," G7 Research Group, University of Toronto, https://v.gd/9atXWv.

of transition" from a fossil fuel to a non-fossil-fuel economy.[136] The Carter administration had recently turned to the accelerated exploitation of coal resources to reduce the nation's dependence on foreign oil, which was, in light of the findings of the 1979 reports on climate change, a particularly concerning prospect, as coal combustion releases more carbon dioxide than either oil or natural gas.[137] Despite those findings, President Carter had been receptive to the findings of another Carroll Wilson exercise, the World Coal Study.

Having recently convened the Workshop on Alternative Energy Strategies, Wilson had turned in 1978 to the promotion of coal. "It was a crusade," wrote a biographer, "an effort by a man, disenchanted with the possibilities of nuclear power, to demonstrate that coal could provide much of the world's 'additional energy needs' during the remaining years of the twentieth century."[138] Supported by the multinational corporation Bechtel, Wilson gathered eighty delegates from the major coal-producing and consuming nations, including Australia, China, West Germany, India, Japan, the UK, and the United States, which together constituted three quarters of the world's energy consumption. Among them were government officials, as well as representatives from companies such as British Petroleum, Broken Hill Proprietary Company Limited, Shell Coal International, and Commonwealth Edison Co. They shared the view that extant energy supplies would not be sufficient to meet the world's growing energy needs, not least as "it is now widely agreed that the availability of oil in international trade is likely to diminish over the next two decades."[139]

The Study's 1980 report acknowledged the findings of the World Climate Conference but suggested that aside from the growing concentration of carbon dioxide in the atmosphere, "there is still great uncertainty about most other aspects of the global carbon cycle." On these grounds, the report concluded that "the present state of knowledge about CO_2 effects on climate does not justify action to limit or reduce the global use of fossil fuels or delay the expansion of coal use even if a mechanism for such concerted actions by all nations were available."[140] Noting the contribution of deforestation to the emission of carbon dioxide, and the difficulties of distinguishing anthropogenic effects from natural climate cycles, the Study's report called for further national and international research on the issue.[141]

---

[136]Edith Brown Weiss, "A Resource Management Approach to Carbon Dioxide during the Century of Transition," *Denver Journal of International Law and Policy* 10, no. 3 (1981): 487.

[137]Weiss, "A Resource Management Approach," 488.

[138]"One of a Kind: Carroll L. Wilson, a Biography by Milton Lomask," in *Carroll L. Wilson, 1910–1983: A Report of the Carroll L. Wilson Awards Committee*, 178. Reproduced from copy T171. M4218.W553.1987 in the Institute Archives and Special Collections, MIT Libraries.

[139]Carroll Wilson (ed.), *Coal: Bridge to the Future, Report of the World Coal Study* (Cambridge, MA: Ballinger, 1980), xvi.

[140]Wilson, *Coal*, 31–2.

[141]Wilson, *Coal*, 149–50.

Counselling a cautious approach, Weiss meanwhile encouraged govern-ments to engage with industry, as they had on the matter of ozone depletion, to examine the prospects for the control or recycling of carbon dioxide emissions, as well as the kind of incentives that might be required.[142] On the climate question, however, industry was already taking a different path. The American Petroleum Institute had in 1979 formed an internal task force to "increase industry's understanding of the CO_2 and climate problem."[143] Despite briefings to the contrary, the Institute suggested in 1980 that the findings of the World Coal Study were "consistent with the authoritative statement on the carbon dioxide question issued by the World Climate Conference in 1979."[144] Gathered in Venice in June 1980, a year after declaring their concern for reducing pollutants in the atmosphere, President Carter and his G7 counterparts agreed to dramatically reduce their reliance on imported oil by doubling coal production by 1990, and developing nuclear power and synthetic fuels.[145]

Conclusion

Writing in *Europe* magazine in 1979, British diplomat Crispin Tickell had echoed the declaration of the World Climate Conference. The time had come, he argued, to embark on negotiations for an international approach to managing climate change. A Foreign and Commonwealth Office diplomat, Tickell had spent a sabbatical year at the Harvard Centre for International Affairs and was now chief of staff for the European Communities Commission president Roy Jenkins. He anticipated that the nature of climate change and variability would demand novel forms of international governance: "The issues raised are very large, and relate to the increasing need for effective international management of the resources of our planet. Such management will require a much greater sense of international responsibility on the part of individual governments than has so far been shown."[146]

---

[142]Weiss, "A Resource Management Approach," 509.
[143]See Neela Banerjee, "Exxon's Oil Industry Peers Knew about Climate Dangers In the 1970s, Too," *Inside Climate News*, December 22, 2015, https://v.gd/QeY6aU.
[144]American Petroleum Institute, *Two Energy Futures: A National Choice for the 80s* (Washington, DC: API, 1980), 80. See Benjamin Franta, "Early Oil Industry Disinformation on Global Warming," *Environmental Politics* 30, no. 4 (2021): 663–8.
[145]Henry Tanner, "US and Allies Vow to Cut Oil Demand and Use More Coal," *New York Times*, June 24, 1980, 1, 6.
[146]Crispen Tickell, "The Constant Inconstant: Climate, Its Study, Its Implications," *Europe* no. 215 (1979): 54–5. See also Crispen Tickell, *Climatic Change and World Affairs* (Cambridge, MA: Harvard University Press, 1978).

The rolling crises of the 1970s had emphatically demonstrated the interdependence of nations and their vulnerability to unanticipated environmental and economic changes. Freshly attuned to the geopolitical implications of fossil fuel reliance and climate variability, scientists, legal observers, and some governments began to recognize that human-induced climate change would warrant the cooperation of many, if not all, nations. Yet the uneven distribution of resource endowments between nations, combined with the mounting claims for resource sovereignty among producing states, complicated the prospects for an international agreement. Any cooperation between governments, as the G7 and industry responses to the 1979 oil shock suggested, would encounter the political imperative to ensure economic growth, which in 1982 was at its lowest since the Second World War.[147]

---

[147]Frank Bösch and Rüdiger Graf, "Reacting to Anticipations: Energy Crises and Energy Policy in the 1970s," *Historical Social Research* 39, no. 4 (2014): 7–21.

4

Global Greenhouse

ı

1988: 351.24 ppm

In November 1981, the World Bank's new president delivered the annual Fairfield Osborn Memorial Lecture in Washington, DC. Appointed by US president Ronald Reagan, the chief executive officer of the Bank of America A. W. "Tom" Clausen had in July replaced the long-serving Robert McNamara, who had been in the role for over a decade, overseeing the institution's expansion and shift toward attending to minimum or basic human needs.[1] In his address, titled "Sustainable Development: The Global Imperative," Clausen spoke of the bank's record on reconciling the "goals of economic development and environmental enhancement" through funding water supply, forestry, and health care programs. Having reminded the international development community of Teddy Roosevelt's dictum "Conservation means development as much as it does protection," Clausen continued, "And I ask you who are ardent defenders of the environment to join efforts with those of us who are trying to assist the developing countries accelerate their economic growth and improve the quality of life of their societies."[2]

Those efforts were undergoing a rapid transformation in the wake of the elections of Prime Minister Thatcher in the UK and President Reagan in the United States. No matter the call of the 1980 Brandt Commission for the redistribution of global wealth, the essence of the emerging Washington Consensus was made clear in Cancún, Mexico, in the month before Clausen's address. Both Thatcher and Reagan resisted any "global idealism" at the meeting between leaders of the eight wealthiest industrial nations and fourteen from what were now described as "less developed countries." In an era of recession, rising interest rates, and unemployment, the North's conservative leaders would only entertain market solutions to the world's

---

[1] See Stephen Macekura, *The Mismeasure of Progress: Economic Growth and Its Critics* (Chicago: University of Chicago Press, 2020), 72–103.
[2] A. W. Clausen, "Sustainable Development: The Global Imperative," *Environmentalist* 2 (1982): 23–8.

inequalities—in short, there would be "trade not aid."[3] Mexico's default on its debt repayments the following year only entrenched this ethos further. As a severe debt crisis engulfed the developing world, the World Bank and the International Monetary Fund embarked on a program of structural adjustment, whereby further loans were contingent on recipients' liberalization of their economies, which advocates argued would encourage economic growth and poverty reduction.[4]

The World Bank's Berg report for sub-Saharan Africa's development had preempted these approaches. Published at the beginning of Clausen's tenure in mid-1981, its authors argued that "a reordering of post-independence priorities" in favor of the world market "is essential if economic growth is to accelerate." They conceded that the prolonged droughts of the 1970s had contributed to "poor performance," noting that "in some locations ... [t]hese were the result of acts of man—a relative overpopulation and semiarid areas under the pressure of human and animal population increases—and not to autonomous changes of climate."[5] Elsewhere, the declining cost of fossil fuels, thanks to the availability of cheap oil and natural gas, prompted renewed interest in the implications of energy use for atmospheric concentrations of carbon dioxide and, thus, the climate. Renewed Cold War tensions, meanwhile, served to provoke fresh concerns about the use of nuclear weapons and their implications for the world's weather. By decade's end, the political and scientific fears of a changing atmosphere had brought climate change firmly into the view of the United Nations (UN).

Environmental Review: Assessing the Stockholm Decade

In May 1982, delegates from over a hundred governments joined representatives of intergovernmental organizations, and nongovernmental organizations at the headquarters of the UN Environment Programme (UNEP) in Nairobi to mark the tenth anniversary of the Stockholm conference.[6] Hoping to "rekindle

---

[3]Walter Goldstein, "Redistributing the World's Wealth: Cancún 'Summit' Discord," *Resources Policy* (1982): 25–40.

[4]John Toye and Richard Toye, *The UN and Global Political Economy: Trade, Finance and Development* (Bloomington: Indiana University Press, 2004), 254–75; Macekura, *The Mismeasure of Progress*, 166–95.

[5]Elliot Berg et al., *Accelerated Development in Sub-Saharan Africa: An Agenda for Action* (Washington, DC: World Bank, 1981), 1.6, 2.7.

[6]On emerging challenges facing the UN Environment Programme in the early 1980s, see Maria Ivanova, *The Untold Story of the World's Leading Environment Institution: UNEP at Fifty* (Cambridge, MA: MIT Press, 2021), 294–7.

the spirit of Stockholm", the programme's director, Mostafa Tolba reflected on the disappointing progress of the past decade.[7] "The choice facing nations in 1982 was an unprecedented one," he observed, "to carry on as they were and face by the year 2000 an environmental catastrophe whose impact would be as devasting and irreversible as that of nuclear war, or to begin a serious co-operative effort to use the world's resources rationally and fairly."[8] Several developing and Scandinavian countries ventured the New International Economic Order as an avenue for such an effort but were met with opposition from the United States delegation.[9]

Guiding the discussions in Nairobi was a 600-page report, *The World Environment, 1972–1982*. Assessing issues from the state of the atmosphere to transport and tourism, the report had been prepared by the chief scientist of the Department of the Environment of the UK, biologist Martin Holdgate; president of the International Union for Conservation of Nature and Natural Resources (IUCN), Egyptian ecologist Mohammed Kassas; and president of the ICSU SCOPE, American geographer Gilbert White. They concluded, "The CO_2 question was undoubtedly the largest outstanding environmental problem confronting the world at the end of the 1970s."[10] Keeling's curve of the rising atmospheric concentrations of carbon dioxide helped convey their concern that "the data that were available in 1980 suggest (establish would be too strong a word) that the implications of the rise in atmospheric carbon dioxide concentrations do need to be taken seriously – especially because most national energy plans assume an increase in carbonaceous fuel consumption."[11]

Accounting for the role of deforestation in changing the atmosphere was another important consideration. In contrast to the study of fossil fuel emissions, carbon dioxide arising from deforestation remained more scientifically unclear. Climate models since the early 1970s had assumed that given that the world's ocean had been found to have a limited capacity to absorb atmospheric carbon dioxide (Chapter 1), forests must be soaking up much of the excess. By the end of the decade, however, the likes of US ecologist George Woodwell and Bert Bolin had each observed that the world's

[7]Robert P. Lamb, "An Interview with UNEP's Chief," *Environmental Conservation* 9, no. 1 (1982): 1–6.
[8]Mostafa Tolba cited in UNEP, *Report of the Governing Council (Session of a Special Character and Tenth Session)* (New York: United Nations, 1982), 10.
[9]Ignacy Sachs, "Environment and Development Revisited: Ten Years after Stockholm Conference," *Alternatives* 8 (1982): 369–78.
[10]Martin W. Holdgate, Mohammed Kassas, and Gilbert F. White, *The World Environment, 1972– 1982: A Report by the United Nations Environment Programme* (Dublin: Tycooly International, 1982), 62.
[11]Holdgate, Kassas and White, *The World Environment, 1972-1982*, 623.

forests, rather than accumulating carbon, were releasing carbon dioxide.[12] With his colleague Richard Houghton, Woodwell argued in 1977 that deforestation was likely an additional source of carbon dioxide entering the atmosphere.[13] Writing in the *Scientific American* the following year, Woodwell warned, "It is difficult to avoid the conclusion that the destruction of the forests of the earth is adding carbon dioxide to the atmosphere at a rate comparable to the rate of release from the combustion of fossil fuels, and if the oxidation of hummus is included, at an appreciably higher rate."[14]

The emerging awareness of the role of forests in the global climate resonated with growing fears that the world's tropical forests were in grave danger. In 1976, a study of Food and Agriculture Organization (FAO) data and land-use and vegetation maps suggested that, once extrapolated to all countries, approximately eleven million hectares of forest was being lost annually, "a figure which will certainly increase progressively in the future, if no firm steps are taken to stop this development."[15] The World Bank likewise warned, "At this rate ... the remaining tropical forests will disappear in about 60–80 years."[16] A month after the publication of the Council on Environmental Quality's report on *The Carbon Dioxide Problem*, of which Woodwell had been lead author (Chapter 3), the deforestation of tropical forests featured as one of President Carter's talking points in an address to Congress in August 1979. Referring to estimates that the world's forests could decline by 20 percent by the year 2000, the president had noted, "Forest loss may adversely alter global climate through the production of CO_2. These changes and their effects are not well understood and are being studied by scientists, but the possibilities are disturbing and warrant caution."[17]

The assessment prepared for the Nairobi meeting observed that such concerns about forest destruction, particularly in the tropics, had rapidly

---

[12] George M. Woodwell, "The Carbon Dioxide Problem," in George M. Woodwell (ed.), *The Role of Terrestrial Vegetation in the Global Carbon Cycle: Measurement by Remote Sensing* (New York: John Wiley & Sons, 1984): 3–17.

[13] George M. Woodwell and Richard A. Houghton, "Biotic Influence on the World Carbon Budget," in Werner Stumm (ed.), *Global Chemical Cycles and Their Alterations by Man* (Berlin: Dahlem Konferenzen, 1977), 61–72; David Schimel, "Forests in the Global Carbon Cycle," in Trevor Fenning (ed.), *Challenges and Opportunities for the World's Forests in the 21st Century* (Dordrecht: Springer, 2014), 231–40.

[14] George M. Woodwell, "The Carbon Dioxide Question," *Scientific American* 238, no. 1 (1978): 34–43.

[15] Adrian Sommer, "Attempt at an Assessment of the World's Tropical Forests," *Unasylva* 28 (1976): 5–25.

[16] Graham Donaldson, *Forestry: Sector Policy Paper* (Washington, DC: World Bank, 1978), 15. See Michael Williams, *Deforesting the Earth: From Prehistory to Global Crisis* (Chicago: University of Chicago Press, [2002] 2006), 417–21.

[17] Jimmy Carter, "Environmental Priorities and Programs: Message to Congress, August 2, 1979," American Presidency Project, https://v.gd/EG0v9I.

increased over the preceding decade. Although the relative contribution of forest clearance to the trend in atmospheric carbon dioxide remained a topic of debate, its editors noted that "tropical forests were converted for agriculture and energy on a scale that was the subject of many conflicting estimates (figures ranged from 7 to 20 million hectares per year), but appeared greatest in Africa."[18] In addition to forest cutting for construction and fuelwood in southern Africa, the editors highlighted the burning of tropical forest in Central and South America for pastures.[19]

India's delegation had reportedly objected to the prominence of carbon dioxide in the discussions in Nairobi and called for greater attention to environmental issues affecting the South, such as soil degradation.[20] This position drew the ire of former diplomat Jayantanuja Bandyopadhyaya, a Gandhian and Marxist scholar of international relations at Jadavpur University in Kolkata. Climate, in his view, was a significant factor in the "origin and the continuance of the North-South dichotomy." In *Climate and World Order*, he argued, "While imperialism is undoubtedly the greatest single, *historical* force responsible for the North–South economic gap, climate is perhaps the greatest *single* factor in the origin and development of the gap."[21] Having canvassed the CIA reports of the mid-1970s, Soviet weather modification endeavors, and various conference proceedings, Bandyopadhyaya concluded that the South should not seek to prevent climatic variability but rather encourage "altering the global climatic status quo, particularly of modifying and ameliorating the tropical climate."[22] Perhaps climate change could redress the South's disadvantage vis-à-vis the North in the ways that the New International Economic Order had yet to realize.[23]

By now the possibility that some developing countries might benefit from climate change had emerged as a potential stumbling block for attempts to craft an international agreement to curb carbon dioxide emissions. Bandyopadhyaya, for instance, had sketched out his position in a 1978 article for the UN Educational, Scientific and Cultural Organization's (UNESCO's) *International Social Science Journal*, while Reid Bryson's suggestion that China, India, and Bangladesh might benefit from a more reliable monsoon

---

[18]Martin Holdgate, Mohammed Kassas, and Gilbert F. White, "World Environmental Trends between 1972 and 1982," *Environmental Conservation* 9, no. 1 (1982): 11–29.

[19]Holdgate, Kassas, and White, *The World Environment, 1972–1982*, 214.

[20]Jayantanuja Bandyopadhyaya, *Climate and World Order: An Inquiry into the Natural Cause of Underdevelopment* (New Delhi: South Asian Publishers, 1983), 30.

[21]Bandyopadhyaya, *Climate and World Order*, 4. Emphasis in original.

[22]Bandyopadhyaya, *Climate and World Order*, 27.

[23]Siddharth Mallavarapu, "The Sociology of International Relations in India: Competing Conceptions of Political Order," in Gunther Hellmann (ed.), *Theorizing Global Order: The International, Culture and Governance* (Frankfurt: Campus Verlag, 2018), 142–71.

had been widely circulated.[24] Energy forecasts anticipated, meanwhile, that the developing countries would increase their contribution to atmospheric concentrations of carbon dioxide from less than 20 percent to some 40 percent by the year 2025.[25] Inferring that warmer temperatures would favor rice production and hinder wheat yields, English science writer John Gribbin surmised that "although Third World countries will produce the greenhouse problem by the early 21st century (because of their economic growth), they will suffer little adverse consequences themselves, and may even benefit as a result."[26]

Gribbin's assessment received a hostile reception from New Delhi's recently established Centre for Science and Environment. Having consulted biologist Suresh K. Sinha, "one of the few Third World scientists who had studied this problem of increasing carbon dioxide," founding director Anil Agarwal dismissed the idea as "totally naïve." Increased temperatures could increase soil evaporation, which would worsen climatic conditions for agriculture in developing countries.[27] Elsewhere, Sinha also warned that should developed countries such as the United States choose to divert farmland to produce biofuels, then developing countries might need to increase their own food production, regardless of the effects of climate change on crop yields.[28] Agarwal expected the North would "argue in the future that the fuel consumption of developing countries ought to be kept in check to control the increase in carbon dioxide, regardless of their own contribution in the past."[29] In contrast to Bandyopadhyaya's welcome vision of climate change for the South, Agarwal hoped that an international agreement could limit the emission of carbon dioxide. "An International Law of the Air, like the International Law

---

[24]Jayantanuja Bandyopadhyaya, "Climate as an Obstacle to Development in the Tropics," *International Social Science Journal* 30, no. 2 (1978): 339–52; Reid Bryson, "A Perspective on Climate Change," *Science* 184 (1974): 753–60; Charles Cooper, "What Might Man-Induced Climate Change Mean?," *Foreign Affairs* 56, no. 3 (1978): 500–20.

[25]G. Marland and R. M. Rotty, "Atmospheric Carbon Dioxide: Implications for World Coal Use," in Michel Grenon (ed.), *Future Coal Supply for the World Energy Balance* (Oxford: Pergamon Press, 1979).

[26]John Gribbin, "The Politics of Carbon Dioxide," *New Scientist*, April 9, 1981, 82–4. Gribbin attributed his assessment to Wilfrid Bach, "The Potential Consequences of Increasing CO_2 Levels in the Atmosphere," in Jill Williams (ed.), *Carbon Dioxide, Climate and Society* (Oxford: Pergamon Press, 1978), 141–68. For a contrasting interpretation to Gribbin's, see Robert Schware and Edward J. Friedman, "Climate Debate Heats Up," *Bulletin of the Atomic Scientists* 37, no. 10 (1981): 31–3.

[27]Anil Agarwal, "Climate and Carbon Dioxide," in Anil Agarwal, Ravi Chopra, and Kalpana Sharma (eds.), *The State of India's Environment 1982: A Citizen's Report* (New Delhi: Centre for Science and Environment, 1982), 85–90. See also Martin Mahony, "The Predictive State: Science, Territory and the Future of the Indian Climate," *Social Studies of Science* 4, no. 1 (2014): 109–33.

[28]Suresh K. Sinha, "The Impact of Carbon Dioxide on Agriculture in Developing Countries," *Impact of Science on Society* 32, no. 3 (1982): 311–23.

[29]Anil Agarwal, "Climate and Carbon Dioxide," 90.

of the Seas, is clearly needed," he observed. "After all, like the oceans, the atmosphere is also a common heritage of humanity."[30]

In preparation for the Nairobi meeting, Tolba had invited Barbara Ward to write a book for the tenth anniversary of the Stockholm conference, which he hoped would emulate the acclaimed *Only One Earth* she had coauthored with René Dubos in 1972. Although her illness prevented her from completing what would become *Down to Earth*, she dictated its foreword before her death in May 1981. In Stockholm, she recalled, "There was a beginning of a sense of shared stewardship for our common planetary home." But that had already begun to fade: "Now, in the 1980s, some Western leaders are starting to abandon the concept of our joint voyage on Spaceship Earth, and to dismiss any concern for the environment or for underdeveloped nations as 'do-goodism'."[31] The challenges facing the UN and its Environment Programme only looked set to continue, as the Reagan administration threatened on the eve of the Nairobi conference to cut its funding to the organization. Although US funding was reinstated and eventually increased, the Programme still faced a budget shortfall, while the episode served to highlight the administration's reluctance to engage in multilateral negotiations to resolve global environmental problems.[32]

Atmospheric Politics: Science and Nuclear Disarmament

Having gathered in New Delhi in January 1985, representatives of Argentina, Greece, India, Mexico, Sweden, and Tanzania sent what they titled the "Delhi Declaration" to the UN secretary-general, Peru's Javier Pérez de Cuéllar. Citing the atomic bombs that had devastated Hiroshima and Nagasaki nearly forty years earlier, the five leaders called for an immediate end to the nuclear arms race and for the transfer of the resources "currently wasted" on military expenditure to social and economic development. Tensions between the superpowers had intensified since the 1979 Soviet invasion of Afghanistan and the North Atlantic Treaty Organization's (NATO's) "dual track" policy to permit US nuclear missiles in Western Europe. Anti-nuclear rallies and diplomatic attempts to broker disarmament, such as the Palme Commission, had yet to move either Moscow

---

[30]Anil Agarwal, "Climate and Carbon Dioxide," 90. At the third United Nations Conference on the Law of the Sea in April 1982, delegates had voted in favor of an international treaty on the use of seabed resources considered as part of the common heritage of mankind.

[31]Barbara Ward, "A Decade of Environmental Action," *Environment* 24, no. 4 (1982): 4–6.

[32]David Struthers, "The United Nations Environment Programme after a Decade: The Nairobi Session of a Special Character," *Denver Journal of International Law & Policy* 12, no. 2 (1983): 269–84.

or Washington, prompting the *Bulletin of the Atomic Scientists* to advance its doomsday clock to three minutes to midnight in early 1984.[33] Hopeful that the renewal of negotiations between the Soviet Union and United States might break the impasse, the New Delhi signatories warned that "nuclear war, even on a limited scale, would trigger an arctic nuclear winter which may transform the earth into a darkened, frozen planet, posing unprecedented peril to all nations, even those far removed from the nuclear explosions."[34]

Concerns as to the atmospheric effects of nuclear weapons had been renewed in the early 1980s. Scientific and popular renderings of what soon became known as "nuclear winter" depicted the planetary aftermath of nuclear war, whereby the accumulation of smoke, soot, and dust in the atmosphere would darken the earth's skies for months, causing temperatures to plummet and thereby crippling human civilization and other life on earth. Thanks to the rise of ecological awareness during the 1970s, the weather-related anxieties that had accompanied atmospheric nuclear weapons testing in the 1950s (Chapter 1) had become reimagined in interconnected and planetary terms that transcended borders and species.[35] Although fears of an imminent nuclear winter faded with the thawing of relations between the two superpowers, the concept's contestation and planetary nature foreshadowed similar framings of anthropogenic climate change by the end of the decade.[36]

In 1980, the American editor of the Royal Swedish Academy of Sciences' journal, *Ambio*, attended a conference where a speaker described the aftermath of nuclear war as "unimaginable." Finding this assessment wanting, journalist Jeannie Peterson took it upon herself to assemble the scientific expertise to determine more specifically the likely results of the outbreak of nuclear war.[37] Hers was not simply an empirical exercise—"a realistic assessment of the possible human and ecological consequences of a nuclear war may

[33]Eckart Conze, Martin Klimke, and Jeremy Varon (eds.), *Nuclear Threats, Nuclear Fear and the Cold War of the 1980s* (Cambridge: Cambridge University Press, 2017). See Palme Commission, *Common Security: A Programme for Disarmament – Report of the Independent Commission on Disarmament and Security Issues* (London: Pan Books, 1982).

[34]"Letter Dated 30 January 1985 from the Representatives of Argentina, Greece, India, Mexico, Sweden and the United Republic of Tanzania to the United Nations addressed to the Secretary General," A/40/114, 1985, United Nations, https://v.gd/qbso4K.

[35]Wilfried Mausbach, "Nuclear Winter: Prophecies of Doom and Images of Desolation during the Second Cold War," in Eckart Conze, Martin Klimke, and Jeremy Varon (eds.), *Nuclear Threats, Nuclear Fear and the Cold War of the 1980s* (New York: Cambridge University Press, 2017), 28. See also John R. McNeill, "The Environment, Environmentalism, and International Society in the Long 1970s," in Niall Ferguson et al., *Shock of the Global: The 1970s in Perspective* (Cambridge, MA: Harvard University Press, 2010), 263–78.

[36]See Fabien Locher, "Neo-Malthusian Environmentalism, World Fisheries Crisis, and the Global Commons, 1950s–1970s," *Historical Journal* 63, no. 1 (2020): 187–207.

[37]Lawrence Badash, *A Nuclear Winter's Tale: Science and Politics in the 1980s* (Cambridge, MA: MIT Press, 2009), 50.

help to deter such a catastrophe," she began the subsequent special issue's editorial.[38] Accordingly, the advisory group she recruited included the special assistant for disarmament to the Swedish minister of defense, a physicist from the Manhattan Project who had founded the Pugwash Conference on Science and World Affairs, and another who had directed the Stockholm International Peace Research Institute.[39]

Having provided a possible scenario of nuclear war to participating scientists, she and her advisors were stunned by findings that fire and smoke, not radioactive fallout, would have the most far-reaching effects.[40] Expecting significant stratospheric ozone depletion, what atmospheric chemist Paul Crutzen and his collaborator John Birks had instead determined for the first time were the quantitative effects of the resultant smoke on sunlight.[41] As a result of the spread of smoke across the northern hemisphere, they estimated that "the average sunlight penetration to the ground will be reduced by a factor between 2 and 150 at noontime in the summer. This would imply that much of the Northern Hemisphere would be darkened in the daytime for an extended period of time following the nuclear exchange." In this "twilight at noon," as they called it, crops and other plants would fail to thrive, just as most of the ocean's phytoplankton and zooplankton would die.[42]

Crutzen and Birks likened the resulting conditions to a grim portrait recently depicted by a US team. In the mid-1970s, Crutzen had worked at the National Center for Atmospheric Research in Boulder, where he had contributed to a study that speculated that the earth's fifth mass extinction some sixty-five million years ago was the result of the depletion of stratospheric ozone.[43] Since then, a new theory had appeared. With his geologist son and two nuclear chemists, retired nuclear physicist and Nobel laureate Luis Alvarez, a veteran of the Manhattan Project, had proposed in 1980 that the impact of an asteroid had caused the mass extinctions of the Cretaceous period. The enormous explosion of dust from such a cosmic impact, they had theorized, would plunge the earth into darkness, suppressing photosynthesis and leading to widespread biological consequences, not least the demise of the dinosaurs.[44]

---

[38]Jeannie Peterson, "About This Issue," *Ambio* 11, no. 2/3 (1982): n.p.

[39]Frank Barnaby et al., "Reference Scenario: How a Nuclear War might be Fought," *Ambio* 11, no. 2/3 (1982): 99.

[40]Badash, *A Nuclear Winter's Tale*, 51.

[41]Paul J. Crutzen and John W. Birks, "The Atmosphere after a Nuclear War: Twilight at Noon," *Ambio* 11, no. 2/3 (1982): 114–25; Badash, *A Nuclear Winter's Tale*, 52.

[42]Crutzen and Birks, "The Atmosphere after a Nuclear War," 120.

[43]G. C. Reid, I. S. A. Isaksen, T. E. Holzer, and P. J. Crutzen, "Influence of Ancient Solar-Proton Events on the Evolution of Life," *Nature* 259 (1976): 177–9.

[44]Luis W. Alvarez et al., "Extraterrestrial Cause for the Cretaceous-Tertiary Extinction," *Science* 208 (1980): 1095–108. On the reception of this theory, see David E. Fastovsky, "Ideas in Dinosaur Paleontology: Resonating to Social and Political Context," in David Sepkoski and Michael Ruse (eds.),

For both the Alvarez team and the *Ambio* authors, the closest analog to such a mass extinction event was the atmospheric effects of the volcanic eruption of Krakatoa in 1883. The consequences of a large asteroid, they stressed, would be of a far greater order of magnitude.[45] Since the 1950s, volcanic activity had continued to provide climate scientists insights into the possible atmospheric effects of nuclear weapons tests.[46] News of the *Ambio* paper, meanwhile, had reached the United States. There, the recent eruption of Mount St. Helens had invited further study of volcanic effects, but it was already clear that such events offered an inadequate atmospheric analog for nuclear war.[47] To this end, a group of atmospheric scientists that would become together known as "TTAPS" (derived from the initials of their surnames) had begun the modeling work from which they would derive their own version of the *Ambio* findings. To describe the atmospheric aftermath of large-scale nuclear war, they coined the term "nuclear winter."[48]

Fronted by scientist Carl Sagan, who had recently hosted the US television series *Cosmos*, and joined by biologist Paul Ehrlich, the TTAPS team quickly became embroiled in the Cold War politics of the early 1980s. This was no accident, however: Sagan saw in nuclear winter a compelling case for disarmament and publicized the concept beyond scientific circles. Just months after President Reagan announced his controversial Strategic Defense Initiative (also known as "Star Wars"), Sagan shared the concept with readers of *Parade* magazine (a popular Sunday supplement to most US newspapers) and *Foreign Affairs* in advance of its publication in the journal *Science*.[49] The latter's publisher took the unusual step of praising the TTAPS authors and Ehrlich's collaborators for their "courage": "It says a good deal for the emergence of the scientific conscience that, in a difficult age of superpower hatreds and technological gusto, the present warning is timely, unvarnished, and stark."[50]

Sagan was from the outset the figurehead of nuclear winter and thus the prime target of its critics. His argument for disarmament won him few

---

The Paleobiological Revolution: Essays on the Growth of Modern Paleontology (Chicago: University of Chicago Press, 2009), 239–53.

[45]Alvarez et al., "Extraterrestrial Cause for the Cretaceous-Tertiary Extinction," 1105; Crutzen and Birks, "The Atmosphere after a Nuclear War," 123.

[46]See Paul N. Edwards, "Entangled Histories: Climate Science and Nuclear Weapons Research," *Bulletin of the Atomic Scientists* 68, no. 4 (2012): 28–40.

[47]Badash, *A Nuclear Winter's Tale*, 35.

[48]R. P. Turco, O. B. Toon, T. P. Ackerman, J. B. Pollack, and C. Sagan, "Nuclear Winter: Global Consequences of Multiple Nuclear Explosions," *Science* 222 (1983): 1283–92.

[49]Cited in Badash, *A Nuclear Winter's Tale*, 89; Carl Sagan, "Nuclear War and Climatic Catastrophe: Some Policy Implications," *Foreign Affairs* 62, no. 2 (1983): 257–92.

[50]William D. Carey, "A Run Worth Taking," *Science* 222, no. 4630 (1983): 1281. See Erik M. Conway, *Atmospheric Science at NASA: A History* (Baltimore, MD: Johns Hopkins University Press, 2008), 206–12.

friends in the defense establishment, whose leading figures condemned the concept on scientific and political grounds, and rallied to defend the Reagan White House.[51] Sagan's engagement with Soviet scientists was further grist for their mill. Naively perhaps, he had invited in 1983 Soviet climate modelers to participate from Moscow in a large public meeting via TV link with Washington. As far as Sagan's critics were concerned, such was the primitive state of Soviet computing that their findings were derivative, at best, and part of Moscow's strategy to exploit Western catastrophism to undermine the United States, at worst.[52] By contrast to their US counterparts, Soviet officials appear to have endorsed the nuclear winter theory, at least initially.[53] At the 1983 World Meteorological Congress, for instance, the delegate for the Soviet Union had "noted with deep satisfaction that all speakers (from the Soviet bloc) were seriously concerned at the danger to our planet, its atmosphere and climate which thermo-nuclear war would present."[54]

Beyond the atmospheric politics of the United States and the Soviet Union, the International Council for Science meanwhile established in 1984 a Scientific Committee on Problems of the Environment (SCOPE) study on the environmental effects of nuclear war. Completed the following year, the authors departed from the term "nuclear winter," which they believed did not adequately convey the complexity of environmental consequences. In a veiled warning to the superpowers, the report's steering committee noted, "Because of the possibility of a tragedy of an unprecedented dimension, any disposition to minimize or ignore the widespread environmental effects of a nuclear war would be a fundamental disservice to the future of global civilization."[55] Led by an Australian atmospheric scientist, the SCOPE-ENUWAR report resonated with the heightened nuclear anxieties pervading the South Pacific in the mid-1980s. Relations between Washington and Wellington had recently soured, after the Aotearoa New Zealand government declined to host the USS *Buchanan* on grounds that it may have been carrying nuclear

---

[51]Lawrence Badash, "Nuclear Winter: Scientists in the Political Arena," *Physics in Perspective* 3 (2001): 76–105; Naomi Oreskes and Erik M. Conway, *Merchants of Doubt: How a Handful of Scientists Obscured the Truth on Issues from Tobacco Smoke to Global Warming* (New York: Bloomsbury Press, 2010), 37–8; Jacob Darwin Hamblin, *Arming Mother Nature: The Birth of Catastrophic Environmentalism* (New York: Oxford University Press, 2013), 237.
[52]Hamblin, *Arming Mother Nature*, 238–41.
[53]Paul Rubinson, "The Global Effects of Nuclear Winter: Science and Antinuclear Protest in the United States and the Soviet Union during the 1980s," *Cold War History* 14, no. 1 (2014): 47–69. Moscow's position shifted in the mid-1980s, as the peace movement's association of nuclear winter with their calls for the Soviet Union's disarmament led to the arrest and imprisonment of activists.
[54]World Meteorological Organization, *Ninth World Meteorological Congress* (Geneva: World Meteorological Organization, 1983), 97–8.
[55]A. Barrie Pittock et al., *Environmental Consequences of Nuclear War: Physical and Atmospheric Effects*, vol. 1, SCOPE 28 (Chichester: John Wiley & Sons, 1986), xxvii.

weapons. Aotearoa had elected in 1984 the Lange Labour government, which had campaigned on an antinuclear platform and had since legislated to prohibit nuclear-armed and powered vessels from the country's ports. As tensions simmered with the Reagan administration, French agents bombed the Greenpeace vessel *Rainbow Warrior* in Auckland Harbor as it prepared to voyage to Mururoa Atoll to protest the Mitterand government's ongoing commitment to underground nuclear testing.[56]

A handful of the scientists engaged in the SCOPE-ENUWAR report subsequently contributed to a study on behalf of the UN secretary-general. A year after the Norwegian Nobel Committee awarded the 1985 Nobel Peace Prize to the International Physicians for the Prevention of Nuclear War, the governments of Bangladesh, India, Mexico, Pakistan, and Sweden had urged Pérez de Cuéllar to assemble an expert panel to assess "the climatic and potential physical effects of nuclear war, including nuclear winter."[57] Although more realistic climate models suggested that the temperature decreases would be smaller than first envisioned, the 1989 report concluded that "a major nuclear war would entail the high risk of a global environmental disruption."[58]

The scientific and political debates over the nature, extent, and likelihood of nuclear winter highlight the convergence of atmospheric anxieties in the 1980s. As Pérez de Cuéllar observed in his foreword to the UN report,

> For all its apparent robustness, the planet on which we live exists in fragile balance. For the first in the history of the human race, humanity is now taking actions that, within the time-span of a single generation, are affecting the global environment in fundamental ways. The effects of acid rain and deforestation are plain to see. The future implications of global warming and ozone depletion are just being fully recognized.[59]

Reactions to the nuclear winter theory not only presaged the divisiveness that would also characterize political responses to anthropogenic climate change, but also set the scene for much closer attention to the earth's changing temperature and vulnerability to human activity.[60]

---

[56]See Lawrence Badash, *A Nuclear Winter's Tale: Science and Politics in the 1980s* (Cambridge, MA: MIT Press, 2009), 256–62.

[57]"Climatic Effects of Nuclear War, Including Nuclear Winter: Draft Resolution," UN General Assembly, October 30, 1986, A/C.1/41/L.36, https://v.gd/n297HQ.

[58]Henry A. Nix (Chair), *Study on the Climatic and Other Global Effects of Nuclear War* (New York: United Nations, 1989), 15.

[59]Cited in *Study on the Climatic and Other Global Effects of Nuclear War*, 5.

[60]See Matthias Dörries, "The Politics of Atmospheric Sciences: 'Nuclear Winter' and Global Climate Change," *Osiris* 26, no. 1 (2011): 198–223; Joshua P. Howe, *Behind the Curve: Science and the Politics of Global Warming* (Seattle: University of Washington Press, 2014), 243–4.

Climate Awakening: An Agenda for Scientific Internationalism

Speaking before his colleagues gathered in the Austrian town of Villach in October 1985, Soviet geophysicist Georgy Golitsyn confessed he could restrain himself no longer. He explained that he could not resist drawing parallels between their discussion of climate change and another issue, that of the environmental consequences of nuclear war. Like others in attendance, he had contributed to the SCOPE report on nuclear war and reflected on their sustained effort to undertake "detailed, authoritative and convincing treatment" to address the problem.[61] They too had grappled with scientific uncertainties, just as they were now in Villach. Encouraged by this experience, he suggested, "We should start hard, consistent and coordinated work without expecting that something will save us."[62]

By the meeting's end, the delegates had produced the first international scientific consensus on the role of increased carbon dioxide and other greenhouse gases on global climate as a result of human activity: "It is now believed that in the first half of the next century a rise of global mean temperature could occur which is greater than any in man's history." Echoing Golitsyn's association of their discussions with the environmental effects of nuclear war, the Villach conference statement highlighted the intersection of climate change with other pressing atmospheric issues: action to ameliorate acid rain, such as reducing coal use, would also reduce greenhouse gas emissions, they argued, just as limiting the release of chlorofluorocarbons would both benefit the ozone layer and slow the rate of anthropogenic climate change (Chapter 3). Although more research was necessary, they agreed that the time had come for governments to limit the emission of greenhouse gases and to consider the "greenhouse question" in their planning.[63]

---

[61] In addition to Golitsyn, the following scientists contributed to both the SCOPE-ENUWAR report and the Villach meeting: Michael C. MacCracken (Lawrence Livermore National Laboratory) and Rafael Herrera (Centro de Ecología, Venezuela), while Thomas F. Malone had chaired the atmospheric and climatic consequences panel of Ehrlich and Sagan's Conference on the Long-Term Worldwide Biological Consequences of Nuclear War. Golitsyn, Herrera, Kassas, and Malone also contributed to the UN secretary-general's report.

[62] Golitsyn, "Structured Response II," in World Meteorological Organization, *Report of the International Conference on the Assessment of the Role of Carbon Dioxide and of Other Greenhouse Gases in Climate Variations and Associated Impacts* (Geneva: World Meteorological Organization, 1986), 36–7.

[63] "Conference Statement," in WMO, *Report of the International Conference on the Assessment of the Role of Carbon Dioxide and of Other Greenhouse Gases in Climate Variations and Associated Impacts*, 1–3.

That week, "the climate changed," New Scientist's Fred Pearce recalled thirty years on.[64] Scientists from twenty-nine countries had met in Villach under the auspices of the World Meteorological Organization's (WMO's) World Climate Programme to undertake a scientific assessment of the climatic impacts of increasing atmospheric concentrations of greenhouse gases. A fraction of the participants had met there five years earlier to launch the World Climate Programme and to discuss the "carbon dioxide/climate issue."[65] Disappointed that the 1980 meeting had not "penetrated the problem" beyond what had been discussed at the World Climate Conference the previous year (Chapter 3), the chair of that meeting, Bert Bolin, was hopeful that this second meeting in Villach would be "more truly international and that it [w]ould go beyond an analysis of the physical aspects of climate change that had dominated the efforts so far." In the intervening years, UNEP's Mostafa Tolba had shared with him his hope to shed further light on the climate change issue, in which he saw a central role for his agency.[66] These discussions led to Bolin's commission to prepare a SCOPE report, which then became the lengthy background paper for the 1985 Villach meeting.[67] That report, Bolin recalled, "brought the issue of human-induced climate change much more to the forefront in the scientific community than earlier assessments had done," including the recent Changing Climate report published by the US National Academy of Science.[68]

For participants at the Villach meeting, two aspects of their discussion stood out. The first was the influence of Bolin's SCOPE report. Studies since Charney's in 1979 (Chapter 3), including this one, had largely agreed that the average global temperature would rise between 1.5 and 5.5°C as a result of doubling the concentration of carbon dioxide (this modeling calculation is called "climate sensitivity").[69] Bolin's SCOPE report found, however, that carbon dioxide was not the only gas of climate concern: there were additional greenhouse gases associated with human activities, particularly

---

[64]Fred Pearce, "Histories: The Week the Climate Changed," New Scientist, October 15, 2005, 52–4.
[65]World Climate Programme, Joint WMO/ICSU/UNEP Meeting of Experts on the Assessment of CO_2 on Climate Variations and Their Impact (Geneva: World Climate Programme, 1981).
[66]Bert Bolin, A History of the Science and Politics of Climate Change: The Role of the Intergovernmental Panel on Climate Change (Cambridge: Cambridge University Press, 2007), 35–6; Mostafa Tolba, "Climate, Environment and Society: Statement to the World Climate Conference, Geneva, Switzerland, February 1979," in Mostafa Tolba, Development without Destruction: Evolving Environmental Perceptions (Dublin: Tycooly International, 1982), 97–100.
[67]Bert Bolin et al., The Greenhouse Effect, Climatic Change and Ecosystems, SCOPE 29 (Chichester: John Wiley & Sons, 1986).
[68]Bolin, A History, 36–7; National Research Council, Changing Climate: Report of the Carbon Dioxide Assessment Committee (Washington, DC: National Academy Press, 1983).
[69]Bolin et al., The Greenhouse Effect, 3–4; Jeroen van der Sluijs et al., "Anchoring Devices in Science for Policy: The Case of Consensus around Climate Sensitivity," Social Studies of Science 28, no. 2 (1998): 195–351.

methane, would also contribute to climate change.[70] This revelation meant that the warming anticipated to arise from a doubling of the atmospheric concentration of carbon dioxide would occur much sooner than previously predicted—"as early as the 2030s."[71] According to Bolin's younger colleague Jill Jaeger, an International Institute for Applied Systems Analysis alumna with whom he had prepared the SCOPE report, this assessment surprised her fellow participants: "Wow, significant change could take place in our lifetimes," she recalled in the late 1990s.[72]

For atmospheric scientist Thomas F. Malone, participating as a representative of one of the meeting's sponsors, the International Council of Scientific Unions (ICSU), this finding prompted him to reassess his position. Having chaired the National Research Council report, *Changing Climate*, he had in early 1984 testified before a joint hearing that the report's findings warranted only "caution, not panic."[73] Now, just over a year later, he realized that "human activity can produce changes on a global scale, comparable to those within a geological time frame—*and to do so within the lifetime of a single member of the human species*." With this fresh outlook, he declared, "I believe it is timely to *start* on the long, tedious and sensitive task of framing a CONVENTION on greenhouse gases, climate change and energy."[74]

Malone's about-face underscored the second key aspect of the Villach meeting. Participants had gathered not as representatives of their governments but rather in their personal capacities. As such, conference chair Canada's Jim Bruce observed, they could "shed their national policy perspectives," which afforded them the scope to make recommendations that were more inclined toward policy than science.[75] Moreover, their participation as private individuals afforded the meeting's sponsors greater influence over the direction of the discussions. Irish hydrologist and politician James Dooge, of the ICSU, urged the participants to provide "a sound foundation and appropriate guidelines for the development of the necessary policies at

---

[70]See Veerabhadran Ramanathan et al., "Trace Gas Trends and Their Potential Role in Climate Change," *Journal of Geophysical Research* 90 (1985): 5547–66; Bernhard Stauffer et al., "Increase of Atmospheric Methane Recorded in Antarctic Ice Core," *Science* 229 (1985): 1386–8.

[71]WMO, *Report of the International Conference on the Assessment of the Role of Carbon Dioxide and of Other Greenhouse Gases in Climate Variations and Associated Impacts*, 2.

[72]Jill Jaeger, cited in Wendy E. Franz, *The Development of an International Agenda for Climate Change: Connecting Science to Policy* (Laxenberg: IIASA, 1997), 18.

[73]Subcommittee on Investigations and Oversight and the Subcommittee on Natural Resources, Agriculture Research and Environment, *Carbon Dioxide and the Greenhouse Effect, February 28, 1984* (Washington, US Government Printing Office: 1984), 104.

[74]World Meteorological Organization, *Report of the International Conference*, 32–3. Emphasis in original.

[75]Jim Bruce, cited in Franz, *The Development of an International Agenda for Climate Change*, 14. See William Kellogg, "Mankind's Impact on Climate: The Evolution of an Awareness," *Climatic Change* 10 (1987): 113–36.

the national and international level."[76] Likewise, Jim Bruce, who was a vice president of the WMO, reminded the group of their task: "to develop sound recommendations for action by countries and by international agencies, based on [their] scientific consensus."[77]

Fresh from his leading role in negotiating the recent Vienna Convention on the Protection of the Ozone Layer, Tolba had great ambitions for the Villach meeting. Since his speech at the 1979 World Climate Conference, Tolba had shared widely his sense of urgency that a "world plan of action for dealing with the carbon dioxide problem" should be developed.[78] From Nairobi, Buenos Aires, and New Delhi to Tokyo and New York City, he argued that such was the interdependence of the global environment that attending immediately to this problem was critical to achieving "development without destruction."[79] That nations had agreed to the Ozone Convention as recently as March 1985, he believed, was "a testament to [their] political maturity because, for the first time, nations are agreeing to take steps to ward off a threat that may be still far in the future."[80] The discovery of the so-called hole in the ozone layer just months later only reinforced the timeliness of that early intervention, and it offered a lodestar for the discussions at Villach.[81]

In recent remarks, Tolba had articulated his case for the more public role of scientists such as those gathered at Villach. Addressing the Sixth General Assembly of SCOPE, he stressed the need for science to inform decision-making and that the likes of SCOPE should "use its reputation for objectivity to hammer home" to governments "the advantages that accrue to all nations from environmentally sound sustainable development." He scolded scientists for what he saw as their reluctance to "commit to how seriously these possibilities should be taken. ... When governments have retreated behind the smoke screen of scientific uncertainty, the scientific community has too often been willing to go along with them."[82] At Villach, he called on the scientists to inform "a wider debate on such issues as the costs and benefits of a radical shift away from fossil fuel consumption."[83]

---

[76]World Meteorological Organization, *Report of the International Conference*, 17.

[77]World Meteorological Organization, *Report of the International Conference*, 7.

[78]Tolba, "Climate, Environment and Society," in Tolba (ed.), *Development without Destruction*, 99.

[79]Tolba, *Development without Destruction*; Mostafa Tolba (ed.), *Sustainable Development: Constraints and Opportunities* (London: Butterworths, 1987).

[80]Tolba, *Sustainable Development*, 175.

[81]J. C. Farman, B. G. Gardiner, and J. Shanklin, "Large Losses of Total Ozone in Antarctica Reveal Seasonal ClO_x/NO_x Interaction," *Nature* 315 (1985): 207–10.

[82]Tolba, *Sustainable Development*, 189, 193.

[83]World Meteorological Organization, *Report of the International Conference*, 12.

Expert Governance: Advocacy and Scientific Internationalism

Following Tolba's recommendation to form an "international co-ordinating committee on greenhouse gases," a group of seven men gathered at the headquarters of the WMO in Geneva in early July 1986.[84] Villach's sponsoring organizations, the WMO, UNEP, and ICSU, had each nominated two scientists for this task force: joining the group's chair, Canada's Ken Hare, were Bolin (Sweden), Golitsyn (USSR), Gordon Goodman (UK), Mohammed Kassas (Egypt), Syukuro Manabe (USA), and Gilbert White (USA), with Villach chair Jim Bruce as their secretary. Their role was twofold: first, to conduct biennial reviews of international and regional studies related to greenhouse gases; and, second, to make periodic assessments of the changing concentrations of greenhouse gases and their effects. To these ends, the group sought to develop a mechanism that would disseminate this information to a wide audience.[85] Tolba's template for these endeavors was the Coordinating Committee on the Ozone Layer (Chapter 3).

The international landscape of ozone diplomacy had undergone significant change since the meeting in Villach in October 1985. Critics had questioned the Vienna Convention as either premature or flimsy. But the Villach report, as well as the publication of the British Antarctic Survey's findings, affirmed the precautionary nature of the Vienna Convention and prompted calls for a stronger international response to ozone depletion.[86] Guiding this process since 1977 had been the scientific expertise of the coordinating committee, led by the director of the National Aeronautics and Space Administration's (NASA's) "Mission to Planet Earth" project, Robert Watson, and reporting directly to Tolba's UNEP (Figure 4.1).[87] Although the committee's insights alone did not overcome the impasse between the parties, US chief negotiator Richard Benedick suggests that scientific consensus and collaboration between scientists and policymakers proved "indispensable" to the adoption

---

[84]World Meteorological Organization, *Report of the International Conference*, 12.

[85]Thomas D. Potter, "Advisory Group on Greenhouse Gases Established Jointly by WMO, UNEP and ICSU," *Environmental Conservation* 13, no. 4 (1986): 365.

[86]Marvis Soroos, *The Endangered Atmosphere: Preserving a Global Commons* (Columbia: University of South Carolina Press, 1997), 158–9.

[87]David Hirst, "Controlling the Agenda: Science, Policy and the Making of the Intergovernmental Panel on Climate Change," in Wolfram Kaiser and Jan-Henrik Meyer (eds.), *International Organizations and Environmental Protection: Conservation and Globalization in the Twentieth Century* (Oxford: Berghahn Books, 2017), 293–316. On NASA's Mission to Planet Earth program, see Naomi Oreskes, "Changing the Mission: From the Cold War to Climate Change," in Naomi Oreskes and John Krige (eds.), *Science and Technology in the Global Cold War* (Cambridge, MA: MIT Press, 2014), 141–88.

FIGURE 4.1 *Dr. Mostafa K. Tolba, executive director of the UN Environment Programme (left), addresses a press conference in New York, October 22, 1991. On the right is Dr. Robert Watson, director of Earth Process Studies at NASA and co-chair of the Coordinating Committee on the Ozone Layer. Dr. Watson later served as the chair of the Intergovernmental Panel on Climate Change (IPCC). UN Photo by John Isaac.*

of the Montreal Protocol in September 1987.[88] Seeing the efficacy of this model firsthand, Tolba anticipated a similar role of political legitimation for the Advisory Group on Greenhouse Gases (AGGG) that emerged from Villach.[89]

Circumstances conspired to limit the work of the original group, however. In the wake of the Chernobyl disaster, Hare had been appointed commissioner of the Ontario Nuclear Safety Review in late 1986, while Bolin was serving as science advisor to the Swedish prime minister. The demands of these roles precluded their further involvement in the group's work. Manabe, meanwhile, felt unsuited to the group's public role and resigned, leaving only White and Kassas for whom climate change was not a primary concern. For his part, Golitsyn was otherwise occupied with his work on nuclear winter.[90]

[88]Richard Benedick, *Ozone Diplomacy: New Directions in Safeguarding the Planet* (Cambridge, MA: Harvard University Press, 1991), 5.
[89]Hirst, "Controlling the Agenda," 301.
[90]Shardul Agrawala, "Early Science-Policy Interactions in Climate Change: Lessons from the Advisory Group on Greenhouse Gases," *Global Environmental Change* 9 (1999): 161.

Consequently, ecologist Goodman, director of Sweden's Beijer Institute, became the group's most active, official member.[91]

In physicist Michael Oppenheimer of the Environmental Defense Fund and ecologist George Woodwell of the Woods Hole Research Center Goodman found partners with whom he could advance this initiative further. Both were veterans of the atmospheric politics of Reagan's United States: Oppenheimer had been a vocal advocate for the regulation of acid rain, while Woodwell had chaired the Conference on the Long-Term Worldwide Biological Consequences of Nuclear War and led the 1979 Council for Environmental Quality report (Chapter 3).[92] With funding from the UNEP and the German Marshall Fund, as well as the Austrian and Swedish governments, they together embarked on the organization of two meetings in 1987 that they hoped would influence the world's policymakers.[93]

The first of these meetings returned to Villach in October, with nearly fifty experts from universities, environmental advocacy groups, and some national environmental agencies. They had much to discuss—scientists at the Climatic Research Unit at the University of East Anglia had recently found that the three warmest years in over a century had occurred in the 1980s.[94] Since then, geochemist Wallace Broecker had warned *Nature*'s readers that, by releasing carbon dioxide and other greenhouse gases into the atmosphere, "the inhabitants of planet Earth are quietly conducting a gigantic environmental experiment." Paleoclimatic evidence showed that the earth's temperature was closely tied to the ocean's circulation and correlated with changes in the atmospheric concentration of carbon dioxide.[95] Broecker urged further scrutiny of this dynamic: "We play Russian roulette with climate, hoping that the future will hold no unpleasant surprises. No one knows what lies in the active chamber of the gun, but I am less optimistic about its contents than many."[96]

---

[91]World Meteorological Organization, *Report of the International Conference*, 43–4.

[92]Paul Ehrlich et al., "Long-Term Biological Consequences of Nuclear War," *Science* 222 (1983): 1293–1300; George Woodwell, "Biotic Effects on the Concentration of Atmospheric Carbon Dioxide: A Review and Projection," in National Research Council, *Changing Climate*, 216–41.

[93]Michael Oppenheimer, "Developing Policies for Responding to Climatic Change," *Climatic Change* 15 (1989): 1–4.

[94]P. D. Jones, T. M. L. Wigley, and P. B. Wright, "Global Temperature Variations between 1861 and 1984," *Nature* 322 (1986): 430–4.

[95]Swiss physicist Hans Oeschger, who had worked with Revelle and Suess at the Scripps Institution of Oceanography in 1959, had in the early 1980s speculated a role for the ocean circulation in what would become understood as 'abrupt climate change'. See Hans Oeschger et al., "Late-glacial Climate History from Ice Cores," in James Hansen and Taro Takahashi (eds), *Climate Processes and Climate Sensitivity* (Washington, DC: American Geophysical Union, 1984), 299–306; Thomas Stocker, "Oeschger, Hans", in Noretta Koertge (ed.), *New Dictionary of Scientific Biography*, vol. 5 (Detroit: Thomson Gale, 2008), 326–31.

[96]Wallace S. Broecker, "Unpleasant Surprises in the Greenhouse?," *Nature* 328 (1987): 123–6.

Having earlier focused on discerning possible climate changes over the next century and the options for limiting or adapting to those changes, the November meeting in Bellagio turned to the policy implications of these findings.[97] Just as generational change influenced the tone of the "Study of Man's Impact on Climate" (Chapter 2), so too a younger group of researchers would make their mark on these proceedings.[98] Joining Jaeger, Oppenheimer, and Bill Clark were new faces, Peter Gleick from the United States, and from the Netherlands, Pier Vellinga, Frank Rijsberman and Rob Swart, who together formed what policy analyst Shardul Agrawala has called a "*de facto* Advisory Group on Greenhouse Gases."[99] Together, they sought to galvanize "a coordinated international response" to the problem of climate change.[100] That response would involve assistance "to pay for anticipatory adaptation in many developing countries." To this end, the Bellagio meeting had attracted a wider array of international representatives. Sweden and the United States, which had been well-represented to date, were now joined by participants from the Environment Agency of Japan, the Commonwealth Secretariat, and the Arab Fund for Economic and Social Development, alongside the European Economic Community.

Among the European countries, the Federal Republic of Germany was especially prominent. West Germany had by now influenced the European Community's ozone policy in favor of the stricter controls that the United States and its allies advocated, leading to the adoption of the Montreal Protocol in September 1987.[101] The strength of its Green Party as well as concerns about acid rain (*Waldsterben*) aligned with the atmospheric alarm that German scientists had raised since the 1970s. Although the likes of Hermann Flohn and Wilfrid Bach had shared their climate concerns widely at home and abroad (Chapter 3), the 1986 publication of the German Physical Society's "Warning of a Threatening Climate Catastrophe" now made the front page of *Der Spiegel*.[102] In the wake of Chernobyl's radioactive plume spreading across West Germany a few months earlier, the August 1986 cover featured an inundated Cologne Cathedral, with the headline "Die Klima-Katastrophe,"

---

[97]Jill Jaeger, *Developing Policies for Responding to Climatic Change: A Summary of the Discussions and Recommendations of the Workshops Held in Villach (28 September – 2 October 1987) and Bellagio (9 – 13 November 1987)* (Geneva: World Climate Programme, 1988).

[98]See Matthias Heymann and Nils Randlev Hundebøl, "From Heuristic to Predictive: Making Climate Models into Political Instruments," in Matthias Heymann, Gabriele Gramelsberger, and Martin Mahony (eds.), *Cultures of Prediction in Atmospheric and Climate Science: Epistemic and Cultural Shifts in Computer-Based Modelling and Simulation* (London: Routledge, 2017), 100–19.

[99]Shardul Agrawala, "Science Advisory Mechanisms in Multilateral Decision Making: Three Models from the Global Climate Change Regime," PhD dissertation (Princeton University, 1999), 102.

[100]Jaeger, *Developing Policies for Responding to Climatic Change*, 38.

[101]Benedick, *Ozone Diplomacy*, 39, 45.

[102]See, for example, Wilfrid Bach, *Our Threatened Climate: Ways of Averting the CO$_2$ Problem through Rational Energy Use*, trans. Jill Jäger (Dordrecht: D. Reidel, 1984).

which brought into a single frame rising atmospheric anxieties about the ozone hole, global warming, and acid rain.[103]

As the West German anti-nuclear movement galvanized in response to Chernobyl and called for an *Ausstieg* or exit from nuclear power, Chancellor Helmut Kohl declared climate change the greatest of all environmental problems when he addressed the Bundestag in March 1987. For over a decade, the greenhouse effect had been among the reasons that supporters of nuclear energy such as Kohl's CDU had used to justify West Germany's rapid expansion of nuclear energy after the 1970s oil crises (Chapter 3).[104] Recognizing the growing political impasse on the "Climate Catastrophe," which informed sociologist Ulrich Beck's formulation of his risk society thesis, the Bundestag established a new Enquete Commission in late 1987 to examine "precautionary measures for protection of the Earth's atmosphere."[105] Comprising politicians and scientific experts (including Crutzen), the Enquete Commission provided an opportunity for the negotiation of science-informed political decisions, which, in this case, led the Commission's first report to conclude that "there was an extraordinary need" to act to avert climate change.[106] Its findings subsequently informed the government's June 1990 agreement to a goal of 25 percent reduction of carbon dioxide emissions in the former West Germany by 2005 based on 1987 levels, which guided the position of the European Community (Chapter 5).[107]

The reorientation of German energy policy aligned with the recommendations of the 1987 meetings of the nongovernmental AGGG. Greater support for the development of alternative (non-fossil-fuel) energy systems was also among the recommendations arising from recent meetings, where participants agreed that "a coordinated international response seems inevitable and rapid movement towards it is urged."[108] Recent scientific findings strengthened their resolve: NASA's James Hansen, for instance, had identified a strong warming trend between 1965 and 1980, which had raised global mean temperature in 1981 to the highest level on record.[109] In the wake of the recent World Climate

---

[103] "Die Klima-Katastrophe," *Der Spiegel*, August 10, 1986, https://v.gd/CxeON5.

[104] See Michael T. Hatch, "The Politics of Global Warming in Germany," *Environmental Politics* 4, no. 3 (1995): 415–40.

[105] Ulrich Beck, *Risk Society: Towards a New Modernity*, trans. Mark Ritter (London: Sage, 1992); Michael T. Hatch, "Corporatism, Pluralism and post-Industrial Politics: Nuclear Energy in West Germany," *West European Politics* 14, no. 1 (1991): 73–97; Matthias Dörries, "Climate Catastrophes and Fear," *WIREs Climate Change* 1 (2010): 885–90.

[106] Cited in Hatch, "The Politics of Global Warming," 424.

[107] Jeannine Cavender and Jill Jäger ,"The History of Germany's Response to Climate Change," *International Environmental Affairs* 5, no. 1 (1993): 4–18.

[108] Jaeger, *Developing Policies for Responding to Climatic Change*, 37.

[109] James Hansen and Sergej Lebedeff, "Global Trends of Measured Surface Air Temperature," *Journal of Geophysical Research* 92, no. D11 (1987): 13345–72.

Conference (Chapter 3), legal scholars had considered in the early 1980s that an international convention to set carbon dioxide emissions standards would be unlikely in the near future. Now, less than a decade later, the advisory group believed it timely for the negotiation of an international agreement on a law of the atmosphere or a convention similar to the Vienna Convention on ozone.[110] To this end, the group hoped its subsequent report would inform the proceedings of the upcoming Changing Atmosphere conference in Toronto in mid-1988 and the Second World Climate Conference in 1990.

Greenhouse Summer: Climate Change and Sustainable Development

Plans for the Toronto conference had emerged from the public hearings for the World Commission on Environment and Development (WCED). Established in 1983 by UN Secretary General Pérez de Cuéllar and chaired by Norwegian prime minister Gro Harlem Brundtland, this commission incorporated these meetings into its deliberations, visiting major cities in Indonesia, Norway, Brazil, Canada, Zimbabwe, Kenya, the Soviet Union, and Japan between 1985 and 1987.[111] At its Ottawa hearing in May 1986, Canadian environment minister Tom McMillan had volunteered his Progressive Conservative government's willingness to host an international conference to consider ways to improve the forecasting of environmental change. Climate change, he suggested, should be the focus of this meeting, which would become the Changing Atmosphere conference held just over two years later in Toronto (Figure 4.2). Environment Canada's Howard Ferguson, who had been involved in the negotiations with the United States to address acid rain and more recently helped organize the 1987 workshops at Villach and Bellagio, was tasked with making McMillan's idea a reality.[112] Neither Ferguson nor McMillan could have anticipated this meeting's influence on establishing climate change at the top of the international agenda, nor on efforts to impose ambitious thresholds to cut greenhouse gas emissions.

Elevating the Toronto conference's scientific and political profile was its connection to both the 1987 meetings that Ferguson had helped convene

---

[110]See, for example, Ved Nanda, "Global Climate Change and International Law and Institutions," in Ved Nanda (ed.), *World Climate Change: The Role of International Law and Institutions* (New York: Westview Press, 1983), 227–39; Jaeger, *Developing Policies for Responding to Climatic Change*, 41.
[111]World Commission on Environment and Development, *Our Common Future* (Oxford: Oxford University Press, 1987), 359.
[112]Howard Ferguson, "Foreword," in *World Conference on the Changing Atmosphere: Implications for Global Security* (Geneva: World Meteorological Organization, 1988), vii–x.

FIGURE 4.2 *Canadian environment minister Tom McMillan (left), Norwegian prime minister Gro Harlem Brundtland, and Canadian prime minister Brian Mulroney at the Changing Atmosphere conference in Toronto, Canada, 1988. Courtesy of Howard Ferguson and the Canadian Meteorological and Oceanographic Society (CMOS) Archives.*

and the Brundtland Commission. To form the conference steering committee, Ferguson called on the expertise of members of the advisory group, Bruce and Goodman, as well as his co-organizers of the more policy-oriented 1987 meetings, Jaeger and Oppenheimer. In the meantime, Goodman had contributed the findings of the 1985 Villach report to Brundtland's *Our Common Future*, which Bolin later believed were "instrumental in bringing the climate change issue to the attention of the UN General Assembly."[113] Although their influence as individuals on the science and politics of climate change would continue, their collective influence as a group declined after the Toronto meeting (Chapter 5).[114]

From the Brundtland Commission, meanwhile, Ferguson recruited its secretary-general Canadian Jim MacNeill who had "helped decisively to forge" its report, *Our Common Future*, with fellow Canadian commissioner, Maurice Strong.[115] MacNeill, who had led Canada's preparations for the 1972 Stockholm conference, had since become the director of the Organisation for Economic Co-operation and Development's (OECD's) Environmental

[113]Bolin, *A History*, 40. See World Commission on Environment and Development, *Our Common Future*, 174–8.
[114]See Shardul Agrawala, "Context and Early Origins of the Intergovernmental Panel on Climate Change," *Climatic Change* 39 (1998): 605–20.
[115]Gro Harlem Brundtland, "Our Common Future – a Climate for Change," in *World Conference on the Changing Atmosphere*, 16.

Directorate, where global economic and ecological interdependence had become the subject of growing interest.[116] After the UN General Assembly endorsed the idea of a special commission that would examine the environmental outlook to the year 2000, its chair Brundtland approached MacNeill in 1984 to join her; she had earlier served as Norway's representative on his OECD Environment Committee. Together, they embarked on an ambitious project that they now named the "World Commission on Environment and Development."[117]

To the commission, MacNeill brought insights from a series of OECD studies and meetings, which had pursued the reciprocal linkages between environmental protection policies and economic growth since the late 1970s, including the recent International Conference on Environment and Economics.[118] That meeting's report had argued that "the environment and the economy, if properly managed, are mutually reinforcing; and are supportive of and supported by technological innovation."[119] The Brundtland Commission's 1987 report *Our Common Future* framed this outlook in terms of "sustainable development," a concept that had undergone a series of transformations since its first use in the 1980 World Conservation Strategy, led by the likes of Stockholm veterans, economist Barbara Ward and Strong.[120] Sustainable development now denoted "a process of change in which the exploitation of resources, the direction of investments, the orientation of technological development, and institutional change are made consistent with future as well as present needs."[121]

In late June 1988, three hundred invited participants, the heads of state of Norway and Canada, and more than a hundred other government officials gathered in Toronto for Ferguson's conference. There, a week earlier, the Group of Seven (G7) industrial nations had endorsed the Brundtland Commission's concept of sustainable development at their Economic Summit, and they encouraged the establishment of an intergovernmental panel on global climate

---

[116]See, for example, OECD, *Economic and Ecological Interdependence: A Report on Selected Environment and Resource Issues* (Paris: OECD, 1982).

[117]For a detailed study of the Brundtland Commission, see Iris Borowy, *Defining Sustainable Development for Our Common Future: A History of the World Commission on Environment and Development (Brundtland Commission)* (London: Routledge, 2014).

[118]Steven Bernstein, *The Compromise of Liberal Environmentalism* (New York: Columbia University Press, 2001), 196–202.

[119]OECD, *Environment and Economics: Results of the International Conference on Environment and Economics* (Paris: OECD, 1985), 10. On the influence of the OECD on the Brundtland Commission and the notion of sustainable development, see Maarten A. Hajer, *The Politics of Environmental Discourse: Ecological Modernization and the Policy Process* (Oxford: Oxford University Press, 1995), 98–9.

[120]See Stephen Macekura, *Of Limits and Growth: The Rise of Global Sustainable Development in the Twentieth Century* (New York: Cambridge University Press, 2015), 219–61.

[121]World Commission on Environment and Development, *Our Common Future*, 9.

change under the auspices of the UNEP and the WMO.[122] Their emphasis on international cooperation in an increasingly interdependent world was shared by those at the Changing Atmosphere conference, not least by Brundtland, who cited the importance of international agreements to the resolution of other atmospheric problems.[123] International cooperation to manage and monitor the "shared resource" of the atmosphere was a key message in the conference's declaration, which urged, among its recommendations, the reduction of carbon dioxide emissions by approximately 20 percent of 1988 levels by the year 2005.[124]

Although this target fell well short of the 66 percent mooted at the 1987 Villach and Bellagio meetings, it was nonetheless an arresting and influential figure that would inform subsequent climate negotiations (see Chapter 5). This conference consensus reflected the momentum that climate change was gathering on the international agenda, which a slew of unusual weather events had only accelerated. "If ever nature could devise a timely demonstration of an environmental problem, then it did so in June 1988," as the UNEP's Peter Usher observed. The conference had "coincided with the severe drought that scorched the midwestern corn belt of the United States. Extreme, often record, weather events were occurring worldwide: heat waves hit New York and Central European capitals; floods in Africa interrupted nearly two decades of drought; and in Britain, summer brought almost continuous rain and cold."[125] The participants had also met shortly after NASA scientist James Hansen had joined Woodwell, Oppenheimer, Manabe, and others in testifying before a US Senate committee on the greenhouse effect and global climate change. In Toronto, Colorado Senator Tim Wirth, who would lead the US negotiation in the climate talks leading up to the Kyoto Protocol (Chapter 6), repeated Hansen's headline-grabbing testimony: "We now can say, with 99% certainty, that the greenhouse effect is upon us and that events such as the North American drought are increasingly likely to occur."[126]

Heightening the urgency of the Toronto recommendations was the conference's framing—that the changing atmosphere had "implications for global security." Stressing the conference subtitle, the meeting's declaration had described the consequences of climate change as "second only to a global

---

[122] "Toronto Economic Summit Economic Declaration," G7 Research Group, University of Toronto, June 21, 1988, https://v.gd/xPkLYc.

[123] Brundtland, "Our Common Future," 16–22.

[124] "The Changing Atmosphere: Implications for Global Security – Conference Statement," in *World Conference on the Changing Atmosphere*, 292–304.

[125] Peter Usher, "World Conference on the Changing Atmosphere: Implications for Global Security," *Environment: Science and Policy for Sustainable Development* 31, no. 1 (1989): 25–7.

[126] Timothy E. Wirth, "Dinner Address," in *World Conference on the Changing Atmosphere*, 40–6. See Senate Committee on Energy and Natural Resources, *Greenhouse Effect and Global Climate Change, Part 2, June 23, 1988* (Washington, DC: US Government Printing Office, 1988).

nuclear war," with implications for "global food security," political stability, and species extinction.[127] Brundtland too made a similar connection, that the impacts of climate change threatened to be more drastic than any other challenges, "with the exception of the threat of nuclear war."[128] These gestures to global security were not only a response to the decade's escalation of Cold War tensions but also derived from the Brundtland report itself. *Our Common Future* argued, "The whole notion of security as traditionally understood ... must be expanded to include the growing impacts of environmental stress."[129] In addition to the anxieties associated with nuclear winter, this emphasis on interdependence also echoed the findings of the 1980 Brandt Commission. Its report had observed, "The world is now a fragile and interlocking system, whether for its people, its economy or its resources ... More and more local problems can be solved only through international solutions—including the environment, energy, and the coordination of economic activity, money and trade."[130]

As chair of the WCED, Brundtland had understood her commission's work as the latest "call for political action," following Willy Brandt's *Programme for Survival and Common Crisis* and Olof Palme's *Common Security*. This heritage imbued her work with an ethos of social-democratic tradition, which she shared with the chairs of these earlier commissions as leaders of Western European social-democratic parties.[131] MacNeill, meanwhile, had introduced the Brandt Commission's outlook to the OECD in 1982, where it became imbued with the imperative of growth and its protection: "The growing scale of [environmental] issues ... and the inability of many Third World countries to deal with them, could have serious and even security consequences for OECD members."[132] So too his Canadian colleague Strong, who in 1984 noted, "The principal environmental battles in the period ahead will undoubtedly be fought in the developing countries and the future security of the global environment will depend on the outcome of these battles."[133] In

[127]"The Changing Atmosphere," 292–304. See also Bentley Allan, "Second Only to Nuclear War: Science and the Making of Existential Threat in Global Climate Governance," *International Studies Quarterly* 61, no. 4 (2017): 809–20.

[128]Brundtland, "Our Common Future," 16–22.

[129]World Commission on Environment and Development, *Our Common Future*, 19.

[130]Independent Commission on International Development Issues, *North-South, a Programme for Survival: A Report of the Independent Commission on International Development Issues* (London: Pan Books, 1980), 33.

[131]Hajer, *The Politics of Environmental Discourse*, 99. On the engagement of European social democrats with the ideas of *The Limits to Growth*, see Elke Seefried, "Rethinking Progress: On the Origin of the Modern Sustainability Discourse, 1970-2000," *Journal of Modern European History* 13, no. 3 (2015): 377–400.

[132]OECD, *Economic and Ecological Interdependence* (Paris: OECD, 1982), 3. See Bernstein, *The Compromise of Liberal Environmentalism*, 201.

[133]Maurice Strong, 'Future Directions for Environmental Policies: Introduction', in OECD, *Environment and Economics*, 212–16.

the Brundtland report, these arguments combined to center the environment with the economy as integrated issues of equal importance and a priority for multilateral cooperation.

In the months after the Toronto meeting, German climate scientist Bach expanded on the conference recommendation to reduce carbon dioxide emissions in an interview with *Der Spiegel*. Although he had helped to draft the meeting's declaration, he believed it had not gone far enough. Drawing on his work for the West German Enquete Commission, he argued that during the 1990s, emissions must be reduced by at least 37 percent compared to 1980 to prevent temperatures rising more than one or two degrees Celsius by the year 2100. "This is a scientifically justifiable value," he added. "Toronto's 20 per cent is a political number."[134] His government would soon host in Hamburg a follow-up conference on climate and development, inviting scientists, politicians, and representatives of organizations such as the WMO, the UNEP, and the Environment Defense Fund. Among the recommendations of this meeting was the call for other industrialized countries to carry out studies "in the nature of the 'Enquete Kommission' " and to implement their recommendations, as Bonn had.[135]

Elsewhere, US geographer William Riebsame noted that the Toronto conference statement had emerged "through in-fighting" among the drafting committee due to "disagreement whether the threat merited immediate action," as well as "the equity issue" between participants from the developed and developing worlds.[136] The Toronto target would nevertheless become an influential but divisive figure for negotiators, scientists, and activists. For the deputy head of the US delegation negotiating the UN Framework Convention on Climate Change in the early 1990s (Chapter 5), where such targets were a sticking point, the Toronto declaration had "grabbed people, and seized them; it's continued to grab them and seize them. They cannot even think of other alternatives or other approaches."[137]

---

[134]Jan Schumann, " 'Die Zeit läuft uns davon': Professor Wilfrid Bach über Programme," *Der Spiegel*, November 6, 1988, https://v.gd/DhglAt. See also Rik Leemans and Pier Vellinga, "The Scientific Motivation of the Internationally Agreed 'Well Below 2°C' Climate Protection Target: A Historical Perspective," *Current Opinion in Environmental Sustainability* 26–27 (2017): 134–42.

[135]"Hamburg Action Plan," in Hans J. Karpe, Dieter Otten, and Sergio C. Trindade (eds.), *Climate and Development: Climate Change and Variability and the Resulting Social, Economic and Technical Implications* (Berlin: Springer-Verlag, 1990), 14.

[136]William E. Riebsame, "Social Perspectives on Global Climate Change," in Steve Rayner, Wolfgang Naegeli, and Patricia Lund (eds.), *Managing the Global Commons: Decision Making and Conflict Resolution in Response to Climate Change* (Oak Ridge: Oak Ridge National Laboratory, 1990), 17.

[137]Daniel Reifsnyder, "Discussion after the Speeches of Daniel A. Reifsnyder and Elizabeth Dowsdesdell," *Canada-United States Law Journal* 18 (1992): 377.

Global Change: Legitimizing an International Response

For Oppenheimer of the Environmental Defense Fund, the formation of the Intergovernmental Panel on Climate Change (IPCC) represented the culmination of the work of the 1987 Villach and Bellagio meetings. Fellow participants at these meetings, he suggested, "should feel that their efforts have born unusual fruit, and not fallen into the usual 'black hole' that often swallows workshop outcomes."[138] Certainly, those meetings had directly informed the 1988 Toronto conference; their role in the IPCC's development, however, less so than Oppenheimer believed. According to US government officials, the panel arose in the wake of the 1985 Villach meeting, when the UNEP's Tolba wrote to Reagan's secretary of state to urge his government to lead the international response to climate change. This correspondence prompted US officials to devise a means to circumvent both Tolba and the advisory group, and in 1987 they employed the diplomatic channels of the WMO to legitimize their agenda.[139]

Tolba's entreaty to the US secretary of state apparently renewed tensions between the Department of Energy and the Environmental Protection Agency. Both now part of the new National Climate Program, each had undertaken climate assessments in 1983 but had arrived at rather different conclusions. As mentioned earlier in this chapter, the former's *Changing the Climate* report did not recommend any changes in energy policy, while the latter's called for prompt action.[140] The Villach report's intervention aligned more closely with the Environmental Protection Agency's position, while the Department of Energy viewed this report as the product of an unaccountable group beyond government oversight. In an intergovernmental mechanism, however, these agencies could find common ground—offering, on the one hand, progress toward an international agreement, and, on the other, delay.[141]

Following the WMO's Tenth Congress in May 1987, its executive council acknowledged the work of the AGGG. To Tolba's disappointment, however, the executive council supported "the establishment of a more broadly representative mechanism" that would provide "objective, balanced,

---

[138]Oppenheimer, "Developing Policies," 3.

[139]Alan Hecht and Dennis Tirpak, "Framework Agreement on Climate Change: A Scientific and Policy History," *Climatic Change* 29 (1995): 371–402; Alan Hecht, "Past, Present and Future: Urgency of Dealing with Climate Change," *Atmospheric and Climate Sciences* 4, no. 5 (2014): 10.4236/acs.2014.45069.

[140]National Research Council, *Changing Climate*; Stephen Seidel and Dale Keyes, *Can We Delay a Greenhouse Warming? The Effectiveness and Feasibility of Options to Slow a Build-Up of Carbon Dioxide in the Atmosphere* (Washington, DC: Environmental Protection Agency, 1983).

[141]Hirst, "Controlling the Agenda," 303–4.

and internationally co-ordinated scientific assessments of the current understanding in terms useful to governments."[142] Although such a mechanism had US roots, the WMO's Australian first vice president John Zillman recalled its emergence in "off-line discussions between the US and Australian delegations to the Congress [that] received general support in subsequent committee and plenary discussions."[143] This support included a call from Botswana's Gladys Ramothwa for a document that would enable her, and others, to answer government questions about climate change.[144] Seeking its cooperation, Bruce, now the acting deputy secretary-general of the WMO, took this idea of an "*ad hoc* intergovernmental mechanism to carry out internationally co-ordinated scientific assessments of the magnitude, timing, and potential impact of climate change" to the governing council of the UNEP. The council responded positively to Bruce's invitation and encouraged Tolba to work with the WMO and the ICSU to "explore, and after appropriate consultation with Governments," establish the mechanism.[145]

Within the US National Climate Program Office, the former director of the World Climate Programme developed a proposal for this new mechanism. As his model, Swedish meteorologist Bo Döös, who had coauthored with Bolin the SCOPE report that informed the 1985 Villach meeting, used the First Global Atmospheric Research Program Global Experiment, which also had an intergovernmental administration (Chapter 2). Working closely with office head Alan Hecht, his draft proposal observed that the decisions to date of the WMO and the UNEP "reflect the need for an *orderly process* to ensure that research, monitoring and impact assessment studies proceed together, and that internationally agreed assessments should be prerequisites for legal or regulatory activities."[146] With the approval of the relevant US agencies, their proposal eventually made its way to the executive council of the WMO, which established the IPCC in June 1988. The status and the role of the AGGG remained uncertain.[147]

Shortly before the IPCC's first plenary session in November 1988, Malta's representative to the UN introduced a draft resolution at the General Assembly

[142]World Meteorological Organization, *Thirty-Ninth Session of the Executive Council, Geneva 1–5 June 1987, Abridged Report with Resolutions* (Geneva: World Meteorological Organization), 6–7.
[143]John Zillman, "Australian Participation in the Intergovernmental Panel on Climate Change," *Energy & Environment* 19, no. 1 (2008): 21–42.
[144]John Zillman, "Atmospheric Science and Public Policy," *Science* 276 (1997): 1084–6.
[145]UN Environment Programme, "14/20 Global Climate Change, 18 June 1987," in *Report of the Governing Council on the Work of its Fourteenth Session* (Nairobi: UN Environment Programme, 1987), 71–2. Emphasis in original.
[146]Bo Döös and Alan Hecht, "Intergovernmental Panel on Climate Change," August 1987, 1–6, cited in Hirst, "Controlling the Agenda," 307.
[147]World Meteorological Organization, *Fortieth Session of the Executive Council, Geneva, 7–16 June 1988* (Geneva: World Meteorological Organization, 1988), 14.

that had a direct bearing on the scope of its activities. Over several sessions, Foreign Minister Vincent Tabone put forward his government's case for the "conservation of climate as part of the common heritage of mankind." Nodding to Malta's leading role in advancing the 1982 UN Convention on the Law of the Sea, he argued that it was "essential that action be taken on a global level to ensure that our planet remains fit to sustain life."[148] In the subsequent resolution, Malta's delegation called on the IPCC to "immediately initiate action" to comprehensively review the state of the science of climate change, the study of its social and economic impacts, possible policy responses to mitigate climate change, relevant treaties and other legal instruments dealing with climate, and to review the "elements for possible inclusion in a future international convention on climate."[149] Taking such a comprehensive approach, as Malta hoped, would require the panel to venture beyond the assessments of physical science that policymakers viewed as authoritative to assess the potential economic costs of damages and responses that were more uncertain.[150]

Foreshadowing the role of the small island nations in the climate negotiations in the decades ahead, Tabone expanded on the unique perspective of his country. Smaller states such as Malta, he argued, could contribute to the UN by ensuring the organization was "constantly attuned to the growing and changing needs of mankind." Tabone's references to Malta's "smallness" invoked a longer conversation on self-determination in the UN Conference on Trade and Development, as well as the Commonwealth Secretariat and the European Union, which had emerged during the wave of decolonization of the 1960s and 1970s. "Smallness," those organizations had worried, represented a challenge to their economic development, which led them to sponsor special programs that would foster regionalism among small states.[151] In his representations to the UN, Tabone also argued that smaller states such as his "can reflect the conscience of mankind," which might have gestured to an address that Maldives president Maumoon Abdul Gayoom had delivered to the General Assembly a year earlier. Speaking after

[148]Malta, "A/43/PV.35 Conservation of Climate as Part of the Common Heritage of Mankind," United Nations General Assembly, October 24, 1988, 11, https://v.gd/8n1VKt.

[149]Malta, "A/43/251 148 Conservation of Climate as Part of the Common Heritage of Mankind: Draft Resolution," United Nations General Assembly, October 26, 1988, 3, https://v.gd/m0MSmp.

[150]See Agrawala, "Context and Early Origins of the Intergovernmental Panel on Climate Change," 616.

[151]Jenny Grote, "The Changing Tides of Small Island States Discourse: A Historical Overview of Small Island States in the International Area," *Law and Politics in Africa, Asia and Latin America* 43, no. 2 (2010): 164–91; Jon Barnett and John Campbell, *Climate Change and Small Island States: Power, Knowledge and the South Pacific* (London: Earthscan, 2010), 155–60; Greg Fry, *Framing the Islands: Power and Diplomatic Agency in Pacific Regionalism* (Canberra: ANU Press, 2019), 191–216.

Brundtland in October 1987, President Gayoom had described the impacts of a recent series of tidal surges on the Maldives, elevating the problem of climate change to the UN General Assembly for the first time since 1968 (Chapter 2). He presented an island perspective on the need for multilateralism to address this challenge: "No one nation, or even group of nations, can alone combat the onset of a global change in the environment." [152]

Conclusion

As the end of the decade neared, US climate scientist and writer Stephen Schneider wondered whether the "the greenhouse century" lay ahead. Already looking set to reprise the positions that nuclear winter had helped to surface, "battle lines were being drawn for what promises to be one of the most important political debates of this—and the next—century: what can or should we do to avert the possibility of an unprecedented threat to the global environment, global warming?" [153] The UN and the IPCC had emerged as the institutions to shape the impending international negotiations that would center on the concerns of governments, rather than those of independently-minded scientists.

Taking shape at the "end of History," these institutions would soon grapple with the geopolitical realignment that followed the collapse of the Soviet Union. [154] Francis Fukuyama marveled at "the end point of mankind's ideological evolution and the universalization of Western liberal democracy as the final form of government," while Bill McKibben mourned the "end of Nature" that anthropogenic climate change represented for the US environmental tradition. [155] From the self-congratulatory to the gloomy, these positions hinted at the Cold War origins of understanding global climate change and at the spectrum of global imaginaries that would guide the negotiations of the emerging regime of international climate governance. [156]

---

[152]Maumoon Abdul Gayoom, Forty-Second Session, UN General Assembly, A/42/PV.41, October 20, 1987, 27, https://v.gd/VA4flq.

[153]Stephen Schneider, *Global Warming: Are We Entering the Greenhouse Century?* (San Francisco: Sierra Club Books, 1989), ix.

[154]Francis Fukuyama, "The End of History?," *National Interest* 16 (1989): 3–18.

[155]Bill McKibben, "The End of Nature," *New Yorker*, September 3, 1989, https://v.gd/39opti.

[156]Donald Worster, *Under Western Skies: Nature and History in the American West* (New York: Oxford University Press, 1992), 238–54; Gregg Mitman, "Hubris or Humility: Genealogies of the Anthropocene," in Gregg Mitman, Marco Armiero, and Robert Emmett (eds.), *Future Remains: A Cabinet of Curiosities for the Anthropocene* (Chicago: University of Chicago Press, 2018), 59–70.

5

Climate Negotiations

1994: 358.71 ppm

Elected to speak on behalf of the world's youth at the 1992 Earth Summit, Wagaki Mwangi of the Kenya-based International Youth Development and Environment Network delivered a scathing assessment of the conference proceedings: "Those of us who have watched the process have said that UNCED have failed. As youth we beg to differ." She continued, "Multinational corporations, the United States, Japan, the World Bank, the International Monetary Fund have got away with what they always wanted, carving out a better and more comfortable future for themselves ... UNCED has ensured increased domination by those who already have power." By this time, UN officials had severed the live transmission of her speech throughout the conference venue, leaving just the delegates assembled inside the Plenary Hall to digest her message.[1]

Contrary to the official line, Mwangi was giving voice to the discontent among those unconvinced that the Earth Summit's spirit of sustainable development heralded anything but a commitment to the status quo. In June, 110 world leaders, 500 nongovernmental groups, 8,000 accredited journalists and 30,000 citizens had converged on Rio de Janeiro for the United Nations Conference on Environment and Development (UNCED). For some, the event was a success that had raised awareness and captured the world's attention to the environmental cause, while others had been left disappointed that new diplomatic and financial mechanisms for solving shared international problems had not emerged, let alone a genuine commitment to an equitable and sustainable new international order.[2] What it did achieve, political scientist Steven Bernstein argues, was an international regime of sustainable development: "the institutionalization of a particular vision or understanding

---

[1] "Rio," *LA Weekly*, July 9, 1992, 22; Wagaki Mwangi, 1992, cited in Pratap Chatterjee and Matthias Finger, *The Earth Brokers: Power, Politics and World Development* (London: Routledge, 1994), 167.
[2] Andrew Jordan, "The International Organisational Machinery for Sustainable Development: Rio and the Road Ahead," *Environmentalist* 14 (1994): 23–33.

of how the international community ought to manage or approach global environmental problems and the norms that would guide future action."[3]

The end of the Cold War and the economic hardship experienced by many developing countries in the 1980s had helped consolidate the influence of Western visions of the superiority of market forces and Western institutions at the "end of History." Jamaica's Michael Manley, for instance, had espoused free market ideology on his return to office in 1989, India had liberalized its economy in 1991, and Guyana's Cheddi Jagan had converted to market economics on the eve of his election victory in 1992.[4] Backed by the International Monetary Fund and the World Bank, the Washington Consensus now prevailed, with its emphasis on economic policies such as fiscal discipline, financial and trade liberalization, privatization, and deregulation. In this market-friendly international economic system where the superpower rivalry had been confined to the past, many governments were optimistic that circumstances would now allow for states to cooperate to resolve global problems that had until recently struggled for attention.[5]

At the 1990 Earth Day rally in Washington, DC, outspoken biologist Paul Ehrlich encouraged the crowd to "realize our problems are absolutely global—a cow breaks wind in Indonesia and your grandchildren could die in food riots in the United States—realize that we're going to have to cooperate with each other."[6] Just a few months earlier, the UN General Assembly had agreed to a resolution to hold a UNCED, which would be held in Rio de Janeiro in June 1992 and headed by Maurice Strong, the secretary-general of the UN Stockholm Conference twenty years earlier.[7] In addition to the global challenges of ozone depletion and anthropogenic climate change, several disasters over the past decade had drawn public attention to the need for international cooperation— the 1984 leak of toxic gas from the Union Carbide plant in Bhopal (India), the 1986 nuclear accident at Chernobyl (Ukraine), and the *Exxon Valdez* oil spill in 1989.[8] Arising from the recommendations of the Brundtland Commission, the UN resolution echoed the issues raised by the newly created South Commission with its emphasis on the responsibility of the developed nations for environmental problems and the need for technology transfers and financial

---

[3]Steven Bernstein, *The Compromise of Liberal Environmentalism* (New York: Columbia University Press, 2001), 100.

[4]Marc Williams, "Re-articulating the Third World Coalition: The Role of the Environmental Agenda," *Third World Quarterly* 14, no. 1 (1993): 7–29.

[5]Bernstein, *The Compromise of Liberal Environmentalism*, 86.

[6]Paul Ehrlich, "Earth Day 1990 Rally," April 22, 1990, C-Span, https://v.gd/9cb7Jj. See Andrew Ross, "Is Global Culture Warming Up?," *Social Text* 28 (1991): 3–30.

[7]"UN Conference on Environment and Development: Resolution," A/RES/4/228, United Nations, https://v.gd/G6WDdL.

[8]Bernstein, The Compromise of Liberal Environmentalism, 85.

transfers to the developing world.[9] The negotiations for this conference would have a significant bearing on the tenor of the climate negotiations that would soon begin. The newly instituted Intergovernmental Negotiating Committee (INC) was charged with ensuring that a climate convention would be ready for signing in Rio—a demanding deadline to negotiate the dynamics between the governments of the North and South that would remain characteristic of international climate diplomacy.

Atmospheric Politics: Establishing the Agenda

Only a few days after representatives from over a hundred nations gathered in London in early 1989 to discuss the state of the ozone layer, France, Norway, and the Netherlands hastily convened a 24-nation summit at The Hague. The European Community's environment ministers had just unexpectedly committed its twelve member states to try banning ozone-threatening chlorofluorocarbon (CFC) gases by the year 2000.[10] By the time the Vienna Convention came into force in early 1988, momentum was growing for a stronger international response to the problem of ozone depletion. The National Aeronautics and Space Administration's (NASA's) Ozone Trends Panel, led by its own Robert Watson, had recently confirmed global losses of ozone of 1 percent since 1979, and that ozone depletion was underway not only over Antarctica but also in the higher latitudes of the northern hemisphere.[11] "The distant threat was clearly becoming nearer," recalled the UN Environment Programme's (UNEP's) Mostafa Tolba.[12] DuPont—the world's leading producer of CFCs—responded quickly, announcing it would halt all production immediately. In the UK, the panel's results combined with political pressure at home led British prime minister Margaret Thatcher to reassess her government's position in favor of international action on both the ozone layer and climate change.[13]

---

[9] Chatterjee and Finger, *The Earth Brokers*, 30–2
[10] "Another Environment Summit Opens, Illustrating Issue's New Currency," *Washington Post*, March 11, 1989, a17.
[11] International Ozone Trends Panel, *Report of the International Ozone Trends Panel, 1988* (Washington, DC: NASA, 1988).
[12] Mostafa Tolba, *Global Environmental Diplomacy: Negotiating Environmental Agreements for the World, 1973–1992* (Cambridge, MA: MIT Press, 1998), 62.
[13] Karen Litfin, *Ozone Discourses: Science and Politics in Global Environmental Cooperation* (New York: Columbia University Press, 1994), 123–7.

Neither the United States nor the Soviet Union were among the governments invited to The Hague, ostensibly to avoid "East-West competition," while Prime Minister Thatcher declined to attend, describing the proceedings as "posturing."[14] For its part, the Bush administration had already expressed its "no regrets" policy to a meeting of the Intergovernmental Panel on Climate Change (IPCC), calling for "prudent steps that are already justified on grounds other than climate change."[15] The declaration adopted at The Hague proposed to create "a new institutional authority" with non-unanimous decision-making rules—that is, a partial renunciation of sovereignty—to protect the atmosphere and prevent global warming.[16] Although this approach did not gain traction ("pathetic," Thatcher noted privately), the engagement of seventeen heads of state from both the developed and developing worlds indicated that the challenge of climate change had risen to the top of the international agenda.[17]

For developing countries, meanwhile, it was evident that "we can not sit on the sidelines anymore," as India's M. S. Swaminathan had declared to colleagues in New Delhi in February 1989. As president of the International Union for the Conservation of Nature and Natural Resources, Swaminathan collaborated with UNEP, the World Resources Institute, George Woodwell's Woods Hole Research Centre, and the Tata Energy Research Institute (led by future IPCC chair Rajendra Pachauri) to convene the first meeting on climate change in a developing country. Attended by some 150 scientists and policymakers, mostly from South Asia, as well as from the United States and the UK, the event sought to raise awareness about climate change beyond the industrialized world. Its report described an "apocalyptic scenario" that "threatens the very survival of civilization, and promises to throw up only losers over the entire international socio-economic fabric."[18] A subsequent meeting of some four hundred delegates from forty nations in Cairo repeated this approach, whereby the US Climate Institute and UNEP co-organized the event with sponsorship from the governments of Egypt, the Netherlands, the

---

[14]Cited in Jim Sheppard, "24 Nations Call for UN Agency to Safeguard the Atmosphere," *Ottawa Citizen*, March 12, 1989, 44.

[15]US Department of State and US Environmental Protection Agency, *US Efforts to Address Global Climate Change: Report to Congress* (Washington, DC: Environmental Protection Agency, 1991), 25.

[16]Daniel Bodansky, "The United Nations Framework Convention on Climate Change: A Commentary," *Yale Journal of International Law* 18 (1993): 466.

[17]Margaret Thatcher, cited in Jon Agar, *Science Policy under Thatcher* (London: University College London, 2019), 245. The attending heads of state at The Hague included Gro Harlem Brundtland (Norway), Helmut Kohl (Federal Republic of Germany), Brian Mulroney (Canada), King Hussein (Jordan), and Robert Mugabe (Zimbabwe). The organizers did not invite representatives from the European Community, China, Belgium, or Greece, while the UK declined to attend.

[18]Sujata Gupta and R. K. Pachauri (eds.), *Proceedings of the International Conference on Global Warming and Climate Change: Perspectives from Developing Countries* (New Delhi: Tata Energy Research Institute, 1989).

United States, and Australia, as well as other nongovernmental organizations and Western philanthropic foundations.[19]

In the wake of July's *Exxon Valdez* oil spill off the Alaskan coast in 1989, the Group of Seven (G7) nations met in Paris, where rising inflation was a cause for concern. Nevertheless, leaders of the seven major industrialized countries gave environmental issues, such as the ozone layer and climate change, new political impetus, devoting eight pages of the meeting's 22-page communique to such questions. Endorsing "the concept of a framework or umbrella convention on climate change," they expressed their support for the work that the IPCC had been undertaking and emphasized their commitment to ensuring the "compatibility of economic growth and development with the protection of the environment."[20]

Climate change likewise entered international conversations among the political leaders of the developing world. At July's South Pacific Forum in Kiribati, delegates drew attention to the implications of rising sea levels for their island countries, which the Commonwealth Secretariat echoed in its October Langkawi Declaration.[21] At the Belgrade meeting of the Non-Aligned Movement in September, leaders agreed to call "for the preparation and adoption of an international convention on the protection and conservation of the global climate on an urgent basis." India's prime minister, for his part, also advocated for a Planet Protection Fund under UN control. By year's end, it had become evident to Woods Hole director Woodwell and his colleague Kilaparti Ramakrishna that "some of the developing nations wish to join the global warming dialogue because they see opportunities for increasing their economic development, while others see a compelling need to find appropriate solutions to arrest global warming because of their fragile geographic and economic situations."[22]

The fault lines emerging between the North and South became clearer at the Noordwijk Ministerial Conference on Atmospheric Pollution and Climate Change in November 1989. Hosted by the Netherlands, a low-lying nation particularly sensitive to the prospect of rising sea levels, and attended by the representatives of nearly seventy states, the event's organization indicated that climate change was yet to be understood as having profound implications

[19]Climate Institute, *World Conference on Preparing for Climate Change, December 17–21 1989* (Washington, DC: Climate Institute, 1990).

[20]"Economic Declaration, Paris, July 16, 1989," *G7 Information Centre*, University of Toronto, https://v.gd/gcl7rP .

[21]"Twentieth South Pacific Forum, Tarawa, Kiribati, 10–11 July 1989," *Pacific Islands Forum*, https://v.gd/1GKwUA ; "Langkawi Declaration on the Environment, 1989," *Commonwealth Heads of Government*, October 21, 1989, https://v.gd/pV3E5w.

[22]George Woodwell and Kilaparti Ramakrishna, "Guest Editorial: The Warming of the Earth: Perspectives and Solutions in the Third World," *Environmental Conservation* 16, no. 4 (1989): 290.

for the world's economies. The conference secretary, coastal engineer Pier Vellinga (of the Bellagio meeting), worked with the Dutch Ministry of Housing, Physical Planning and Environment to dispatch invitations to other environmentally oriented ministries.[23] The United States, however, where climate change had already been subject to domestic scrutiny for over a decade, led opposition to any quantitative limits on greenhouse gas emissions and joined with Japan and the Soviet Union to counter European efforts to establish a date and level of emissions at which the stabilization of greenhouse gases might occur.

In the resulting Noordwijk Declaration, the delegates reiterated many of the aspirations of the 1988 Toronto Statement such as its emissions reduction target, with much greater attention to the interests of the developing countries.[24] These interests were amplified by the Cairo Compact the following month, which called for "affluent nations" to provide financial and technical assistance for developing countries. By late December 1989, a revised version of Malta's resolution had been adopted in the UN General Assembly (Chapter 4). Now heeding the perspectives of the majority world, this resolution duly observed the messages of Noordwijk and Cairo—that the developed world, having produced the pollutants, must take the "main responsibility" for combatting climate change and provide technical and financial assistance to developing nations. The UN General Assembly reaffirmed Principles 21 and 22 of the Stockholm Declaration and restated the urgent need for governments to prepare a framework convention on climate that would take into account both scientific knowledge and the specific development needs of developing countries. Most importantly for the negotiations ahead, the General Assembly itself was affirmed as the "appropriate forum" for political action on "global environmental problems."[25]

Scientific Diplomacy: Debating Global Climate Change

With the renewed support of the UN, the IPCC continued to prepare its report to inform a series of international meetings in late 1990 and early

---

[23]Piero Morseletto, Frank Biermann, and Philipp Pattberg, "Governing by Targets: *Reductio ad unum* and Evolution of the Two-Degree Climate Target," *International Environment Agreements* 17 (2016): 655–76.

[24]Netherlands Ministry of Housing, Physical Planning and Environment, *Ministerial Conference on Atmospheric Pollution and Climatic Change* (The Hague: Ministry, 1989).

[25]"Protection of Global Climate for Present and Future Generations of Mankind: Resolution," December 22, 1989, A/RES/44/207, United Nations, https://v.gd/NaktUD.

1991, including the Second World Climate Congress and the Eleventh World Meteorological Congress. In August 1990, the IPCC plenary met in Sundsvall, Sweden, to approve the First Assessment Reports of the three working groups and to develop an overview statement of their findings. Focused on the scientific assessment of climate change, the report of Working Group I commanded the most attention among policymakers and would shape the convention negotiations to come. Led by UK meteorologist John Houghton, this group had earlier agreed to prepare a short policy document that would contextualize their findings specifically for this audience, in which it advised, should current trends continue (a "business as usual" scenario), the global mean temperature would increase by about 0.3°C per decade over the course of the next century—"greater than that seen over the past 10,000 years." Echoing Wallace Broecker's 1987 *Science* commentary, the summary advised that such was the complexity of the climate system and scientific uncertainty as to the patterns and processes of climate change that "we cannot rule out surprises."[26] Carbon dioxide emissions, the group concluded, would need to be reduced immediately by over 60 percent if atmospheric concentrations were to be stabilized at 1990 levels.[27]

Arriving at a conclusion was especially challenging for the panel's Working Group III, which had been charged with assessing response strategies. According to UK diplomat and negotiator Tony Brenton, the divisions that had emerged since early 1989 came to the fore in Sundsvall. Proceedings there, he recalled, "finished at 4 o'clock in the morning, one day late, with most of the delegates having abandoned their chairs to gather on the front podium and shout at each other."[28] What would become familiar positions were asserted—the United States sought to emphasize uncertainties, while European nations insisted on targets for carbon dioxide emissions and deforestation. Renewing their position forged at Stockholm nearly twenty years earlier, the developing countries, led by India, resisted mention of deforestation and their own potential action, arguing instead for the responsibility of the developed world and the need to transfer finance and technology to the majority world.

As Group of 77 (G77) negotiator Tariq Hyder put it shortly afterwards, "many people in the South are skeptical about some of the environmental studies pertaining to developing country emissions that have been prepared by various

---

[26]Shardul Agrawala, "Structural and Process History of the Intergovernmental Panel on Climate Change," *Climatic Change* 39 (1998): 633.
[27]John Houghton et al., "Policymakers Summary," in John Houghton, Geoffrey Jenkins, and J. Ephraums (eds.), *Climate Change: The IPCC Scientific Assessment* (Geneva: Intergovernmental Panel on Climate Change, 1990), xi.
[28]Tony Brenton, *Greening Machiavelli: The Evolution of International Environmental Politics* (Milton Park: Routledge, [1994] 2019), 182.

private organizations in the North."[29] After the IPCC report's publication in 1990, Indian scientist Jyoti Parikh argued in *Nature* that Working Group III's report contained "a lot of subjective assumptions, many of which do not do justice to the developing countries." The report assumed, she argued, "that the present inequalities among different world regions will increase considerably," leading to stabilization scenarios that "stabilize the lifestyles of the rich and adversely affect the development of the poor."[30]

Parikh's intervention spurred further reflection within the IPCC, as did those of Brazil's science officials, who had also shared their concerns about representations of the scale of their country's emissions. The IPCC subsequently recommended more inclusive modeling approaches and exploring different economic development pathways, such as closing the income gap between industrialized and developing regions.[31]. The panel also adjusted its structure to replace the single chairs of each working group with two cochairs, which would be elected from a developed country and a developing country. During preparations for the Second Assessment Report, for instance, the former director of the Brazilian space agency, climatologist Luiz Gylvan Meira Filho, joined Working Group I as Houghton's cochair.[32] It also mandated financial support for at least one developing country expert to attend every writing team meeting of every chapter.[33]

Even as the Soviet Union was grappling with an existential crisis of another kind, its scientists had been engaged in their own battles. At a Working Group II meeting in Moscow in early 1989, the gathered scientists debated the use of paleoclimate analogues as a forecasting technique. This method, strongly advocated by climatologist Mikhail Budyko, allowed for interpretations of climate change as a welcome phenomenon for the Soviet Union and other parts of the northern hemisphere. Such prospects invited a rather different outlook on the climate impacts this group was tasked with assessing: as

[29]Tariq Osman Hyder, "Climate Negotiations: The North/South Perspective," in Irving Mintzer (ed.), *Confronting Climate Change: Risks, Implications and Responses* (New York: Cambridge University Press, 1992), 326.

[30]Jyoti Parikh, "IPCC Strategies Unfair to the South," *Nature* 360 (1992): 507–8.

[31]Joseph Alcamo et al., "An Evaluation of the IPCC IS92 Emission Scenarios," in John Houghton et al. (eds.), *Climate Change 1994: Radiative Forcing of Climate Change and an Evaluation of the IPCC IS92 Emission Scenarios* (Cambridge: Cambridge University Press, 1995), 247–304. See Jiesper Pederson et al., "IPCC Emission Scenarios: How Did Critiques Affect Their Quality and Relevance 1990–2022," *Global Environmental Change* 75 (2022): 10.1016/j.gloenvcha.2022.102538.

[32]Myanna Lahsen, *Brazilian Climate Epistemers' Multiple Epistemes: An Exploration of Shared Meaning, Diverse Identities and Geopolitics in Global Change Science* (Cambridge, MA: Belfer Center for Science and International Affairs, 2002), 14.

[33]John Zillman, "Australian Participation in the Intergovernmental Panel on Climate Change," *Energy & Environment* 19, no. 1 (2008): 21–42; Agrawala, "Structural and Process History," 630.

Budyko told a British interviewer, "global warming is a good thing, it will increase harvests everywhere."[34] Backed by the group's Soviet chair Yuri Izrael, Budyko criticized Houghton's Working Group I for failing to consider this technique in their scientific assessment. For the panel's assessment, as Houghton put it, "this is hot just an academic issue: the forecasts from the two techniques are different and may have different policy implications, and so it is important that the issue is resolved amongst those who are best equipped to deal with it."[35]

Resolving these divergent approaches was a delicate matter not only for the IPCC but also for a joint research project underway between the Soviet Union and the United States since the 1987 Washington Summit. Given the overlapping expertise and contents of the reports, the US scientists were especially sensitive to placating their Soviet colleagues and attempted to persuade the panel accordingly: "If they (the non-Soviet scientists) are too dismissive of paleo-analogs they run the risk of sabotaging the whole US-USSR operation. On the other hand they (the non-Soviet scientists) recognize that, if they leave the door open on paleo-analogs as predictions, they will undermine the conclusions of the [IPCC's Working Group I] report."[36] After interventions by Houghton and IPCC chair Bert Bolin, the Working Group I report concluded, "The paleo-analogue approach is unable to give reliable estimates of the equilibrium climatic effect of increases in greenhouse gases." It conceded, however, stating that "information on past climates will provide useful data against which to test the performance of climate models when run with appropriate forcing and boundary conditions."[37] The final report of the US-USSR exercise—the result of nearly twenty years of "steadily intensifying, enriching and rewarding interactions"—was more conciliatory: "It is possible to combine insights from both numerical models and empirical methods to provide some projections of general climate trends for the future."[38] As for the IPCC's Working Group II, concerns about the Soviet dominance of

[34]Fred Pearce and Jack Miller, "Soviet Climatologist Predicts Greenhouse 'Paradise'," *New Scientist*, August 26, 1989, https://v.gd/kgNFof. Budyko had made similar arguments at the Hamburg meeting in 1988 (Chapter 4).

[35]John Houghton, November 1989, cited in Tora Skodvin, *Structure and Agent in the Scientific Diplomacy of Climate Change: An Empirical Case Study of Science Policy Interactions in the Intergovernmental Panel on Climate Change* (New York: Kluwer, 2000), 140.

[36]Internal letter, Working Group I, May 1990, cited in Skodvin, *Structure and Agent*, 140–1. See also Katja Doose, "A Global Problem in a Divided World: Climate Change Research during the late Cold War, 1972–91," *Cold War History* 21 (2021): 469–89.

[37]J. F. B Mitchell et al., "Equilibrium Climate Change – and Its Implications for the Future," in Houghton, Jenkins, and Ephraums (eds.), *Climate Change*, 159.

[38]"Summary," in Michael MacCracken et al., *Prospects for Future Climate: A Special US/USSR Report on Climate and Climate Change* (Chelsea, MI: Lewis, 1990), xii–xiii.

its authorship led to the Australian co-vice-chair assuming responsibility for compiling and editing its final report.[39]

The Advisory Group on Greenhouse Gases (AGGG), meanwhile, had continued its own work. Like the IPCC, its three working groups were also preparing for the upcoming World Climate Conference in Geneva. Since the Toronto meeting, the nongovernmental group had increasingly come under the auspices of Gordon Goodman's Stockholm Environment Institute, with which the Beijer Institute had merged during the recent Swedish economic crisis. Just a few months later in Geneva, however, only one paper resulted from the advisory group's efforts: a proposal for climate policy based on "risk assessment and maximum tolerable climate change."[40] Now in the shadows of the IPCC, this event would be the group's last intervention in the negotiations of climate diplomacy.[41] Having helped to craft the scientific understanding of global climate change, and advocated for an international response, the group would cease to affect the emerging climate regime that would be centered on the negotiations between governments.

By the time scientists and government representatives assembled for the Second World Climate Conference in November 1990, some parts of the developed world had already taken steps to limit their own greenhouse gas emissions. Sweden led the way in this regard, having undertaken in 1988 to stabilize its carbon dioxide emissions at 1988 levels by the year 2000; Norway and the Netherlands followed suit in 1989, and others joined them in the ensuing months: Denmark, Italy, and the UK in May; Austria, Canada, Germany, and the Netherlands (with an updated target) in Aotearoa New Zealand in July; France in September; Australia, the European Community, and Japan in October; and, finally, Iceland and Switzerland in November.[42] The United States, constituting over 20 percent of global carbon dioxide emissions in 1990, was not among them, although the Bush administration had by now conceded that it would engage in the negotiation of a framework convention on climate change.[43] Although the conference's Ministerial Declaration "welcome[d]" these

---

[39]Bert Bolin, *A History of the Science and Politics of Climate Change: The Role of the Intergovernmental Panel on Climate Change* (Cambridge: Cambridge University Press, 2007), 65. See W. J. McG. Tegart, G. W. Sheldon, and D. C. Griffiths (eds.), *Climate Change: The IPCC Impacts Assessment* (Canberra: Australian Government Publishing Service, 1990).

[40]Pier Vellinga and Rob Swart, "The Greenhouse Marathon: Proposal for a Global Strategy," in Jill Jäger and Howard Ferguson (eds.), *Climate Change: Science, Impacts and Policy – Proceedings of the Second World Climate Conference* (New York: Cambridge University Press, 1991), 129–34. This paper was based on Frank R. Rijsberman and Robert J. Swart (eds.), *Targets and Indicators of Climatic Change* (Stockholm: Stockholm Environment Institute, 1990).

[41]See Shardul Agrawala, "Science Advisory Mechanisms in Multilateral Decision Making: Three Models from the Global Climate Change Regime," PhD dissertation (Princeton University, 1999), 129.

[42]Mathew Paterson, *Global Warming and Global Politics* (London: Routledge, 1996), 40.

[43]"United States: What Share of Global CO_2 Emissions Are Emitted by the Country?," *Our World in Data*, https://v.gd/TjLyAq.

decisions, the resulting text had "the non-committal blandness that we have come to expect from such documents," as Canada's Ken Hare later put it.[44]

The meeting's Canadian organizer, Howard Ferguson, who had been responsible for the Toronto Conference in 1988, successfully attracted several heads of state to Geneva—including King Hussein of Jordan, Prime Minister Margaret Thatcher of the UK, Prime Minister Michel Rocard of France, and Prime Minister Bikenibeu Paeniu of Tuvalu. Joining them were some nine hundred participants from nearly 140 countries, representing 80 percent of the UN. The scientific program, meanwhile, involved nearly 750 participants from over 115 nations. Ferguson's planning for the event had commenced in 1986, originally to review the progress of the World Meteorological Organization's (WMO's) World Climate Programme, but in light of the work of the IPCC and the rising political import of climate change, the meeting's schedule had been expanded to include the ministerial component.

In Geneva, the dominance of delegates from the developing world helped to ensure these nations made their own mark on the proceedings, as they reprised their stance from the 1972 Stockholm meeting. Again, they pressed their case for the recognition of the "special needs of developing countries ... with various recommendations stressing the need for technical and financial support to encourage sustainable economic development along environmentally-beneficial pathways."[45] Delegates from the majority world had earlier affirmed their shared cause in New Delhi, where India's Singh government had invited representatives from eighteen developing countries in Asia, Africa, South America, and Europe. At this conference on global environmental issues, they agreed on the position: (1) that the responsibility for climate change lay with the North; (2) that they required technical and financial assistance, (3) that they would accept no impediments to their economic development, and (4) that they would accept no emission reduction targets.[46]

For a small group of island nations, the climate conference in Geneva marked a key moment in sharing their cause with the world. In 1989, the South Pacific island nations had met in Majuro, in the Marshall Islands, where they discussed the "potentially catastrophic" changes that global warming represented for their countries.[47] Another fourteen island nations gathered

---

[44]"Ministerial Declaration," in Jäger and Ferguson (eds.), *Climate Change*, 536; F. Kenneth Hare, "Review – Climate Change: Science, Impacts and Policy," *Environmental Conservation* 19 (1992): 189.

[45]G. O. P. Obasi et al., "Foreword," in Jäger and Ferguson (eds.), *Climate Change*, xi.

[46]Anil Agarwal, Sunita Narain, and Anju Sharma, "Boiling Point: The United Nations Framework Convention on Climate Change/Kyoto Protocol," in Anil Agrawal, Sunita Narain, and Anju Sharma (eds.), *Green Politics* (New Delhi: Centre for Science and Environment, 1999), 15–122.

[47]South Pacific Commission, *Intergovernmental Meeting on Climatic Change and Sea Level Rise in the South Pacific* (Noumea: South Pacific Commission, 1989).

in the Maldives, but their call for action was overshadowed by the fall of the Berlin Wall just the week before.[48] Vanuatu's Ernest Bani, meanwhile, appealed to industrial nations at an IPCC Working Group III meeting in Geneva "to prevent us from becoming endangered species or the dinosaurs of the next century."[49] Now, in his opening address to the Second World Climate Conference, Tuvaluan prime minister Paeniu seized the opportunity to remind delegates, "We in the Pacific, the Caribbean and elsewhere had done the least to create these hazards (of the Greenhouse Effect and sea level rise) but now stand the most to lose."[50]

Although these island nations were ultimately disappointed with the climate conference's pared-back ministerial statement, they came away from Geneva having formally organized themselves as the Alliance of Small Island States (AOSIS).[51] With the legal support of the recently formed British group, the Foundation for International Environment Law and Development, the island nations understood that their interests might be better served collectively as a UN bloc in the upcoming negotiations of a convention on climate change. Led by Robert van Lierop, a US civil rights lawyer and Vanuatu's ambassador to the UN, AOSIS would enter those negotiations armed with the newly salient precautionary principle, which was a device increasingly invoked to counsel decision-makers to take caution in advance of scientific certainty to protect the environment.[52]

Environmental groups that had followed the proceedings in Geneva likewise gestured to this concept in their own conference statement. Coordinated by Annie Roncerel of the newly established consortium, Climate Action Network, Europe, the statement declared that "the world had embarked on an unprecedented global experiment with the climate system." These groups also stressed that scientific uncertainties "should neither be ignored, nor should they be allowed to serve as an excuse for inaction." Pointing to the work of the AGGG, the statement urged governments adopt temperature and sea level

---

[48]"Threatened Islands Demand Urgent Actions on Global Warming," New Scientist, December 2, 1989; "Letter Dated 89/11/20 from the Permanent Representative of Maldives to the United Nations Addressed to the Secretary-General," A/C.2/44/7, United Nations, https://v.gd/hJdGvX.

[49]Futa Helu, "The Endangered Species," Pacific Islands Monthly 61, no. 8 (1991): 13.

[50]Bikenibeu Paeniu, "Address," in Jäger and Ferguson (eds.), Climate Change, 527.

[51]Peter Aldous, "Dissent Hits Climate Accord," Nature 188 (1990): 188. On the views of the island nations expressed in the negotiations of the ministerial statement, see Jeremy Leggett, The Carbon War: Global Warming and the End of the Oil Era (New York: Routledge, 2001), 23–7.

[52]Timothy O'Riordan and Andrew Jordan, "The Precautionary Principle in Contemporary Environmental Politics," Environmental Values 4, no. 3 (1995): 191–212; John Ashe, Robert Van Lierop, and Anilla Cherian, "The Role of the Alliance of Small Island States (AOSIS) in the Negotiation of the United Nations Framework Convention on Climate Change (UNFCCC)," Natural Resources Forum 23 (1999): 209–20.

targets.[53] The larger of these nongovernmental environmental groups, such as Greenpeace, the Friends of the Earth, and the World Wildlife Fund for Nature, had been actively engaged in the intergovernmental negotiations process from the outset. Through that process, they shifted away from their earlier focus on the "South's population problem" to appreciate the concerns and priorities of their developing world colleagues, particularly those of "potential victim states."[54]

Negotiating the UN Framework Convention on Climate Change

Just as King Hussein of Jordan had warned at the climate conference in Geneva, smoke from Kuwait's burning oil fields filled the sky, thus marking the dawn of US president George Bush's "new world order."[55] Following the conclusion of the climate conference, the UN General Assembly had adopted a new resolution that initiated a single intergovernmental negotiating practice under the auspices of the General Assembly. In doing so, the General Assembly relieved the UNEP and the IPCC of responsibility for overseeing the negotiations. Despite Tolba's efforts to replicate the successful process his Environmental Programme had led for the Vienna Convention, developing countries had argued otherwise. Having felt excluded first from the negotiations of the ozone regime and then from the IPCC's assessment process, they had advocated for the UN General Assembly to oversee the negotiations of the nascent climate regime. The December 1990 resolution established an INC for the development of a framework convention on climate change prior to the Earth Summit in June 1992. Meeting for the first time in February 1991, the negotiations would be open to all state members of the UN, with the participation of observers according to the established practices of the General Assembly.[56]

At the first INC meeting in February 1991, delegates elected France's Jean Ripert to the role of chair, with representatives from Algeria, Romania, Argentina, and India as vice-chairs. A veteran of the UN, Ripert had served

---

[53] "Statement of Environmental Non-governmental Organizations," in Jäger and Ferguson (eds.), *Climate Change*, 547.

[54] Atiq Rahman and Annie Roncerel, "A View from the Ground Up," in Irving Mintzer and J. Amber Leonard (eds.), *Negotiating Climate Change: The Inside Story of the Rio Convention* (New York: Cambridge University Press, 1994), 246–7.

[55] Hussein bin Talal, "Address," in Jäger and Ferguson (eds.), *Climate Change*, 511.

[56] "Protection of Global Climate for Present and Future Generations of Mankind: Resolution," UN General Assembly, 45th Session, December 21, 1990, https://v.gd/eCdHAp.

FIGURE 5.1 *Intergovernmental Negotiating Committee (INC) executive secretary Michael Zammit Cutajar, chair Jean Ripert, and secretary Erik Jensen at the second session of the INC for a Framework Convention on Climate Change at the Palais des Nations, Geneva, June 28, 1991. UN Photo.*

as UN director general for Development and International Economic Cooperation, and, more recently, as the chair of the IPCC's Special Committee on Developing Countries, which had been created after the first session in November 1988 to enhance the participation of developing countries. To serve as the INC's executive secretary, the UN secretary-general appointed Malta's Michael Zammit Cutajar, who had served the UN Conference on Trade and Development as well as the UNEP (Figure 5.1).[57] Ripert's committee had a daunting task ahead to finalize the text of a convention on climate change by June 1992. In contrast to the negotiations of the Convention on Long-Range Transboundary Air Pollution (LRTAP) (1979) and the Vienna Convention (1985), which had involved fewer than fifty states, Ripert's committee had just eighteen months to navigate the diverse interests of some 150 nations and to arrive at a consensus agreement that promised to have vastly more far-reaching economic and political implications.[58] This consensus approach, a

---

[57]Kilaparti Ramakrishna and Oran R. Young, "International Organizations in a Warming World: Building a Global Climate Regime," in Irving M. Mintzer (ed.), *Confronting Climate Change: Risks, Implications and Responses* (Cambridge: Cambridge University Press, 1992), 258.
[58]Bodansky, "The United Nations Framework Convention," 474–5, 482.

UN tradition, meant that further agreements would very likely be settled at the level of the least ambitious of the parties.[59]

Negotiating a framework convention had not been the only option mooted in response to Malta's UN resolution. In the wake of the 1988 Toronto Conference (Chapter 4), the Canadian government had convened the International Meeting of Legal and Policy Experts on the Protection of the Atmosphere in Ottawa in February 1989. A nongovernmental meeting, participants from twenty-five countries and eight international bodies discussed different ways of devising a convention on climate change, including a "law of the atmosphere," akin to the Law of the Sea, which would address the interdependence of atmospheric processes and problems. For his part, Tolba saw this approach as far from politically pragmatic, not least as the Law of the Sea had still yet to come into force seven years after its agreement.[60] Such a path would be a "Mission Impossible," he told the meeting. Given Tolba's stature at this time, his assessment dampened enthusiasm for pursuing further such an approach, and primed participants to acquiesce to the adoption of the model established for the Vienna Convention for addressing another anthropogenic atmospheric problem, climate change.[61]

It would be difficult to overstate the influence of the emerging ozone regime on the ensuing negotiations of the UN Framework Convention on Climate Change (UNFCCC). In technical terms, the latter followed the Vienna approach of negotiating a framework convention by establishing general obligations and a rough legal and institutional framework for future action, after which states then develop specific measures and more detailed mechanisms in protocols. This approach, advocates believed, allowed for negotiations to proceed in an incremental manner that would accommodate emerging scientific evidence and encourage ongoing conversations between different countries.[62] In terms of the substance and dynamics of the climate negotiations, the developing countries would not be latecomers to the negotiations, as they had been in the ozone process. For many of these countries, ozone depletion had not been a priority concern, not least as the effects of ozone-depleting gases appeared to be centered at the poles and higher latitudes where fears of skin cancer among fair-skinned populations had prompted government concern.[63]

---

[59]Delphine Borione and Jean Ripert, "Exercising Common but Differentiated Responsibility," in Mintzer and Leonard (eds.), Negotiating Climate Change, 77–96.
[60]Mostafa Tolba, "A Step-by-Step Approach to Protection of the Atmosphere," International Environmental Affairs 1, no. 4 (1989): 307. The 1982 UN Convention on the Law of the Sea entered into force in 1994.
[61]Bodansky, "The United Nations Framework Convention," 472.
[62]Bodansky, "The United Nations Framework Convention," 493–4.
[63]Litfin, Ozone Discourses, 142–3.

Recognizing that their own prospects for CFC production might be curtailed by the ozone negotiations, developing countries began to engage more actively in what Malaysia's negotiator likened to a "trade war by environmental decree."[64] At Thatcher's London meeting in early 1989 (as discussed earlier), developing countries such as China and India demanded assurances that financial aid and technology transfer would be forthcoming to assist their implementation of the ozone treaty. During the negotiations, Indian environment minister Ziuk Rahman Ansari observed pointedly, "Lest someone in this conference think of this as charity, I would like to remind them of the excellent principle of 'polluter pays' adopted in the developed world" in the early 1970s (Chapter 2).[65] By now, Southern delegations had become closely attuned to the inequitable nature of the Montreal Protocol: not only would it serve to increase the dependence of developing countries on the North, but it also allowed just a few states effective veto over decision-making.[66] Negotiations turned to ensuring that a fund would be established, operating under the principle of "additionality," which finally won over the developing countries and ensured that the Montreal Protocol would become global.[67]

Ten months and three sessions after first meeting in Chantilly, Virginia, a draft version of the negotiating climate text emerged in December 1991. Unable to secure the support of the Bush administration for targets and timelines, UK officials took it upon themselves to fly to Washington in April 1992 to negotiate a compromise text with their US counterparts.[68] On May 9, the UNFCCC was finally adopted in time for the Earth Summit in June and entered force on March 21, 1994, having been ratified by fifty states.[69] Like the LRTAP convention (Chapter 3) and the Vienna Convention, the convention sets out a legal and institutional plan or "framework" to guide the future work of its "parties," including the negotiation of detailed obligations in the form of protocols. The convention's preamble restates that the adverse effects of the changing climate are "a common concern of humankind," warranting

[64]Cited in Richard Benedick, *Ozone Diplomacy: New Directions in Safeguarding the Planet* (Cambridge, MA: Harvard University Press, 1998), 100.
[65]Cited in Jonathan Randal, "Third World Seeks Aid before Joining Ozone Pact," *Washington Post*, March 7, 1989, 16.
[66]Ian Rowlands, *The Politics of Global Atmospheric Change* (Manchester: Manchester University Press, 1995), 171.
[67]Litfin, *Ozone Discourses*, 154–5.
[68]Michael T. Hatch, "Domestic Politics and International Negotiations: The Politics of Global Warming in the United States," *Journal of Environment & Development* 2, no. 2 (1993): 26–7.
[69]Daniel Bodansky, "The History of the Global Climate Change Regime," in Urs Luterbacher and Detlef Sprinz (eds.), *International Relations and Global Climate Change* (Cambridge, MA: MIT Press, 2001), 32.

FIGURE 5.2 *Brazilian president Fernando Collor De Mello signs the UN Framework Convention on Climate Change (UNFCCC) as UN secretary-general Boutros Boutros-Ghali (second from right) and INC executive secretary Michael Zammit Cutajar applaud, June 4, 1992, Rio de Janeiro, Brazil. UN Photo by Michos Tzovaras.*

international cooperation to "protect the climate system for present and future generations" (Figure 5.2).[70]

Influenced by the statements issued at Noordwijk and the Second World Climate Conference, the main objective of the climate convention is to achieve the "stabilization of greenhouse gas concentrations in the atmosphere at a level that would prevent dangerous anthropogenic interference with the climate system," leaving in question the "dangerous" threshold and how this may be averted.[71] Although the convention sets out commitments for all states to follow, such as devising and publishing their greenhouse gas emissions and removals by sinks, the notion of "differentiated responsibilities" underscores further commitments in Annex I specific to the members of the Organisation for Economic Co-operation and Development (OECD) and the countries of the former Eastern bloc ("economies in transition"), and others in Annex II for the OECD nations alone. Annex I nations must limit the anthropogenic sources

[70]"Preamble," in United Nations, *United Nations Framework Convention on Climate Change*, New York, May 9, 1992, 1, 3.
[71]Michael Oppenheimer and Annie Petsonk, "Article 2 of the UNFCCC: Historical Origins, Recent Interpretations," *Climatic Change* 73 (2005): 195–226.

of greenhouse gas emissions and protect and enhance their "sinks," while Annex II nations must provide new and additional financial resources to assist developing countries.[72] Presiding over the convention is the Conference of the Parties, which is the mechanism for ensuring its implementation and to steer nations toward the achievement of its objective.[73]

Contested Climate: Geopolitics of Climate Diplomacy

Little, if any, of the convention's content went uncontested among the 140-odd nations that eventually engaged in the process of the negotiations. As the discussions leading up to the Second World Climate Conference had already indicated, too much was at stake. Three areas of the convention were especially disputed, cleaving the parties beyond the old East-West divide and configuring new alliances. The first pertained to the matter of establishing quantitative targets and timetables for limiting the emissions of the developed countries. European countries and AOSIS were in favor of such an approach, while the United States and the oil-producing nations were staunchly opposed. Other developing countries generally favored these tools, adamant that these would only pertain to the developed world. In the Nairobi Declaration of May 1990, for instance, African countries had urged "the industrialized nations to act on this matter (climate change) with the same vigor with which they expect the developing nations to reduce deforestation and population growth rates."[74]

Next on the agenda was the matter of financial assistance and technology transfer from the developed nations to the developing world. Developing countries sought a new fund, while developed countries favored the Global Environmental Facility, a joint project of the World Bank, UNEP, and the UN Development Program that had been established in 1991 to assist developing countries in addressing major environmental problems.[75] Furthermore, developing countries argued for a commitment from developed countries that they would provide "new and additional" financial resources over and above existing aid flows to assist their efforts to implement the convention. Finally, the question of institutions and implementation mechanisms: the developed world

---

[72]"Article 4," in United Nations, *United Nations Framework Convention on Climate Change*, 6–8.

[73]"Article 7," in United Nations, *United Nations Framework Convention on Climate Change*, 10–11.

[74]African Centre for Technology Studies and the Woods Hole Research Center, *The Nairobi Declaration on Climatic Change: International Conference on Global Warming and Climatic Change, African Perspectives* (Nairobi: African Centre for Technology Studies, 1990), 15.

[75]On the contested nature of the Global Environmental Facility, see Joyeeta Gupta, "The Global Environmental Facility in Its North-South Context," *Environmental Politics* 4 (1995): 19–43.

(including the United States) advocated for strong implementation machinery (including detailed reporting requirements and a noncompliance procedure modeled on the Montreal Protocol), while developing countries resisted what they perceived to be both an affront and a threat to their sovereignty.[76]

These concerns had their foundations in questions of energy dependence, the prevailing geopolitical and economic status quo, and vulnerability to the likely consequences of climate change.[77] Given the implications of the convention for energy use and energy sources, which are directly related to greenhouse gas emissions, reliance on particular forms of energy closely informed a nation's engagement in the negotiating process. Oil-producing nations, for example, had everything to lose and acted accordingly once negotiations began in earnest.[78] Extremely dependent on oil exports, these nations were largely opposed to such a convention and deliberately tried to slow down the proceedings, with the support of the Climate Council, a US fossil fuel lobby group.[79] In the shadow of Chernobyl, the likes of Saudi Arabia and Kuwait successfully pressed their case to insert "safe" anywhere near the mention of nuclear energy, so as to diminish the range of options for abatement. They likewise attempted to erase the specific reference to carbon dioxide as a greenhouse gas on the grounds of scientific uncertainty, and they advocated instead for a greater focus on sinks, such as forests and oceans.[80]

Customers of the Organisation of the Petroleum Exporting Countries (OPEC), those nations largely dependent on imported fossil fuels, took the opposite approach. Japan and most European countries (excluding oil-producers UK and Norway), formerly heavily reliant on heavily on imported oil, had developed less-energy-dependent cultures in the wake of the 1970s oil crises. France, for instance, had expanded its reliance on nuclear energy nearly tenfold, from 8 percent in 1973 to 70 percent by the mid-1980s, while Japan had diversified its sources of energy and promoted its more efficient use.[81] Reducing emissions, therefore, was in the economic interest of oil importers, as doing so would improve their overall balance of payments. Germany, meanwhile, had demonstrated the economic benefits of acting first

---

[76]Bodansky, "The History of the Global Climate Change Regime," 33–4.

[77]Paterson, *Global Warming and Global Politics*, 77–90.

[78]Earlier in Noordwijk, Saudi Arabia's delegation had described climate change as a "life or death issue for considerable areas of the earth" and necessitated the reduction of greenhouse gas emissions. See Bodansky, "The United Nations Framework Convention," 467.

[79]Joanna Depledge, "Striving for No: Saudi Arabia in the Climate Change Regime," *Global Environmental Politics* 8, no. 4 (2008): 9–35.

[80]Paterson, *Global Warming and Global Politics*, 79.

[81]Howard Geller, "Policies for Increasing Energy Efficiency: Thirty Years of Experience in OECD Countries," *Energy Policy* 34 (2006): 556–73; Jonas Meckling et al., "Why Nations Lead or Lag in Energy Transitions," *Science* 378 (2022): 31–3.

on environmental issues, as it had on acid rain, which the likes of Japan hoped to emulate in the climate negotiations.[82]

By contrast, energy-exporting nations such as the UK, Norway, and Australia were more lukewarm on the question of targets. Least enthusiastic were the large energy-independent nations, such as the United States and China, as well as India and Mexico, where abundant coal and oil reserves ensured cheap supplies and had fostered (or would likely soon enable) profligate energy cultures. With the Soviet Union, these countries collectively accounted for about half of the world's greenhouse gas emissions in 1991.[83] It followed that those countries, particularly the United States and Australia, were home to proactive political lobbies that agitated for the protection of their fossil fuel industries from efforts to curb greenhouse gas emissions. The influence of these lobby groups extended to negotiations of the climate convention: for instance, the World Coal Institute (with Australia's support) helped to draft Article 4, which provides for special consideration to those countries whose economies are "highly dependent" on producing or consuming fossil fuels.[84] Their greater dependence on fossil fuels also encouraged a perception (well-founded or otherwise) that reducing emissions and seeking alternative energy sources would be prohibitively costly, whereas Japan and European countries were likely to interpret those policies in more optimistic terms.[85]

During the 1988 US presidential election campaign, Republican nominee Bush had declared, "Those who think we're powerless to do anything about the greenhouse effect are forgetting about the White House effect." As the Rio Earth Summit loomed, however, he made it clear that if a climate convention with specific commitments for signatories was agreed, he would not attend.[86] Key figures in the Bush White House, such as John Sununu (chief of staff) and Dan Quayle (vice president), were certainly important to framing the administration's position—each staunchly opposed to any measure that might constrain economic growth, as demonstrated during both the ozone and climate talks.[87] But these individuals only gave voice to a wider ideological disposition in US politics, which would continue to shape the nation's climate diplomacy under the Clinton administration (Chapter 6). After the Earth Summit, to which Bush had acquiesced under the pressure of the 1992 presidential

---

[82]Matthew Paterson and Michael Grubb, "The International Politics of Climate Change," *International Affairs* 68, no. 2 (1992): 293–310.
[83]"Cabinet Memorandum 8486 – Negotiations for a Convention on Climate Change – Progress Report – Decisions 16125/SD and 16144," A14039, NAA, Canberra.
[84]See Peter Newell and Matthew Paterson, "A Climate for Business: Global Warming, the State and Capital," *Review of International Political Economy* 5 (1998): 693–703.
[85]Paterson, *Global Warming and Global Politics*, 80–1.
[86]Rowlands, *The Politics of Global Atmospheric Change*, 134–7.
[87]Benedick, *Ozone Diplomacy*, 159–61; William Nitze, "A Failure of Presidential Leadership," in Mintzer and Leonard (eds.), *Negotiating Climate Change*, 187–200.

campaign, a seasoned US negotiator characterized this general position as such: "suspicious of multilateral institutions and proposed measures of a non-market nature, hostile to actions that might seem to kowtow to a demanding Third World, and negative on perceptions of 'environmental activism'."[88]

In the United States, the White House and the fossil fuel lobby proved to be very adept in their strategic deployment of scientific "gaps" and "uncertainty."[89] From the outset, the scientific deliberations of the IPCC provided a useful device for some negotiators to justify their reluctance to endorse any commitment to reducing greenhouse gas emissions. American negotiators of the climate convention were likewise encouraged "to raise the many uncertainties that need to be understood on this issue."[90] Like-minded scientists and commentators such as Fred Singer and William Nierenberg, for whom policy action on climate change was anathema, extended these critiques of scientific uncertainty to the authority and legitimacy of climate science. Having attempted to discredit the nuclear winter theory in the United States (Chapter 4), for instance, associates of the George C. Marshall Institute and the recently established Global Climate Coalition focused their attentions on disputing both the processes and conclusions of the IPCC and the work of its affiliated scientists (Chapter 6).

The Global Climate Coalition had been formed in the United States in 1989 by the National Association of Manufacturers in response to the emerging climate regime. Composed chiefly of coal and oil companies, particularly from the United States, its founding principle held that as climate change is a global issue, "actions should be taken in a global context"—that is, developing countries should also undertake emissions reductions.[91] Meanwhile, Exxon and the American Petroleum Institute had already mobilized a trans-Atlantic network of the world's major oil companies through the London-based International Petroleum Industry Environmental Conservation Association to mount a sustained campaign of disinformation about climate change and climate science.[92] As a former French oil executive recalled, "What we

---

[88]James Sebenius, "Towards a Winning Climate Coalition," in Mintzer and Leonard (eds.), *Negotiating Climate Change*, 288.

[89]See, for example, Rowlands, *The Politics of Global Atmospheric Change*, 76–7; Hatch, "Domestic Politics and International Negotiations," 1–33.

[90]"Talking Points," cited in Nicholas Hildyard, "Foxes in Charge of the Chickens," in Wolfgang Sachs, *Global Ecology: A New Arena of Political Conflict* (London: Zed Books, 1993), 29.

[91]Cited in Simone Pulver, "Power in the Public Sphere: The Battles between Oil Companies and Environmental Groups in the UN Climate Change Negotiations, 1991–2003," PhD dissertation (University of California, Berkeley, 2004), 105.

[92]Robert J. Brulle, "Advocating Inaction: A Historical Analysis of the Global Climate Coalition," *Environmental Politics* 32, no. 2 (2023): 185–206; Christophe Bonneuil, Pierre-Louis Choquet, and Benjamin Franta, "Early Warnings and Emerging Accountability: Total's Responses to Global Warming, 1971–2021," *Global Environmental Change* 71 (2021): 102386, https://doi.org/10.1016/j.gloenvcha.2021.102386.

feared was that in this kind of conference (Rio), for reasons of diplomacy and communication, the world would take measures harmful to the industry."[93]

Climate Coalitions: Development, Vulnerability, and Security

Guiding much of the developing world's approach to the convention negotiations were questions of equity.[94] Asserting from the outset that the historical responsibility for the climate change problem lay solely with the industrialized nations, developing countries resisted any approach that might encroach on their own prospects for development. As the Rio Summit approached, the South Centre had stressed that Brundtland's concept of sustainable development meant "that the needs of the North should be met in ways that do not compromise the satisfaction of the present and future needs of the South." Moreover, developing states should ensure they negotiate for adequate "environmental space for [their] future development."[95] Maintaining this stance throughout the negotiations, this message was also at the heart of the Beijing Ministerial Declaration on Environment and Development that resulted from a meeting of the representatives of over forty developing countries in China in mid-1991.[96] In the climate negotiations, both scientific and intergovernmental, China's government had found a useful means to overcome its diplomatic isolation in the wake of the events in Tiananmen Square in 1989.[97]

Climate diplomacy, therefore, became a matter of staking claims to the remaining atmospheric space. Prior to the climate negotiations getting underway, the World Resources Institute, a US nonprofit organization, had published an assessment that pointed to the high greenhouse gas emissions of developing countries on account of deforestation and agricultural activities.[98] Brazil, China, and India ranked in its top five largest contributors to annual greenhouse gas emissions, with the United States and Soviet Union. The report drew swift

[93] Bernard Tramier, November 24, 2020, cited in Bonneuil, Choquet, and Franta, "Early Warnings and Emerging Accountability," *Global Environmental Change* (2021): 5.

[94] Hyder, "Climate Negotiations," 328.

[95] South Centre, *Environment and Development: Towards a Common Strategy of the South in the UNCED Negotiations and Beyond* (Geneva: South Centre, 1992).

[96] "Letter dated 91/07/01 from the Chargé d'affaires a.i. of the Permanent Mission of China to the United Nations Addressed to the Secretary-General," A/46/293, United Nations, https://v.gd/tJeHh9.

[97] Gang Chen, *China's Climate Policy* (London: Routledge, 2012), 6.

[98] World Resources Institute, *World Resources, 1990–1991* (New York: Oxford University Press, 1990).

condemnation from New Delhi's Centre for Science and Environment. In *Global Warming in an Unequal World: A Case of Environmental Colonialism*, which Southern environmental groups circulated at the first Negotiating Committee meeting, the centre's Anil Agarwal and Sunita Narain reprised the position on emissions that Agarwal had asserted in the early 1980s (Chapter 4).[99] They argued, "[the World Resources Institute's] main intention seems to be to blame developing countries for global warming and perpetuate the current global inequality in the use of the earth's environment and its resources."[100] Their articulation of the differences between the luxury emissions of the North and the survival emissions of the South, with its calculus of a per capita allowance of greenhouse gas emissions, would be a defining message for the negotiators representing the governments of the developing world.

An ethical approach to calculating each nation's share of responsibility for climate change, Agarwal and Narain argued, should take into account national sources of emissions and sinks, "further matched with each nation's just and fair share of the oceanic and tropospheric sinks—a common heritage of humankind."[101] In addition to India's treatment, their report also questioned the emphasis of the World Resources Institute's assessment on the emissions from deforestation in Brazil. During the 1980s, the enormous toll of large development projects on the forests and local peoples of the Amazon had attracted growing international concern. Earlier in the decade, environmental organizations in the South and North had brought the attentions of the European Commission and the US Congress to the World Bank's support for these schemes, particularly the Polonoroeste and Carajás projects. Since then, images of burning forests, rising deforestation figures, reports of human rights abuses, and, most recently, the murder of grassroots leader Chico Mendes in December 1988, had only caused further alarm.[102]

Reports that the contribution of Amazon deforestation to global carbon dioxide emissions had increased during the 1980s renewed the region's importance to international discussions about climate change, including at the 1989 G7 Summit in Paris.[103] Having earlier resisted the "internationalization" of

[99] Anil Agarwal and Sunita Narain, *Global Warming in an Unequal World: A Case for Environmental Colonialism* (New Delhi: Centre for Science and Environment, 1991).

[100] Agarwal and Narain, *Global Warming*, 1.

[101] Agarwal and Narain, *Global Warming*, 7. See also Sheila Jasanoff, "India at the Crossroads in Global Environmental Policy," *Global Environmental Change* 3, no. 1 (1993): 32–52.

[102] Ans Kolk, *Forests in International Environmental Politics: International Organisations, NGOs and the Brazilian Amazon* (Utrecht: International Books, 1996), 87–171. See also, Jochen Kemner, "Fourth World activism in the First World: The Rise and Consolidation of European Solidarity with Indigenous Peoples," *Journal of Modern European History* 12, no. 2 (2014): 262–79.

[103] Eustáquio J. Reis and Sérgio Margulis, "Options for Slowing Amazon Jungle Clearing," in Rudiger Dornbush and James M. Poterba (eds.), *Global Warming: Economic Policy Responses* (Cambridge, MA: MIT, 1991), 335–74.

the Amazon, historian Seth Garfield observes that "facing massive foreign debt, hyperinflation, and neoliberal restructuring following the return to democracy," Brazil's government had become "increasingly sensitive to foreign censure and incentives towards environmental policymaking."[104] French president François Mitterand, for instance, had gestured to the Amazon when he commented at the 1988 ozone conference in The Hague that some countries should renounce some portion of their sovereignty concerning issues of global significance. The contribution of European governments to development schemes in Brazil also came under close scrutiny, particularly in Germany, which had been closely involved in the Carajás project. Chancellor Helmut Kohl successfully proposed to the 1990 meeting of the G7 in Houston an aid program to assist Brazil in the preservation of the Amazon.[105] President Fernando Collor de Mello's offer to host the 1992 Earth Summit had been partly motivated to offer a more positive depiction of Brazil's environmental record in light of these concerns.

Although the forest question was negotiated separately from climate change in the lead up to the Rio conference, the climate convention nevertheless made several provisions for countries to monitor, conserve, and enhance sinks and reservoirs of carbon in forest ecosystems.[106] The convention's reduction of such ecosystems to their carbon storage was at odds, however, with the wider recognition at Rio of their significance to Indigenous peoples. In contrast to the other international agreements adopted there in 1992—the Rio Declaration, Agenda 21, and the Forest Principles—and the negotiation of the Convention on Biological Diversity, only the climate convention failed to acknowledge the importance of Traditional Knowledge, Local Knowledge, and Indigenous Knowledge.[107] A decade later, the Convention on Biological Diversity would offer a template for Indigenous peoples to engage directly in the negotiations of the climate regime (Chapter 6).

On the road to Rio, some developed countries had interpreted the majority world's position as the renewal of calls for a New International Economic Order of the 1970s (Chapter 3). For instance, Australia's negotiators reported after

---

[104]Seth Garfield, *In Search of the Amazon: Brazil, the United States, and the Nature of a Region* (Duke University Press, 2014), 218.

[105]Matt McDonald, "Environment and Security: Global Eco-Politics and Brazilian Deforestation," *Contemporary Security Policy* 24, no. 2 (2003): 69–94.

[106]See Karin Bäckstrand and Eva Lövbrand, "Planting Trees to Mitigate Climate Change: Contested Discourses of Ecological Modernization, Green Governmentality and Civic Environmentalism," *Global Environmental Politics* 6 (2006): 57–8.

[107]Andrés López-Rivera, *Blurring Global Epistemic Boundaries: The Emergence of Traditional Knowledge in Environmental Governance*, Global Cooperation Research Papers, no. 25 (Duisburg: Käte Hamburger Kolleg/Centre for Global Cooperation Research, 2020), 23. On the contribution of Indigenous peoples in the Amazon to this recognition, see Sonja K. Pieck, "Opportunities for Transnational Indigenous Eco-Politics: The Changing Landscape in the New Millennium," *Global Networks* 6, no. 3 (2006): 309–29.

the fourth meeting of the INC in late 1991 that the objectives of the "dominant ideologists of the South—India, China, Malaysia and Mexico ... relate less to the protection of the global environment, than to the reversal of the imbalance of wealth between developed and developing countries, through *inter alia* the acquisition of Western technology and financial assistance."[108] Pakistan's Hyder, who had led the G77 during both the Rio and climate negotiations, understood the situation rather differently. The developed countries, in his view, "sought to give primacy to (global) environmental protection at the cost of the universal right to development." Relatedly, he continued, the developed countries feared "their own high per capita emission levels and consumption and quality of life would suffer if the developing countries were allowed unlimited economic growth."[109]

The notion of "common but differentiated responsibility" that eventually emerged (first as Principle 7 of the Rio Declaration and then in the climate convention) was a compromise between the developed states, which sought to avoid any notion of legal responsibility, and the developing states. The OECD had already established a precedent for this distribution, to which the developed countries had agreed in 1991—to acknowledge "the responsibility that they bear in the international pursuit of sustainable development."[110] The nascent environmental justice movement viewed these outcomes with great skepticism. The editorial of UK journal *Ecologist* warned that the South's calls for additionality and funding played directly into the North's hand, "since it effectively frames environmental problems in terms of 'solutions' which only the North provide." The journal's editors continued, "Casting environmental problems in the language of development diverts attention from the policies, values and knowledge systems that have led to the crisis—and the interest groups that have promoted them."[111]

For all the solidarity the G77 demonstrated in Beijing and elsewhere, there were nevertheless stark differences between the developing nations. By 1992, China alone accounted for about 11 percent of total energy-related carbon dioxide emissions, while all the other developing countries amounted to about 14 percent.[112] Beijing also anticipated that coal would account for almost three-quarters of the nation's electricity generation by the year 2000.[113]

---

[108]"Cabinet Memorandum 8486." Emphasis in original.

[109]Hyder, "Climate Negotiations," 206–8.

[110]Duncan French, "Developing States and International Environmental Law: The Importance of Differentiated Responsibilities," *International and Comparative Law Quarterly* 49 (2000): 36–7.

[111]Oliver Tickell and Nicholas Hildyard, "Editorial: Green Dollars, Green Menace," *Ecologist* 22, no. 3 (1992): 82–3.

[112]Helge Ole Bergesen and Anne Kristin Sydnes, "Protection of the Global Climate – Ecological Utopia or Just a Long Way to Go?," in Helge Ole Bergesen, Magnar Norderhaug and Georg Parmann (eds.), *Green Globe Yearbook 1992* (Oxford: Oxford University Press, 1992), 41.

[113]Henry Shue, "Subsistence Emissions and Luxury Emissions," *Law & Policy* 15 (1993): 42.

Other nations in the bloc, such as AOSIS and the drought-prone Sahelian states, were "genuinely worried about the adverse consequences for them of climate change," as Australia's negotiators observed.[114] In contrast to China's stance, these nations called loudly for immediate cuts to greenhouse gas emissions and advocated a stabilization target at 1990 levels by 1995.[115] The work of the Alliance earned the INC's recognition, which invited Vanuatu (as the Alliance's chair) to join its bureau and to co-chair one of its working groups.[116] Other developing countries, Australia's negotiators noted, "are concerned that the tough conditions proposed by India and other hard-liners will prove unacceptably costly for the developed countries and deprive the poorer countries of the more modest financial and technology transfer likely to be produced by the Convention."[117]

Among the developed countries, vulnerability to the potential impacts of climate change also influenced the debates that unfolded during the convention's negotiations. As far as the United States was concerned, the economic costs of reducing greenhouse gas emissions would be prohibitive relative to the likely consequences that climate change posed to the nation.[118] What the end of the Cold War had taught its more conservative thinkers was that the flexibility of their own system would afford the United States the means to adapt to any change and that, ultimately, the market-driven West would prevail.[119] The European Community was more wary, however. In the prospect of rising sea levels, its member states saw not only their own coastlines inundated but also the likelihood of migrants from other inundated parts of the world seeking safe haven within their borders.[120] In a 1990 lecture to the Royal Geographical Society, for instance, the UK's Sir Crispin Tickell (Chapter 3) had advocated a program of international action to address climate change in order to avoid its human consequences:

> If only one per cent of a world population of 8 billion in 2020 were affected by such events, that would still mean some 80 million migrants or

[114] "Cabinet Memorandum 8486."

[115] Paterson, *Global Warming and Global Politics*, 85. The 1990 baseline originates in part from the decision of some European states in 1990 to stabilize their carbon dioxide emissions, see Susan Suback, "National Greenhouse Gas Emissions," in Timothy O'Riordan and Jill Jäger (eds.), *Politics of Climate Change: A European Perspective* (London: Routledge: 1996), 51–64.

[116] Djoghlaf, 'The Beginnings of an International Climate Law," in Mintzer and Leonard (eds.), *Negotiating Climate Change*, 105–6.

[117] "Cabinet Memorandum 8486."

[118] Konrad von Moltke and Atiq Rahman, "External Perspectives on Climate Change: A View from the United States and the Third World," in O'Riordan and Jäger (eds.), *Politics of Climate Change*, 330–45.

[119] Jacob Darwin Hamblin, *Arming Mother Nature: The Birth of Catastrophic Environmentalism* (New York: Oxford University Press, 2013), 246–50.

[120] Paterson, *Global Warming and Global Politics*, 86.

environmental refugees; and five per cent would produce 400 million. Even 80 million would represent a problem of an order of magnitude which no one has ever had to face. Yet the flooding of a quarter of Bangladesh alone could displace over 30 million people.

Human displacement on such scales, he warned, "would be a challenge which many governments could not hope to meet."[121] The UN and concerned governments would reprise such associations of climate change with migration and conflict in the decades ahead.

Conclusion

Writing as the climate convention was about to come into force in 1994, Kenyan activist Grace Akumu noted, "In our opinion, the North has a hidden agenda. They would like to continue unchecked their economic development at the expense of the South forever. The South should carry the burden of the profligate lifestyle of the North. However, our memories are still fresh with George Bush's revelation at Rio … that the American lifestyle is not negotiable."[122] Having flown to Brazil reluctantly, US president Bush added his signature to the climate convention that his administration had worked hard to craft in its favor. From the perspective of countries in the majority world, as Akumu's comments suggest, the convention that arose from these difficult negotiations would not redress the inequalities of the global economic system.

Although the global environment had emerged as a common ground for countries to negotiate the post-Cold War order, the negotiations to forge the 1992 climate convention revealed the extent to which such environmental challenges were fundamentally economic in nature. It became clear that vulnerability to the material effects of climate change and to the economic effects of mitigation policies would prove decisive in the negotiations of the decade ahead. Devising the multilateral means to address climate change would require governments, international organizations, and activists to seek out measures in the interests of the atmospheric commons that would challenge neither the territorial nor the development sovereignty of participating states.

---

[121]Crispin Tickell, "Human Effects of Climate Change: Excerpts from a Lecture Given to the Society on 26 March 1990," *Geographical Journal* 156 (1990): 325–9.
[122]Cited in Joyeeta Gupta, *The Climate Change Convention and Developing Countries: From Conflict to Consensus?* (Dordrecht: Kluwer Academic Publishers, 1998), 126.

6

Allocated Atmosphere

2001: 370.93 ppm

In the Swiss ski resort town of Davos, international business and political leaders have met each year for the annual meeting of the World Economic Forum since the late 1980s. At their 1999 event, UN secretary-general Kofi Annan shared his vision of a global compact between his organization and the private sector "to underpin the new global economy." Focusing on human rights, labor standards and environmental practices (areas already defined by international agreements, including the 1992 Rio Declaration), such a compact "would lay the foundation for an age of global prosperity, comparable to that enjoyed by the industrialized countries in the decades after the Second World War." Pressure to improve standards in these areas was "a threat to the open global market, and especially to the multilateral trade regime," he warned. "These are legitimate concerns," Annan continued, "but restrictions on trade and impediments to investment flows are not the means to use when tackling them."[1]

The secretary-general's proposal reflected his wider effort to engage industry in global environmental governance in the context of accelerating economic integration and international institutions, such as the North America Free Trade Agreement and the World Trade Organization. The private sector had been warmly welcomed at Rio in 1992, as Maurice Strong reprised his role as secretary-general twenty years after the Stockholm meeting (Chapter 2). Joining him was Swiss industrialist Stephan Schmidheiny, whom Strong had invited to act as his special advisor on business and government, and who laid the foundations for the Business Council for Sustainable Development. The International Chamber of Commerce, which had also participated in the 1972 Stockholm meeting, similarly pursued the sustainable development agenda, forming the World Industry Council for the Environment in 1991.[2]

---

[1]Kofi Annan, "Kofi Annan's Address to World Economic Forum in Davos," *United Nations Secretary-General*, February 1, 1999, https://v.gd/XR4p8C.
[2]Adil Najam, "World Business Council for Sustainable Development: The Greening of Business or a Greenwash?," in Helge Ole Bergesen, Georg Parmann, and Øystein B. Thommessen

At Rio, such organizations could engage directly with governments and other organizations to advocate for the compatibility of business and environmental interests. Among their favored tools were market-oriented and industry-based regulation instruments, which aligned with the consolidation of liberal environmentalism at the Earth Summit.[3] With such business advocacy groups permitted to attend its meetings as observers, the UN Framework Convention on Climate Change (UNFCCC) was imbued with much of this spirit, supporting an open economic system and the maintenance of economic growth.[4] From the outset, Business and Industry Non-governmental Organizations were recognized as constituencies with observer status in the UNFCCC process, along with Environmental Non-governmental Organizations. When the convention came into force in 1994, negotiations turned to translating these principles into a climate regime that would avoid dangerous climate change.[5]

Island Initiative: Making the Berlin Mandate

When countries gathered in Berlin for the first Conference of the Parties in 1995, prospects for their arrival at a protocol looked bleak (Figure 6.1). Concern over climate change had begun to wane, as environmental issues competed for public attention. The majority of the Organisation for Economic Co-operation and Development (OECD) countries were not on track to meet their commitments under the convention (to return their emissions to 1990 levels by 2000), and no industrialized nation had called for a review of the adequacy those commitments.[6] Meanwhile, the developing countries had splintered between those that continued to question the need for a convention at all, namely the members of the Organisation of the Petroleum Exporting Countries (OPEC), and those that campaigned for a binding target and timeline, the Alliance of Small Island States (AOSIS). Already, Saudi Arabia

(eds.), *Yearbook of International Cooperation on Environment and Development, 1999/2000* (London: Earthscan, 1999), 65–75.

[3]José Célio Silveira Andrade and José Antônio Puppim de Oliveria, "The Role of the Private Sector in Global Climate and Energy Governance," *Journal of Business Ethics* 130 (2015): 375–87; Stephan Schmidheiny, *Changing Course: A Global Business Perspective on Development and the Environment* (Cambridge, MA: MIT Press, 1992).

[4]Marten Boon, "A Climate of Change? The Oil Industry and Decarbonization in Historical Perspective," *Business History Review* 93 (2019): 101–25.

[5]Robyn Eckersley, "Understanding the Interplay between Climate and Trade Regimes," in Benjamin Simmons, Harro van Asselt, and Fariboz Zelli (eds.), *Climate and Trade Policies in a Post-2012 World* (Geneva: UNEP, 2009), 11–18.

[6]Ian H. Rowlands, *The Politics of Global Atmospheric Change* (Manchester: Manchester University Press, 1995), 157; Michael Grubb, Christiaan Vrolijk, and Duncan Brack, *The Kyoto Protocol: A Guide and Assessment* (London: Earthscan, 1999), 45.

FIGURE 6.1 *German federal minister for the environment, nature conservation and nuclear safety, and conference president Angela Merkel (foreground) opens the first Conference of the Parties of the UN Framework Convention on Climate Change (UNFCCC) at the Internationalen Congress Centrum, Berlin, March 28, 1995. Bundesarchiv, B145 Bild-00075830 / Photographer: Bernd Kühler.*

(with the help of fossil fuel lobbyists) had submitted a proposal that protocols to the convention be adopted by three-quarters of the parties, instead of the existing two-thirds. Other countries interpreted this proposal as an attempt to effectively gain veto power over the whole process, which left all decisions to be taken by consensus.[7] Other members of the Group of 77 (G77), meanwhile, remained steadfastly opposed to any commitments on their part, much to the chagrin of the United States and its allies in what became the "Umbrella Group": Japan, Canada, Australia, and Aotearoa New Zealand.[8]

Led by Annette des Iles of Trinidad and Tobago, the island nations had prepared a proposal for a protocol—the only one formally submitted for consideration at Berlin.[9] Drawing on the 1988 Toronto target (Chapter 4), the ambitious proposal called for a basic commitment of all parties, "taking into account their common but differentiated responsibilities and their specific national and regional development priorities." Industrialized nations, the proposal outlined, should commit to a legally binding reduction of their carbon emissions by 2005 to a level at least 20 percent below 1990 levels.[10] German observers interpreted the intervention as a move to "raise the expectations around the first review of the adequacy of commitments."[11] Reluctant to support a proposal they saw as politically unrealistic both domestically and internationally, the German hosts sought to broker an agreement between the parties.[12]

In the meantime, India's lead negotiator T. P. Sreenivasan had formed a "green" group of "like-minded States" among the developing countries (including AOSIS, but not OPEC) and worked with the World Wildlife Fund for Nature (WWF) to prepare a "green draft" of what would become the Berlin Mandate. India's negotiators and their allies sought to press upon Germany and the European Union (EU) their case for a strong protocol, with no additional commitments for the industrializing states.[13] Environmental groups such as the WWF constituted about a third of the 165 nongovernmental organizations

---

[7] Harriet Bulkeley and Peter Newell, *Governing Climate Change* (London: Routledge, 2010), 91.

[8] Bas Arts and Wolfgang Rüdig, "Negotiating the 'Berlin Mandate': Reflections on the First 'Conference of the Parties' to the UN Framework Convention on Climate Change," *Environmental Politics* 4, no. 3 (1995): 481–7. After the Berlin conference, the Umbrella Group expanded to include Switzerland and Norway.

[9] Trinidad and Tobago, "Statement Introducing a Draft Protocol to the UN Framework Convention on Climate Change on Greenhouse Gas Emissions Reduction: Submitted by the Alliance of Small Island States," February 8, 1995, https://v.gd/6I709P /.

[10] Sebastian Oberthür and Hermann Ott, *The Kyoto Protocol: International Climate Policy for the 21st Century* (Berlin: Springer-Verlag, 1999), 227.

[11] Oberthür and Ott, *The Kyoto Protocol*, 45.

[12] Oberthür and Ott, *The Kyoto Protocol*, 46.

[13] Sandeep Sengupta, "India's Engagement in Global Climate Negotiations from Rio to Paris," in Navroz K. Dubash (ed.), *India in a Warming World: Integrating Climate Change and Development* (New Delhi: Oxford University Press, 2019), 114–41.

at the Berlin conference, which joined together under the umbrella of the Climate Action Network (Chapter 5). Their engagement in the proceedings was especially important to facilitate the participation of delegations from the South, for whom the demands of multilateral environmental diplomacy were (and remain) prohibitive.[14] Observers at the meeting also noted the Network's efforts to "embarrass" the delegations of the developed world by reporting on their lack of commitment to the Framework Convention and to making the negotiations more transparent through reporting in daily newspapers, such as *Eco* and the *Earth Negotiations Bulletin*.[15]

On the last night of the meeting, Germany's Angela Merkel, the conference president, embarked on shuttle diplomacy between the separate groups to forge the final compromise.[16] Acknowledging the proposal of the small island nations, the final document maintained the convention's focus on the emissions of the developed countries alone and emphasized that their current nonbinding agreements were inadequate. These countries were obliged to act first before any binding obligations could be negotiated for the developing countries. Finally, it established a deadline for the negotiation of a protocol in time for the third Conference of the Parties in 1997. What targets the protocol would prescribe the industrialized countries, and what means they could use to achieve them, dogged discussions leading up to Kyoto and only intensified during the 1997 conference.

Contesting Science: The Geopolitics of Knowledge

As the Ad-hoc Working Group on the Berlin Mandate continued to meet during 1995, the Intergovernmental Panel on Climate Change (IPCC) presented its Second Assessment Report. Its authors concluded that "the balance of evidence suggests a discernible human influence on global climate."[17] Scientists had not only detected climatic change, but could now attribute that change to human activity. Since its First Assessment Report in 1990, the IPCC had also produced two interim reports, the first in 1992 to coincide with the final negotiations of the Climate Convention,

---

[14]See Pamela S. Chasek, "NGOs and State Capacity in International Environmental Negotiations: The Experience of the *Earth Negotiations Bulletin*," *Review of European, Comparative & International Environmental Law* 10, no. 2 (2001): 168–76.

[15]Arts and Rüdig, "Negotiating the 'Berlin Mandate'," 481–7.

[16]T. P. Sreenivasan, *Words, Words, Words: Adventures in Diplomacy* (Delhi: Pearson Longman, 2008), 110–11; Chad Carpenter, Pamela Chasek, Anilla Cherian, and Steve Wise, "Adequacy of Commitments," *Earth Negotiations Bulletin* 12, no. 21 (1995): 10–11.

[17]"Summary for Policymakers," in John Houghton et al. (eds.), *Climate Change 1995: The Science of Climate Change* (Cambridge: Cambridge University Press, 1995), 4.

and the second, two years later, on the radiative forcing of climate change. The IPCC's Working Groups II and III had, meanwhile, undergone some adjustment. Working Group II now assessed responses or mitigation options, as well as the impacts of climate change and adaptation. After the difficulties during the previous round of assessment, the collapse of the Soviet Union made it possible for the IPCC to quietly move Yuri Izrael from leading this group to the role of panel co-vice-chair.[18] The new Working Group III was focused on the economics and social dimensions of climate change, which reflected the growing field of economic analyses of climate change, as well as the Climate Convention itself, which stated that "policies and measures to deal with climate change should be cost-effective so as to ensure global benefits at the lowest possible cost."[19]

In the IPCC's First Assessment, Working Group III had been responsible for "formulating response strategies." With no mandate to assess the extant scholarship or undergo peer review, the group had largely focused on developing explicit policy advice. As negotiations for the Climate Convention had got underway, this group became "the cockpit for much of the climate change politics of the subsequent 18 months," the UK's Tony Brenton recalled.[20] Aside from questions about the group's use of scenarios, its work had also become too closely entangled with the US camp: its chair Robert Reinstein was also the chief US negotiator for the Climate Convention.[21] Wary of this situation's implications for the credibility of the other working groups, IPCC chair Bert Bolin had welcomed the establishment of the Intergovernmental Negotiating Committee, as it allowed the panel to focus solely on the development of peer-reviewed assessments.[22]

Regarding the work of the new Working Group III, Bolin told an International Institute for Applied Systems Analysis (IIASA) meeting in 1992, "In the future the IPCC would undertake economic analyses of climate change with the same vigor that it has demonstrated in its other scientific assessments."[23] Assessing the costs and benefits of action and inaction, the group's report

---

[18]Bert Bolin, *A History of the Science and Politics of Climate Change: The Role of the Intergovernmental Panel on Climate Change* (Cambridge: Cambridge University Press, 2007), 83.
[19]Samuel Randalls, "Optimal Climate Change: Economics and Climate Science Policy Histories (from Heuristic to Normative)," *Osiris* 26 (2011): 224–42.
[20]Tony Brenton, *The Greening of Machiavelli: The Evolution of International Environmental Politics* (London: Earthscan, 1994), 179.
[21]Bolin, *A History*, 83; Robert Reinstein, "Climate Negotiations," *Washington Quarterly* 16 (1993): 79–95.
[22]Tora Skodvin, *Structure and Agent in the Scientific Diplomacy of Climate Change: An Empirical Case Study of Science Policy Interaction in the Intergovernmental Panel on Climate Change* (New York: Kluwer, 2000), 120–1.
[23]Y. Kaya et al., "Introduction," in Y. Kaya et al. (eds.), *Costs, Impacts, and Benefits of CO_2 Mitigation: Proceedings of a Workshop Held on 28–30 September 1992, at IIASA, Laxenburg, Austria* (Laxenburg: IIASA, 1993), iv.

counseled the flexibility that early mitigation offered policymakers—that is, "balancing the economic risks of rapid abatement now (that premature capital stock retirement will later be proved unnecessary) against the corresponding risk of delay (that more rapid reduction will then be required, necessitating premature retirement of future capital stock)."[24] To help lower emissions cost-effectively, the report continued, policymakers could address "imperfections and institutional barriers in markets through policy instruments based on voluntary agreements, energy efficiency incentives, product efficiency standards, and energy efficiency procurement programmes."[25]

With much of its assessment broadly aligned with the prevailing ethos of market-led sustainable development, the working group encountered resistance as it prepared to finalize its report. Delegates representing developing countries objected to the methods and assumptions one chapter's authors had used to calculate "the statistical value of life."[26] The chapter had assessed differential values for lives in the developed and majority worlds to calculate a global cost-benefit analysis of reducing greenhouse gas emissions. Measured by the willingness to pay to avoid an environmental harm, these values are based on the assumption that a life's value was roughly proportional to national per capita GDP. According to this chapter's arithmetic, life in the developed countries was valued fifteen times more highly than that in developing countries, or, "15 dead Chinamen equal to one dead Englishman," as critics put it.[27]

Although concerns had been raised about this approach a couple of years earlier, drafts of Working Group III's report suggested that the authors had not altered course.[28] As delegations gathered in Berlin for the first Conference of the Parties, Indian environment and forests minister, Kamal Nath, wrote to them to denounce

the absurd and discriminatory global cost/benefit analysis procedures propounded by economists in the work of IPCC WGIII. ... [W]e unequivocally reject the theory that the monetary value of people's lives around the world is different because the value imputed should be proportional to the disparate income levels of the potential victims. ... [I]t is

---

[24]"Summary for Policymakers," in James Bruce, Hoesung Lee, and Erik Haites (eds.), *Climate Change 1995: Economic and Social Dimensions of Climate Change* (Cambridge: Cambridge University Press, 1996), 5.

[25]"Summary for Policymakers," in Bruce, Lee, and Haites (eds.), *Climate Change 1995*, 13.

[26]Ehsan Masood, "Developing Countries Dispute Use of Figures on Climate Change Impacts," *Nature* 376 (1995): 374.

[27]Tom Wakeford and Aubrey Meyer, "Valuing the Environment and Valuing Lives," *Environmental Politics* 5, no. 2 (1996): 363.

[28]Richard Douthwaite, "Who Says That Life Is Cheap?," *Guardian*, November 1, 1995, 49–50.

impossible for us to accept that which is not ethically justifiable, technically accurate or politically conducive to the interests of poor people as well as the global common good.[29]

By now, the London-based Global Commons Institute had shared its concerns with readers of the *Times of India*, the *Guardian*, and *Nature*, collecting signatures for a petition to have the offending chapter withdrawn from the report.[30] If lives were to be valued, the institute's researchers argued, then they should be valued equally or, at the value of the industrial countries, as they were historically responsible for climate change.[31]

The lead author of the chapter in question, economist David Pearce, described these concerns as "a matter of scientific correctness versus political correctness."[32] In the late 1980s, he had coauthored *Blueprint for a Green Economy*, which had been commissioned by the UK Department of Environment and advocated the use of market incentives in environmental policymaking.[33] Determining environmental values, he maintained, was a means to present "the arguments in terms of units that politicians understand … [as] the record of decision-making in the absence of such valuations is hardly encouraging for the environment."[34] Later defending his chapter's approach, Pearce argued that his critics were attempting "to hijack an essentially scientific process for political and ideological ends that appear to have little, and perhaps nothing, to do with climate change itself."[35] As the IPCC faced a deadlock over the Working Group III report, he suggested the removal of the chapter entirely.[36]

Ultimately, the chapter remained in the report, but not before a special meeting to resolve the formal protests that delegates of Cuba and Brazil had lodged.[37] The Summary for Policymakers, however, made clear what Pearce

---

[29]Kamal Nath, March 24, 1995, cited in Michael Grubb, "Seeking Fair Weather: Ethics and the International Debate on Climate Change," *International Affairs* 71 (1995): 471.

[30]Aubrey Meyer, "Commons People Given Green Light," *Guardian*, July 29, 1994, 19; Aubrey Meyer, "Correspondence: Economics of Climate Change," *Nature* 378 (1995): 433.

[31]Aubrey Meyer and Tony Cooper, "A Recalculation of the Social Costs of Climate Change," Global Commons Institute, September 1995, https://v.gd/Pq3RBE.

[32]David Pearce, cited in Fred Pearce, "Global Row over Value of Human Life," *New Scientist*, August 19, 1995, https://v.gd/1IFTrQ.

[33]David Pearce, A. Markandya, and E. Barbier, *Blueprint for a Green Economy* (London: Earthscan, 1989). See Peter Newell and Matthew Paterson, *Climate Capitalism: Global Warming and the Transformation of the Global Economy* (Cambridge: Cambridge University Press, 2010), 24–5.

[34]David Pearce, "Green Economics," *Environmental Values* 1 (1992): 3–13.

[35]David Pearce, "Climate Confusion," *Environment and Planning A* 28, no. 1 (1996): 8.

[36]Pearce, cited in Pearce, "Global Row."

[37]Douthwaite, "Who Says That Life Is Cheap?," 49–50; David Pearce et al., "The Social Costs of Climate Change: Greenhouse Damage and the Benefits of Control," in Bruce, Lee, and Haites (eds.), *Climate Change 1995*, 179–224.

and his supporters had been unwilling to countenance, that economists were engaged in making political choices and those choices were not value-neutral. Their refusal to engage with the concerns of environmental groups and other economists only fueled suspicion and distrust among developing countries toward the IPCC.[38] The summary, which governments must approve, distanced itself from the chapter's assessment: "There is no consensus about how to value statistical lives or how to aggregate statistical lives across countries."[39] Although the chapter's authors may have been following the conventions of their discipline, they had failed to appreciate the implications of their approach in a deeply divided world. The very attempt to monetize human lives was at the heart of the moral outrage of developing countries and their advocates.[40]

This episode reflected a wider geopolitics of knowledge that favored Northern scientists and, thus, presented Northern values and perspectives as global.[41] Developing countries remained poorly represented in the area of climate research and in the workings of the IPCC, which diminished the prospect of Southern interests informing global scientific inquiries. Social scientists Milind Kandlikar and Ambuj Sagar pointed out, for instance, that the Working Group II report for the IPCC's Second Assessment had made no reference to the impacts of climate change on the Asian monsoon, which had a "strong bearing on the lives of about a third of the world's population."[42] Such issues seemed to confirm the value of the Subsidiary Body for Scientific and Technological Advice, which met for the first time at the Berlin climate conference in 1995. This body forms a scientific wing of the Climate Convention's secretariat, which acts as an intermediary between the IPCC and the delegations from member states.[43] Following similar terms to the procedures of the Climate Convention,

---

[38]Simon Shackley, "The Intergovernmental Panel on Climate Change: Consensual Knowledge and Global Politics," *Global Environmental Change* 7 (1997): 77–9.

[39]"Summary for Policymakers," in Bruce, Lee, and Haites (eds.), *Climate Change 1995*, 9–10. See David Demeritt and Dale Rothman, "Figuring the Costs of Climate Change: An Assessment and Critique," *Environment and Planning A* 31 (1999): 389–408.

[40]Shinichiro Asayama, Kari de Pryck, and Mike Hulme, "Controversies," in Kari de Pryck and Mike Hulme (eds.), *A Critical Assessment of the Intergovernmental Panel on Climate Change* (Cambridge: Cambridge University Press, 2022), 152.

[41]See Jean Carlos Hochsprung Miguel, Martin Mahony, and Marko Synésio Alves Monteiro, "'Infrastructural Geopolitics' of Climate Knowledge: The Brazilian Earth System Model and the North-South Knowledge idvide," *Sociologias* 21, no. 51 (2019): 44–74.

[42]Milind Kandlikar and Ambuj Sagar, "Climate Change Research and Analysis in India: An Integrated Assessment of a South-North Divide," *Global Environmental Change* 9 (1999): 119–38; Frank Biermann, "Big Science, Small Impacts – in the South? The Influence of Global Environmental Assessments on Expert Communities in India," *Global Environmental Change* 11, no. 4 (2001): 297–307.

[43]Amy Dahan-Dalmedico, "Climate Expertise: Between Scientific Credibility and Geopolitical Imperatives," *Interdisciplinary Science Reviews* 33, no. 1 (2008): 71–81.

this body operates by consensus, which affords negotiators from developing countries greater opportunity to solicit and shape scientific advice, and thus enhance the credibility of the climate regime.[44]

Fossil fuel lobbyists based in the United States, meanwhile, had also found fault with the IPCC's Second Assessment Report. Working Group I's attribution of observed climate change to human activity for the first time posed a danger to their business interests, as it would strengthen the case for reducing greenhouse gas emissions from fossil fuels. Continuing their campaign of discrediting climate science, associates of the Global Climate Coalition and George C. Marshall Institute targeted the IPCC's processes and the scientific conduct of two of its authors, claiming in the *New York Times* and *Wall Street Journal* that they had improperly altered its findings.[45] Acquiring the support of a handful of US senators, they appealed to both the IPCC and US State Department to challenge the report's findings. Although their clients among the OPEC nations also raised concerns at the next climate conference in Geneva in mid-1996, they failed to turn the international negotiations.[46] "These concerns were raised not by scientists involved in the IPCC; not by participating governments, but rather by naysayers and special interests," declared Tim Wirth, the undersecretary for global affairs and US lead negotiator. "Let's take a false issue off the table; there can be no question but that the IPCC's findings meet the highest standards of scientific integrity."[47] The IPCC nevertheless moved to prevent such criticism, by formalizing its rules of procedure for peer review and appointing review editors to oversee the process.[48] After the United States withdrew from the Kyoto Protocol in 2001, the Global Climate Coalition retreated from the international scene. Fearing reputational risk, as this chapter will discuss, some US companies began to shift away from the Coalition's intransigence, in favor of the new business alliances championed by the Pew Center on Global Climate Change, a thinktank founded in 1998 by former Assistant Secretary of State for Oceans and International Environmental and Scientific Affairs, Eileen Claussen.

---

[44]Clark N. Miller, "Challenges in the Application of Science to Global Affairs: Contingency, Trust and Moral Order," in Clark A. Miller and Paul N. Edwards (eds.), *Changing the Atmosphere: Expert Knowledge and Environmental Governance* (Cambridge, MA: MIT Press, 2001), 247–86.

[45]Bolin, *A History*, 127–32; Naomi Oreskes et al., "The Denial of Global Warming," in Sam White, Christian Pfister, and Franz Mauelshagen (eds.), *The Palgrave Handbook of Climate History* (New York: Palgrave, 2018), 149–71.

[46]Chad Carpenter et al., "Other Plenary Meetings," *Earth Negotiations Bulletin* 12, no. 38 (1996): 3.

[47]Timothy Wirth, "Climate Change Speech by Under Secretary of State Timothy Wirth," *Clean Air Report* 7, no. 15 (1996): 22–4.

[48]Paul N. Edwards and Stephen H. Schneider, "Self-Governance and Peer Review in Science-for-Policy: The Case of the IPCC Second Assessment Report," in Miller and Edwards (eds.), *Changing the Atmosphere*, 219–46; Paul N. Edwards, "Peer Review," in Kari de Pryck and Hulme (eds.), *A Critical Assessment*, 96–104.

FIGURE 6.2 *Conference chair Raúl Estrada-Oyuela (right) and UNFCCC executive secretary Michael Zammit Cutajar embrace each other after the Kyoto Protocol is adopted at the Kyoto International Conference Center, Japan, December 11, 1997. Photo 691569582, The* Asahi Shimbun *via Getty Images.*

Allocating Emissions: Negotiating the Kyoto Protocol

Meeting for the third climate conference in late 1997, the warmest year yet on record, marathon negotiations concluded with the adoption of the Kyoto Protocol (Figure 6.2).[49] As signs of a global economic downturn loomed, about

---

[49]National Oceanic and Atmospheric Administration, "The Climate of 1997 Annual Global Temperature Index," January 1998, https://v.gd/luPvSH.

ten thousand participants gathered at Kyoto's International Conference Center for the event, where US vice president Al Gore made a brief appearance to reignite the talks. Chaired by Argentina's ambassador to China, Raúl Estrada-Ouyela, who also oversaw the Ad-hoc Working Group on the Berlin Mandate, the conference agreed to the commitment of developed countries and "economies in transition" (collectively termed the Annex B countries in the Kyoto Protocol) to reduce their overall greenhouse gas emissions to at least 5 percent below 1990 levels during the five-year period from 2008 to 2012 (the first commitment period).

Based on their "national circumstances," each party committed to their own legally binding target: the European Community agreed to an 8 percent reduction, the United States 7 percent, and Japan 6 percent, while Russia and Ukraine did not have to make any changes. As neither Russia nor Ukraine was likely to reach 1990 levels on account of their ongoing economic transition, they could sell their assigned amounts to other Annex B countries—a windfall that cynics dubbed "hot air." Of the remaining Annex B countries, only four could increase or stabilize their emissions: Iceland (10 percent), Australia (8 percent), Norway (1 percent), and Aotearoa New Zealand (0 percent). To facilitate the achievement of those commitments, the protocol provided for market-based instruments as well as emissions sinks, such as land-use and forestry changes.[50] The parties agreed to defer to future negotiations the question of just how the market-based instruments would work.

This agreement bore only a modest resemblance to the ambitions of those gathered in Toronto nearly a decade earlier that had inspired the small island states' proposal tabled in Berlin. At the Changing Atmosphere conference in Toronto in 1988, scientists, activists, and international organizations had attempted to craft a politically realistic goal of an overall 20 percent reduction in greenhouse gas emissions by 2005 for both developed and developing countries (Chapter 4). At the conclusion of the Kyoto negotiations over a decade later, the parties had wrangled a compromise: Developing countries had no commitments, the member nations of the European Community could work together as a "bubble," and countries could use sinks to meet their emissions targets. Further still, the Annex B parties were committed to both different baseline years and different targets, with a new deadline of 2012.

As far as critics were concerned, the developed world had shirked its moral obligation at Kyoto to accept historical responsibility for climate change and to dramatically reduce its greenhouse gas emissions. The selection of 1990 as the baseline year for the Kyoto Protocol was no arbitrary decision, they argued: the developed nations had sought the most flattering year, and 1990 happened to

---

[50] "Kyoto Protocol to the United Nations Framework Convention on Climate Change," FCCC/CP/1997/L.7/Add.1, December 10, 1997, UNFCCC.

favor several in particular—the UK, Germany, and Russia.[51] Emissions in these nations had inadvertently peaked that year and fallen since. On account of industrial decline since the dissolution of the Soviet Union in 1991, Russia, Ukraine, and other "economies in transition" could take advantage of their higher emissions prior to the fall of the Iron Curtain. So too reunified Germany, where the collapse of industry in the east brought "wall-fall-profits" as emissions fell 12 percent between 1990 and 1995. The UK, meanwhile, could reap the rewards of the Thatcher government's privatization of the electricity sector and switch to natural gas, which produces lower emissions of carbon dioxide per unit of electricity. There, carbon dioxide emissions were 7 percent lower in 1995, compared to 1990. Under the European Community's burden-sharing or "bubble" approach of a single reduction target, these windfalls for the UK and Germany could offset the higher emissions of other member nations, namely, the less-developed "cohesion countries," Greece, Ireland, Spain, and Portugal.[52]

For its part, Japan had favored a baseline year of 1995 for one of the greenhouse gases under review, hydrofluorocarbons. Their production had increased in both Japan and the United States during the early 1990s as a substitute for the chlorofluorocarbons (CFCs) prohibited under the Montreal Protocol. The EU had also argued for their separate regulation, favoring limits on just three gases. To reduce the cost of lowering their emissions, however, the United States insisted on a "comprehensive approach" or bundle of greenhouse gases in addition to carbon dioxide: methane, nitrous oxide, hydrofluorocarbons, perfluorocarbons, and sulfur hexafluoride. These gases were bundled together and each weighted according to their "global warming potentials," as per the model adopted by the IPCC.[53] Much to the satisfaction of the chemical industry, which had lobbied for this outcome, nations could then choose how to concentrate their reduction efforts according to their own bundle of emissions.[54] In short, these were not exactly moral undertakings: Japan's chief negotiator Toshiaki Tanabe put it plainly, "At the core of the negotiation process was the issue of whether each country would be able to maintain its international [economic] competitiveness."[55] Developed nations were reluctant to support any accord that might bring them economic disadvantage.

[51] Diana Liverman, "Conventions of Climate Change: Constructions of Danger and the Dispossession of the Atmosphere," *Journal of Historical Geography* 35 (2009): 292.
[52] Scott Barrett, "Political Economy of the Kyoto Protocol," *Oxford Review of Economic Policy* 14, no. 4 (1998): 34.
[53] Liverman, "Conventions of Climate Change," 289.
[54] Oberthür and Ott, *The Kyoto Protocol*, 125–6.
[55] Toshiaki Tanabe, "Reflections on the Kyoto Conference on Global Warming," *Asia-Pacific Review* 5, no. 2 (1998): 40.

As for the targets themselves, negotiations since Berlin had indicated that a "flat rate" target would simply not eventuate. Neither would the idea of emissions entitlements allocated on a per capita basis, or a convergence toward this, which had become a popular principle among developing countries.[56] Different targets, or differentiation, for the industrialized countries became the means for Chair Estrada-Oyuela to ensure the Kyoto meeting would culminate in a protocol. Nevertheless, given their implications, legally binding, quantified targets were the subject of tough negotiations and bargaining with the EU (–8 percent), the United States (–7 percent), Japan (–6 percent), and Canada (–6 percent). The wide proliferation of mobile phones had allowed delegates to coordinate their positions with their both governments and nongovernmental groups, including the some 250-odd environmental organizations that were undertaking intensive lobbying in Kyoto.[57] Targets for the remaining countries largely resulted from voluntary pledges based on their "willingness to pay." As a result of this voluntary approach, German analysts Sebastian Oberthür and Hermann Ott observed, "Their targets do not rely on any particular rationale and reflect basically the degree of intransigence and *chutzpah* employed by these countries."[58]

In this regard, some countries had taken a tougher stance at Kyoto than others. The president of the Coal Association of Canada argued as much: "We're dumbfounded how Australia came out as a clear winner. [Prime Minister] Mr Chrétien should learn how Australia went about protecting their interests, while Canada didn't."[59] Unlike most of the developed world, Australia had emerged from Kyoto with a target of *higher* emissions relative to 1990. At the eleventh hour of the conference, its delegation had also negotiated the so-called "Australia clause" (Article 3.7), which allows any country with net land clearing in 1990 to include the equivalent emissions in its baseline. As the only developed nation still clearing land, Australia was the only Annex I country to benefit from this clause—land clearing had peaked in 1990 and fallen thereafter. Consequently, Australia's emissions of industrial greenhouse

[56]Grubb, Vrolijk, and Brack, *The Kyoto Protocol*, 95.

[57]William Sweet, *Climate Diplomacy from Rio to Paris* (New Haven: Yale University Press, 2016), 130; Michele Betsill, "Environmental NGOs and the Kyoto Protocol Negotiations: 1995 to 2007," in Michele Betsill and Elisabeth Corell (eds.), *NGO Diplomacy: The Influence of Nongovernmental Organizations in International Environmental Negotiations* (Cambridge, MA: MIT Press, 2007), 43–66.

[58]Oberthür and Ott, *The Kyoto Protocol*, 120.

[59]Sydney Sharpe, "A Theatre of the Absurd," *Calgary Herald*, December 13, 1997, 62. On Canada's position at Kyoto, see Kathryn Harrison, "The Struggle of Ideas and Self-interest in Canadian Climate Policy," in Kathryn Harrison and Lisa McIntosh Sundstrom (eds.), *Global Commons, Domestic Decisions: The Comparative Politics of Climate Change* (Cambridge, MA: MIT Press, 2010), 169–200.

gases could *increase* to at least 120 percent of 1990 levels, while still meeting its overall Kyoto target.

These peculiar arrangements were the culmination of a sustained campaign to establish Australia as a "special-case," which relied on dubious economic analysis showing the country was especially vulnerable to emissions reduction as a carbon-intensive nation.[60] During the Kyoto negotiations, it became widely known that the sponsors of the government's economic analysis were the likes of Exxon, Mobil, and Texaco, as well as coal and aluminum companies, and other representatives of Australian industry.[61] Besides, Australia had already flagged its position in 1996 in Geneva, having sided with Russia and the OPEC nations against an internationally binding treaty with targets and timetables.[62] Having successfully tested the desire for consensus among the 160-odd governments at Kyoto, Australia "had got away with it," observed Ritt Bjerregaard, Europe's commissioner for the environment.[63] The following year, the Australian government made the decision not to ratify the Kyoto Protocol until the United States had done so.

Market Mechanisms: Sharing the Responsibility for Mitigation

Closely attuned to the potential economic burden of meeting their Kyoto commitments, the developed nations turned to the market. Market-based, liberal policies, their negotiators believed, would allow these states to "distribute mitigation costs," as the IPCC's Second Assessment Report put it.[64] That market-based instruments could afford the developed world more efficient (lower cost) ways to reduce emissions had been anticipated in the Framework Convention. Industrialized countries, according to its Article 4.2b, could stabilize their emissions "individually or jointly." Soon understood as "joint implementation," such measures permitted OECD nations and "economies in transition" (Annex B countries) to offset their domestic emissions by investing

---

[60] Kate Crowley, "Climate Clever? Kyoto and Australia's Decade of Recalcitrance," in Harrison and Sundstrom (eds.), *Global Commons, Domestic Decisions*, 201–28. On Australia's climate policy during this period, see Clive Hamilton, *Running from the Storm: The Development of Climate Change Policy in Australia* (Sydney: UNSW Press, 2001).

[61] William Hare, "Australia and Kyoto: In or Out?," *UNSW Law Journal* 24, no. 2 (2001): 556–64.

[62] Oberthür and Ott, *The Kyoto Protocol*, 54.

[63] Lenore Taylor, "Australia's Greenhouse Triumph," *Australian Financial Review*, December 12, 1997, https://v.gd/DnsQ6W.

[64] Bert Bolin et al. (eds.), *Climate Change 1995: IPCC Second Assessment – A Report of the Intergovernmental Panel on Climate Change* (Cambridge: Cambridge University Press, 1996), 15.

in abatement measures elsewhere in the developed world, that is, with other Annex B countries.[65]

Leading up to the Kyoto meeting, the EU had argued on this basis for its "bubble" or burden-sharing approach to juggling the diverse interests of its member states. Not only did this position allow for other parties to form their own bubbles, but the EU's advocacy for joint implementation also gave the United States and its allies in the Umbrella Group cause to renew their calls for approaches that would afford them the most flexibility in meeting their Kyoto commitments.[66] The head of the US delegation, Tim Wirth, had tabled his government's support for this approach at the Geneva conference in 1996. There he had welcomed the possibility of meeting commitments "through maximum flexibility in the selection of implementation measures, including the use of reliable activities jointly, and trading mechanisms around the world."[67] Pointing to the cost-effectiveness of the US tradeable permit system to control sulfur emissions from power plants, which had begun in the late 1980s, the US delegation argued that such market instruments provided the most efficient way to manage greenhouse gases.[68]

Proposals to apply emissions trading to climate diplomacy had emerged in the late 1980s. Attuned to the zeitgeist of sustainable development, these early plans aimed to facilitate wealth and technology transfers between the developed and developing countries, and assumed that per capita emissions would be the most equitable approach to distributing permits.[69] Although that version quickly faded, the appeal of emissions trading grew rapidly, thanks to the success of sulfur dioxide trading in the United States, and found an audience at the UN Conference on Trade and Development.[70] Having helped to orchestrate the US sulfur dioxide program, the Environmental Defense Fund mobilized industry support for the scheme, with BP, DuPont, and the International Climate Change Partnership. Such an alliance would have been

---

[65]Mathew Paterson, *Global Warming and Global Politics* (London: Routledge, 1996), 110.

[66]Peter Newell, "Who 'CoPed' Out in Kyoto? An Assessment of the Third Conference of the Parties to the Framework Convention on Climate Change," *Environmental Politics* 7, no. 2 (1998): 153–9. See, for example, Timothy Wirth, "Press Conference on the Kyoto Conference on Climate Change, Washington, DC," US Department of State, November 12, 1997, https://v.gd/DIOOQN.

[67]Timothy Wirth, "Making the International Climate Change Process Work," *US Department of State Despatch* 7, no. 30 (1996): 376–78. See Robert N. Stavins, *Project 88: Harnessing Market Forces to Protect the Environment – Initiatives for the New President* (Washington, DC: Environmental Policy Institute, 1988). Note that *Project 88* was a bipartisan initiative, cosponsored by Tim Wirth, then Democrat senator for Colorado, and John Heinz, Republican senator for Pennsylvania.

[68]Barry Solomon and Hugh Gorman, "The Origins, Practice, and Limits of Emissions Trading," *Journal of Policy History* 14, no. 3 (2002): 293–320.

[69]Michael Grubb, *The Greenhouse Effect: Negotiating Targets* (London: Royal Institute of International Affairs, 1989).

[70]Scott Barrett et al., *Combating Global Warming: Study on a Global System of Tradeable Carbon Emission Entitlements* (New York: UNCTAD, 1992).

unthinkable at the first climate meeting in Berlin. Since the US government had signaled its support for a binding international treaty in Geneva, however, advocates argued that emissions trading was efficient and market-friendly—this was the means to "unleash forces that are desperately needed to solve a problem like global warming," as the head of the Environmental Defense Fund declared at Kyoto.[71]

Having resulted in the departure of the Environmental Defense Fund from the Climate Action Network, emissions trading also helped to unsettle the international coalition of business interests that had been staunchly opposed to regulation. As Kyoto approached, only the insurance industry appeared to have the resources to counter the influence of the fossil fuel industry.[72] On the eve of the Berlin meeting in 1995, Greenpeace had arranged a seminar for the insurance industry, which the organization's Jeremy Leggett had been courting since the Rio Earth Summit. Although cynics suggested that insurers were merely seeking government support for large insurance claims, some companies were already undertaking their own research, and two of the world's largest reinsurers, Swiss Re and Munich Re, had expressed fears that climate change could bankrupt the industry.[73] Leggett had called on insurers to lobby industry and government to cut their greenhouse gas emissions, which representatives of Munich Re, Swiss Re and Lloyd's of London pursued in Berlin.[74] Concerned insurers such as General Accident and Swiss Re also joined with the UN Environment Programme in 1995 to form the Insurance Industry Initiative and set out a Statement of Environmental Commitment by the Insurance Industry, which some sixty companies had joined by the Kyoto conference in 1997.[75] Unlike the industry representatives, however, they were "the new kids on the block" and yet to wield their collective power in the negotiations.[76]

Siding with the Environmental Defense Fund led BP to break ties with the Global Climate Coalition in late 1996, as the company's CEO John Browne took the counsel of the IPCC's John Houghton and sought to exercise

[71]Fred Krupp, 1997, cited in Jonas Meckling, *Carbon Coalitions: Business, Climate Politics, and the Rise of Emissions Trading* (Cambridge, MA: MIT Press, 2011), 82.

[72]Michael Tucker, "Climate Change and the Insurance Industry: The Cost of Increased Risk and the Impetus for Action," *Ecological Economics* 22, no. 2 (1997): 85–96.

[73]Fred Pearce, "Greenpeace: Storm-Tossed on the High Seas," in *Green Globe Yearbook 1996* (Oxford: Oxford University Press, 1996), 73–80; Matthew Paterson, "Risky Business: Insurance Companies in Global Warming Politics," *Global Environmental Politics* 1 (2001): 18–42.

[74]Virginia Haufler, "Insurance and Reinsurance in a Changing Climate," in Henrik Selin and Stacey D. VanDeveer (eds.), *Changing Climates in North American politics: Institutions, Policymaking and Multilevel Governance* (Cambridge, MA: MIT Press, 2009), 241–62.

[75]Sverker C. Jagers and Johannes Stripple, "Climate Governance beyond the State," *Global Governance* 9, no. 3 (2003): 385–99.

[76]Julian E. Salt, "Kyoto and the Insurance Industry: An Insider's Perspective," *Environmental Politics* 7, no. 2 (1998): 160–5.

first-mover advantage. DuPont similarly saw benefits in taking early action, as it had in the ozone regime (Chapter 4).[77] Although the EU shared with the Climate Action Network suspicions of emissions trading, this position softened as the challenges of meeting their own emissions targets without a carbon tax became clear.[78] Attempts in the early 1990s to introduce a tax had spurred what the *Economist* described in 1992 as "the massed ranks of Europe's industrialists to mount what is probably their most powerful offensive against an EC proposal."[79] The March 1997 Economists' Statement on Climate Change added another supporting voice to emissions trading, which signatories argued would allow "the world to achieve its climatic objectives at minimum cost."[80]

Ambivalence toward the Kyoto Protocol had also begun to abate somewhat as reports emerged in 1998 that the hole in the ozone layer was shrinking as a consequence of the 1987 Montreal Protocol. The ozone regime had similarly set phase-down targets for CFCs, and in the United States, the Environmental Protection Agency had developed an emissions trading system to comply with the protocol.[81] Although the problem of ozone depletion was different from the challenges of climate change, the effectiveness of the Montreal Protocol helped to show that the multilateral negotiations of atmospheric diplomacy could yield promising results.[82]

For the Clinton administration, securing the "meaningful participation of developing countries" at Kyoto had become of paramount importance in the wake of the 1997 Byrd–Hagel resolution. Backed by the fossil fuel lobby and passed unanimously, this resolution prevented the United States from becoming a signatory to a protocol that did not mandate specific commitments for developing countries or that might result in "serious harm" to the nation's economy.[83] The issue of emissions reduction in developing countries had dogged negotiations since the early 1990s, but now a proposal from the Brazilian delegation looked set to break the impasse. On the eve of the Kyoto meeting, Brazil's delegation had mooted the creation of a "green development fund." Premised on the accumulation of penalties for noncompliance by the industrialized countries, this proposal quickly gained the support of other G77

[77]Meckling, *Carbon Coalitions*, 82–3.

[78]Grubb, Vrolijk, and Brack, *The Kyoto Protocol*, 93–6; Meckling, *Carbon Coalitions*, 88–9.

[79]"Europe's Industries Play Dirty," *Economist* 323, no. 7758 (1992): 85.

[80]"The Economists' Statement on Climate Change," in Stephen J. Decanio, *The Economics of Climate Change* (San Francisco: Redefining Progress, 1997), 2.

[81]Arno Simons and Jan-Peter Voß, "Politics by Other Means: The Making of the Emissions Trading Instrument as a 'Pre-history' of Carbon Trading," in Benjamin Stephan and Richard Lane (eds.), *The Politics of Carbon Markets* (London: Routledge, 2014), 51–68.

[82]Grubb, Vrolijk, and Brack, *The Kyoto Protocol*, 25–6.

[83]Christian Downie, "Transnational Actors in Environmental Politics: Strategies and Influence in Long Negotiations," *Environmental Politics* 23 (2014): 376–94.

nations, which had long pressed for the means of financial and technology transfer to the developing world.[84] For instance, the idea of a World Atmosphere Fund to facilitate such wealth transfers had been one of the recommendations of the 1988 Toronto conference (Chapter 4). The following year, Indian prime minister Rajiv Gandhi had leant his support for a Planet Protection Fund at the Non-Aligned Movement meeting in Belgrade, which his successor echoed at the Earth Summit in Rio, joining similar proposals made there by Argentina and Brazil (Chapter 5).[85]

By calculating emissions reductions based on historical responsibility, Brazil's new proposal further aligned with the G77's established position. Prior to the Kyoto meeting, both the US delegation's Jonathan Pershing and IPCC chair Bolin had separately raised the likelihood of majority world emissions surpassing those of the developed world around the year 2025.[86] Concerned that these estimates presaged a renewed attempt to impose emissions limits on the developing world, Brazil's delegation mounted an expressly scientific counterattack. If the baseline year for calculating the accumulation of emissions was instead 1840, Brazil argued, then the developing world's contribution to increasing global temperature would not equal those of the developed world until the year 2162. In light of these calculations, they argued, only developed nations were obliged to limit greenhouse gas emissions and should be penalized if they failed to meet their commitments. The accumulation of these penalties would then be dispersed to developing countries according to their relative contribution to climate change, ranging from China at 32 percent to Niue at just 0.00005 percent.[87] This calculation was especially favorable for Brazil: some 80 percent of its emissions derived from agriculture, half of which arose from deforestation in the Amazon and Cerrado, while the ongoing financial crisis there made the prospect of international finance for sustainable development especially appealing.[88]

As the proposal's notion of penalties quickly receded from the negotiations, the US delegation welcomed the flexibility that the idea appeared to offer to

---

[84]Grubb, Vrolijk, and Brack, *The Kyoto Protocol*, 58.

[85]"Statement by H.E. Mr P.V. Narasimha Rao, Prime Minister of the Republic of India," in *Report of the United Nations Conference on Environment and Development, Rio de Janeiro, 3–14 June 1992*, vol. 3 (New York: United Nations, 1993), 1.

[86]John Cole, "Genesis of the CDM: The Original Policymaking Goals of the 1997 Brazilian Proposal and Their Evolution in the Kyoto Protocol Negotiations into the CDM," *International Environmental Agreements* 12 (2012): 43.

[87]Brazil, "Proposed Elements of a Protocol to the United Nations Framework Convention on Climate Change, Presented by Brazil in Response to the Berlin Mandate," in *Implementation of the Berlin Mandate – Additional Proposals from Parties, Addendum*, Ad Hoc Group on the Berlin Mandate, UNFCCC, May 30, 1997, FCCC/AGBM/1997/MISC.1/Add.3, https://v.gd/NNVdPn.

[88]Cole, "Genesis of the CDM," 47.

industrialized countries. As the *Earth Negotiations Bulletin* reported in Kyoto, "the US has come to view the initiative as the key to a neat fix, linking a number of their interests, including emissions trading, and engaging some developing countries in meaningful participation with the promise of generating funds for technology."[89] Brazilian and US negotiators reworked the plan to facilitate a form of joint implementation between the developed and developing worlds. This approach resembled a trial underway since the Berlin Mandate, whereby industrial countries (or the private sector) could invest in emissions reductions projects in developing nations or in countries with "economies in transition." So far, only a handful of projects had commenced, mostly in the Baltic States and South America, where Costa Rica had moved especially quickly to take advantage of the program.[90]

What became the Clean Development Mechanism (CDM) now allowed countries in the North to invest in sustainable development projects in the South in return for credits toward their Kyoto commitments. This "Kyoto Surprise," as Chair Estrada-Oyuela called it, appeared to be a winning formula: the United States and Umbrella nations had achieved their desired flexibility, while the developing countries could be assured of additional funds from the North.[91] Brazil's intervention reflected a wider shift in the nation's approach to foreign policy. Engaging, rather than resisting, the climate negotiations emerged from a desire to not only improve relations with the United States but also move away from "an outdated and counter-productive 'third-worldism'."[92] The CDM also represented an opportunity for India, which had been initially hostile to such transnational mitigation programs.[93] In Berlin, Environment and Forests Minister Kamal Nath had only assented to voluntary joint implementation programs with the South on the grounds that they would not be "an excuse by the North to continue with their present profligate consumption patterns which are at the root of the unsustainable mess we find ourselves in."[94] As negotiations of the Kyoto mechanisms continued in the

---

[89]Paola Bettelli, Chad Carpenter, Deborah Davenport, Peter Doran and Steve Wise, "In the Corridors," *Earth Negotiations Bulletin* 12, no. 71 (1997): 2.

[90]Ana V. Rojas, "Costa Rica and Its Climate Change Policies: Five Years after Kyoto," in Velma I. Grover (ed.), *Climate Change: Five Years after Kyoto* (Boca Raton, FL: CRC Press, 2004), 297–308.

[91]David Ciplet, J. Timmons Roberts, and Mizan Khan, *Power in a Warming World: The New Global Politics of Climate Change and the Remaking of Environmental Inequality* (Cambridge, MA: MIT Press, 2015), 60.

[92]Myanna Lahsen, *Brazilian Climate Epistemers' Multiple Epistemes: An Exploration of Shared Meaning, Diverse Identities and Geopolitics in Global Change Science* (Cambridge, MA: Belfer Center for Science and International Affairs, 2002), 8.

[93]Grubb, Vrolijk, and Brack, *The Kyoto Protocol*, 100–1.

[94]"Statement by Kamal Nath, Minister for Environment and Forests India," COP1, Berlin, April 6, 1995, Global Commons Institute, https://v.gd/FAH4Mu.

late 1990s, a "rapprochement in Indo-American relations" encouraged closer engagement between US and Indian officials and businesses, which led the Indian government to warm to the idea.[95]

Of the flexible mechanisms that arose from Kyoto, only the CDM involved developing countries. In addition to providing a means to facilitate investment in sustainable development projects in the South, a levy on those North–South projects would contribute to the newly established Adaptation Fund. The IPCC's Third Assessment Report had recently emphasized that given the inertia of the climate system, climate change would continue even once countries had curbed their greenhouse gas emissions below "dangerous levels." Adaptation, the report continued, was "inevitable" and would be generally "cheaper, if taken earlier than later." As the adverse impacts of climate change were likely to "fall disproportionately upon developing countries," where populations had "lesser adaptive capacity relative to developed countries," the negotiations at Bonn in mid-2001 turned to supporting the enhancement of their "adaptive capabilities."[96] With the creation of the Adaptation Fund, developing countries could now access additional financial resources to support "concrete adaptation projects and programmes," to which Canada declared it would provide an immediate "jump-start" grant.[97] Although the CDM provided this fund a source of assured regular income, analysts Joyeeta Gupta and Alison Lobsinger wondered why "only North-South cooperation is taxed." They continued, "This increases the price of North-South cooperation vis-à-vis North-North cooperation, while arbitrarily exempting other flexibility mechanisms from such a tax."[98] According to some calculations, the protocol's flexible mechanisms would leave 450 million metric tons more carbon in the atmosphere by 2012 than if emission reductions were made domestically. As a comparison, the US exit from Kyoto in 2001 had the equivalent effect.[99]

[95]Hayley Stevenson, "India and International Norms of Climate Governance: A Constructivist Analysis of Normative Congruence Building," *Review of International Studies* 37 (2011): 997–1019.
[96]Robert Watson (ed.), *Climate Change 2001: Synthesis Report* (Cambridge: Cambridge University Press, 2001), 12.
[97]Saleemul Huq, "The Bonn–Marrakech Agreements on Funding," *Climate Policy* 2 (2002): 243.
[98]Joyeeta Gupta and Alison Lobsinger, "Climate negotiations from Rio to Marrakech: An Assessment," in Velma I. Grover (ed.), *Climate Change: Five Years after Kyoto* (New York: CRC Press, 2004), 81.
[99]Suraje Dessai, Nuno S. Lacasta, and Katharine Vincent, "International Political History of the Kyoto Protocol: From The Hague to Marrakech and Beyond," *International Review for Environmental Strategies* 4, no. 2 (2003): 197; Liverman, "Conventions of Climate Change," 279–96.

Sinking Kyoto? Forests, Power, and Climate Mitigation

Just months into the presidency of George W. Bush, his administration announced in March 2001 the withdrawal of the United States from the Kyoto Protocol. Without the world's largest emitter of carbon dioxide (23 percent that year), commentators from all sides saw this move as the death knell for the climate treaty. Only time would tell whether other industrialized nations would follow Washington's lead, namely, its allies in the preceding negotiations—Japan, Canada, Australia, and Russia. So far, negotiations in Buenos Aires and The Hague had yet to finalize the details of the Kyoto mechanisms, as familiar flashpoints of developing country commitments and market mechanisms resurfaced. The use of carbon sinks such as forests to achieve emissions reductions had also become a source of tension, stalling talks until an unlikely breakthrough in Bonn in July 2001.[100] Negotiations turned, then, to the Marrakech Accords at year's end, which would pave the way for the protocol's entry into force.

In the wake of President Bush's announcement, the EU rallied to ensure the negotiations would prevail. The findings of the IPCC's Third Assessment Report in 2001 had also reinforced the urgency of their climate diplomacy. The IPCC confirmed that there was "new and stronger evidence that most of the observed warming over the last 50 years is attributable to human activities," and that the effects of climate change "are expected to be greatest in the developing countries."[101] Even without the United States, however, the final negotiating text of the Kyoto Protocol appeared to have been "Made in the USA," as the former executive secretary of the Framework Convention, Michael Zammit Cutajar, later put it.[102] This result was largely because the so-called Umbrella Group had gained even greater bargaining power in the wake of the exit of the United States. The protocol would only enter into force once it had been ratified by fifty-five parties accounting for 55 percent of 1990 emissions by Annex I countries.[103] Without Russia (17.4 percent) or Japan (8.5 percent), the 55 percent rule could not be met.

[100]Dessai, Lacasta, and Vincent, "International Political History of the Kyoto Protocol," 183–205.
[101]Robert Watson, John Houghton, and Ding Yuhui, "Preface," in John Houghton et al. (eds.), *Climate Change 2001: The Scientific Basis – Contribution of Working Group I to the Third Assessment Report of the Intergovernmental Panel on Climate Change* (Cambridge: Cambridge University Press, 2001), ix.
[102]Michael Zammit Cutajar, "Reflections on the Kyoto Protocol – Looking Back to See Ahead," *International Review for Environmental Strategies* 5, no. 1 (2004): 61–70.
[103]Suraje Dessai and Lisa Schipper, "The Marrakech Accords to the Kyoto Protocol: Analysis and Future Prospects," *Global Environmental Change* 13, no. 2 (2003): 150.

As far as the European delegations were concerned, the CDM resembled another loophole for the Umbrella nations to avoid domestic action. To diminish this risk, they demanded that emissions reductions arising from such projects should be real, measurable, and additional to any that would have occurred in their absence.[104] European negotiators were similarly lukewarm toward another US-led initiative for Annex B industrialized nations: to ensure the most generous reading of the Kyoto Protocol's text with regard to the use of forest and cropland management ("terrestrial carbon sinks") to offset their greenhouse gas emissions. The rationale for the US-led position lay in the scientific interpretation of the global carbon cycle: the greater the biomass or terrestrial sink, the lower the amount of carbon dioxide that reaches the atmosphere.

Harnessing biomass for the mitigation of climate change had been investigated since at least the 1970s.[105] As Bolin and Woodwell were attempting to quantify the global carbon cycle (Chapter 3), some enterprising scientists were surveying ways to arrest the trend of rising levels of atmospheric carbon dioxide due to fossil fuel use. Physicist Freeman Dyson was among them, setting out in 1976 a "proposal for purging CO_2 from the atmosphere biologically" by planting fast-growing trees.[106] His was not the only such plan to emerge from the Oak Ridge National Laboratory in Tennessee, for which the "global carbon dioxide problem" offered new possibilities for nuclear research.[107] A group led by chemist Charles Baes Jr., for instance, speculated that converting "more land to woods" could "counterbalance the current annual production of CO_2 from fossil fuel."[108]

During the Kyoto negotiations over two decades later, these ideas became newly relevant. Scientific efforts to quantitatively trace the whereabouts of carbon emissions indicated a "missing carbon sink" that was slowing the atmospheric uptake of carbon dioxide, a sign that scientists still had much to learn about the carbon cycle.[109] A study led by physicist Song-Miao Fan at

[104]Eva Lövbrand, Teresia Rindefjäll, and Joakim Nordqvist, "Closing the Legitimacy Gap in Global Environmental Governance? Lessons from the Emerging CDM Market," *Global Environmental Politics* 9, no. 2 (2009): 74–100.

[105]Eva Lövbrand, "Pure Science or Policy Involvement? Ambiguous Boundary-Work for Swedish Carbon Cycle Science," *Environmental Science and Policy* 10, no. 1 (2007): 39–47.

[106]Freeman J. Dyson, "Can We Control the Carbon Dioxide in the Atmosphere?," *Energy* 2, no. 3 (1977): 287–91.

[107]Jacob Darwin Hamblin, *The Wretched Atom: America's Global Gamble with Peaceful Nuclear Technology* (New York: Oxford University Press, 2021), 496–97. See, C. F. Baes Jr. et al., *The Global Carbon Dioxide Problem* (Oak Ridge: Oak Ridge National Laboratory, 1976).

[108]C. F. Baes Jr. et al., "Carbon Dioxide and Climate: The Uncontrolled Experiment," *American Scientist* 65, no. 3 (1977): 311.

[109]W. Neil Adger and Katrina Brown, "Policy Implications of the Missing Global Carbon Sink," *Area* 27 (1995): 311–17.

Princeton University's Carbon Modelling Consortium had sought to estimate the geographic distribution of biomass absorption or "uptake" of carbon dioxide.[110] Their 1998 findings suggested that "North America may have drawn the winning ticket in the carbon sink sweepstakes," as a *Science* journalist reported. According to their calculations, this sink "sops up ... enough to suck up every ton of carbon discharged annually by fossil fuel burning in Canada and the United States."[111] For negotiators in the United States as well as their forested allies in Canada, Australia, and Russia, these results were a boon leading into the negotiations that followed Kyoto.[112] As far as Australia's delegation was concerned, including such sinks "maximises the cost effectiveness of mitigation action through allowing individual countries to tailor their approach to their own emissions profile."[113] Most European nations, on the other hand, had more limited forest areas from which to benefit from carbon sink accounting, putting them at a distinct economic disadvantage relative to their chief competitor, the United States.[114]

Tensions over "the missing sink" had yet to be resolved when the IPCC concluded its special report on *Land Use, Land-use Change, Forestry* in late 2000.[115] Requested by the Framework Convention secretariat's Subsidiary Body for Scientific and Technological Advice (SBSTA) in 1998 to clarify the implications of the protocol's approach to carbon sinks, this report was quickly assembled to inform the ongoing negotiations. Over a hundred researchers contributed during the two-year period, mostly from the United States and Europe, with few representatives from developing countries, and none from the island states—those nations most opposed to the inclusion of sinks. The stakes were high—at the plenary session to approve the Summary for Policymakers, government representatives argued over the report's findings: Canada sought particular definitions of afforestation and reforestation, the United States negotiated for the broad inclusion of forestry activities, the UK argued for greater clarification of the uncertainties to limit forestry activities, and Germany wanted to exclude biotic carbon sequestration

---

[110]Song-Miao Fan et al., "A Large Terrestrial Carbon Sink in North America Implied by Atmospheric and Oceanic Carbon Dioxide Data and Models," *Science* 282, no. 5388 (1998): 442–6.

[111]Jocelyn Kaiser, "Possibly Vast Greenhouse Gas Sponge Ignites Controversy," *Science* 282, no. 5388 (1998): 386–7.

[112]Lövbrand, "Pure Science or Policy Involvement?," 453.

[113]"Paper 1: Australia," in UNFCCC, *Response from Parties on Issues Related to Sinks: Comments from Parties*, November 30 1997, 4, FCCC/AGBM/1997/MISC.4/Add.1, https://v.gd/0K5rzE.

[114]Myanna Lahsen, "A Science–Policy Interface in the Global South: The Politics of Carbon Sinks and Science in Brazil," *Climatic Change* 97 (2009): 346; Philip M. Fearnside, "Saving Tropical Forests as a Global Warming Countermeasure: An Issue That Divides the Environmental Movement," *Ecological Economics* 39 (2001): 167–84.

[115]Robert Watson et al., *Land Use, Land-use Change, and Forestry: A Special Report of the IPCC* (Cambridge: Cambridge University Press, 2000).

activities under the CDM.[116] Such was the state of the negotiations, the panel's Robert Watson scolded the delegates: "There is no chance that we will finish the document if we continue at this rate. Everyone is trying to position themselves for COP-6" in the Netherlands.[117]

The IPCC chair was proven correct: this report became the primary reference for all negotiators as they gathered to resume talks in The Hague in December 2000. Bolin, a coordinating lead author of the special report, had already expressed his reservations about the Kyoto Protocol's inclusion of sinks. He suspected that they might "well supply loopholes for countries that were unable to fulfil their commitments."[118] The European countries shared this position and were unwilling to countenance the US proposal that countries receive credits for carbon absorption or sinks from all managed land.[119] In contrast to the domestic scene in the United States, European governments were exploring more radical ideas at home, such as the "contraction and convergence" idea that the Global Commons Institute had raised in the wake of the "statistical lives" episode, discussed earlier.[120] The UK's independent Royal Commission on Environmental Pollution, for instance, had recommended in 2000 that the Blair Labour government "press for a future global climate agreement based on the contraction and convergence approach, combined with international trading in emission permits."[121] At The Hague climate conference in 2000, French president Jacques Chirac had similarly declared, "France proposes that we set as our ultimate objective the convergence of per capita emissions."[122]

Whether, or how, sinks might be included in the CDM was another source of tension, with its implications for potential projects dividing the developing countries that would be their hosts. Brazil, China, India, and the small island states were especially opposed to the inclusion of forests under the Mechanism. Their concerns included the potential encroachment on their sovereignty; the transfer of Mechanism credits away from other projects, such as renewable

---

[116]Cathleen Fogel, "Biotic Carbon Sequestration and the Kyoto Protocol," *International Environmental Agreements* 5 (2005): 198–9.

[117]Watson, cited in Fogel, "Biotic Carbon Sequestration," 200.

[118]Bolin, *A History*, 160.

[119]Michael Grubb and Farhana Yamin, "Climatic Collapse at The Hague: What Happened, Why, and Where Do We Go from Here?," *International Affairs* 77 (2001): 261–76.

[120]"Contraction and convergence" refers to the idea that once states agree to a particular level of greenhouse gas emissions to avoid "dangerous climate change" (contraction), states would allocate per capita emissions such that over time, the per capita emissions of all countries would eventually converge.

[121]Royal Commission on Environmental Pollution, *Energy – The Changing Climate: Summary of the Royal Commission on Environmental Pollution's Report* (London: Royal Commission on Environmental Pollution, 2000), 28.

[122]Cited in Alex Kirby, "Climate Treaty 'Almost Irrelevant'," *BBC News* November 23, 2000, https://v.gd/NmHkAp.

energy; and the displacement of deforestation elsewhere.[123] Under pressure to sustain the Kyoto Protocol without the United States after The Hague talks collapsed, the parties finally struck an agreement in Bonn in mid-2001. They agreed in favor of generous sink provisions, limiting eligible carbon uptake activities to afforestation, reforestation, and deforestation, where deliberate human activity could be more easily assessed.[124] During the first commitment period of the Kyoto Protocol (2005–12), industrialized countries would be permitted to acquire emissions credits under the Mechanism for tree planting but not for forest conservation.[125]

During the negotiations about sinks in the SBSTA at Lyon in 2000, AOSIS had attempted unsuccessfully to raise the concerns of Indigenous Peoples. According to Ian Fry, Tuvalu's ambassador for climate change, advocates had circulated a recent declaration that argued that sinks in the Clean Development Mechanism would "constitute a worldwide strategy for expropriating our lands and territories, and violating our fundamental rights that would culminate in a new form of colonialism."[126] This declaration had arisen from the First International Forum on Indigenous Peoples on Climate Change, which had gathered delegates from Latin America, South Asia, Southeast Asia, Africa, and Oceania to coincide with the climate negotiations in Lyon.[127] As the Framework Convention does not refer to Indigenous Peoples, the declaration gestured to other international conventions, such as the Convention on Biological Diversity and the International Labour Organization. The delegations for Samoa and Tuvalu attempted to elevate these concerns in the SBSTA negotiations, proposing that "activities associated afforestation,

---

[123]Ian Fry, "More Twists, Turns and Stumbles in the Jungle: A Further Exploration of Land Use, Land-Use Change and Forestry Decisions within the Kyoto Protocol," *Review of European Community & International Environmental Law* 16, no. 3 (2007): 341–55.

[124]Matthew Paterson and Johannes Stripple, "Singing Climate Change into Existence: On the Territorialization of Climate Policymaking," in Mary Pettenger (ed.), *The Social Construction of Climate Change: Power, Knowledge, Norms, Discourses* (Aldershot: Ashgate, 2007), 161. See also Eva Lövbrand, "Bridging Political Expectations and Scientific Limitations in Climate Risk Management – on the Uncertain Effects of International Carbon Sink Policies," *Climatic Change* 67 (2004): 449–60.

[125]Cathleen Fogel, "The Local, the Global, and the Kyoto Protocol," in Sheila Jasanoff and Marybeth Long Martello (eds.), *Earthly Politics: Local and Global in Environmental Governance* (Cambridge, MA: MIT Press, 2004), 103–25.

[126]"Declaration of the First International Forum of Indigenous Peoples on Climate Change, Lyon, France, September 4–6, 2000," Amazon Watch, September 6, 2000, https://v.gd/3lozsc; Ian Fry, "Twists and Turns in the Jungle: Exploring the Evolution of Land Use, Land-Use Change and Forestry Decisions within the Kyoto Protocol," *Review of European, Comparative and International Environmental Law* 11 (2002): 159–68.

[127]This declaration shares elements with the Quito Declaration, signed in Ecuador in May 2000, by representatives of Indigenous organizations and local communities from Ecuador, Bolivia, Panamá, Indonesia, Venezuela, South Africa, Kenya, Colombia, and Mexico; see "Quito Declaration," Center for International Environmental Law, c. 2000, https://v.gd/xcRv5s.

reforestation and deforestation shall be consistent with and supportive of the principles of sustainable development ... and the protection of social, cultural and community rights."[128] Fry recalls that members of the Umbrella Group rejected these proposals on the grounds that they were not relevant in the context of the discussion. As a result, the official negotiating texts do not include any recognition of the concerns of Indigenous Peoples about land-use change, forestry, or the Mechanism.[129]

The 2000 Declaration of the Indigenous Peoples had called for the Climate Convention to formally recognize Indigenous Peoples and their organizations in their processes at the upcoming climate conference at The Hague. This recommendation echoed a similar call made in 1998 in the Albuquerque Declaration, prepared by Indigenous Peoples from across North America and presented at the climate conference in Buenos Aires in November 1998.[130] Following the example already set in the Convention on Biological Diversity, the Declaration called for the establishment of an "Inter-Sessional Open-Ended Working Group For Indigenous Peoples" under the UNFCCC.[131] The experiences of Indigenous peoples and local peoples in Southern countries were also at the forefront of the climate justice meeting that CorpWatch organized at The Hague for the climate conference in late 2000, at which a documentary film showed UNFCCC delegates the experiences of Inuvialuit in the changing Arctic climate.[132]

In response to these growing calls for Indigenous recognition in the intergovernmental climate negotiations, the UNFCCC Secretariat recognized the Indigenous Peoples Organizations as an official constituency in 2001, which allows for speaking rights and office space.[133] The following year, a coalition of organizations came together as the International Climate Justice Network at the World Summit on Sustainable Development in Johannesburg, where they developed and shared the Bali Principles of Climate Justice.[134] Advocating for

---

[128]For example, Samoa (on behalf of the AOSIS), "Textual Proposals Associated with Land Use, Land Use Change and Forestry," Methodological Issues, Land-Use, Land-use Change and Forestry, UNFCCC SBSTA, August 23, 2000, 6, https://v.gd/dQ5Tzl.

[129]Fry, "Twists and Turns in the Jungle," 159–68.

[130]"Albuquerque Declaration," in Nancy G. Maynard (ed.), *Circles of Wisdom: Native Peoples–Native Homelands Climate Change Workshop* (Washington, DC: US Global Change Research Program, 1998), 69–74.

[131]Andrés López-Rivera, "Diversifying Boundary Organizations: The Making of a Global Platform for Indigenous (and Local) Knolwedge in the UNFCCC," *Global Environmental Politics* (2023): https://v.gd/caYATu.

[132]"Alternative Summit Opens with Call for Climate Justice," *CorpWatch*, November 19, 2000, https://v.gd/yJtz1x; Marybeth Long Martello, "Arctic Indigenous Peoples as Representations and Representatives of Climate Change," *Social Studies of Science* 38, no. 3 (2008): 351–76.

[133]Ella Belfer et al., "Pursuing an Indigenous Platform: Exploring Opportunities and Constraints for Indigenous Participation in the UNFCCC," *Global Environmental Politics* 19, no. 1 (2019): 12–33.

[134]David Schlosberg and Lisette B. Collins, "From Environmental to Climate Justice: Climate Change and the Discourse of Environmental Justice," *WIREs Climate Change* 5, no. 3 (2014): 359–74.

the rights of Indigenous peoples was part of the objectives of this growing movement: "Climate Justice affirms the right of Indigenous Peoples and local communities to participate effectively at every level of decision-making, including needs assessment, planning, implementation, enforcement and evaluation, the strict enforcement of principles of prior informed consent, and the right to say 'No'."[135]

Framing this mobilization was the First International Decade of the World's Indigenous People (1995–2004), which had been proclaimed by the UN General Assembly, and it marked the drafting of the Declaration on the Rights of Indigenous Peoples. As Andrés López-Rivera observes, "The unresponsiveness of states in the climate field contrasted with the breakthrough of Indigenous issues at the United Nations through the establishment of the Permanent Forum on Indigenous Issues in 2000."[136] Indigenous peoples' calls for official recognition in the UN's climate negotiations reflected, then, a wider Indigenous internationalism that stemmed at least as far back as George Manuel's "Fourth World" of the 1970s (Chapter 2). In the legal documents arising from the 1992 Rio Summit (Chapter 5), which had stressed the vital importance of Indigenous peoples and Indigenous knowledge to global environmental governance, Indigenous peoples identified new strategic fronts to assert their calls for self-determination.[137]

Conclusion

Two weeks after the parties agreed to the final components of the Kyoto Protocol in Marrakech in 2001, the chair of the G77 reflected on the outcome of the decade-long process. Before the UN General Assembly, Iran's Bagher Asadi observed, "Apart from substance and speculation on what could have been, the long, tortuous process ending with the Marrakech Accords carries a very important lesson and message; multilateralism and international cooperation work." He followed this thinly veiled critique of the United States by adding "that the developing countries, who are the most vulnerable to the adverse effects of climate change, will continue their active and proactive engagement in the on-going multilateral process in this field with the very

---

[135]"Bali Principles of Climate Justice," *Energy Justice Network*, August 29, 2002, https://v.gd/C5IN2X.

[136]López-Rivera, "Diversifying Boundary Organizations," 7.

[137]Miranda Johnson, "Indigenizing Self-Determination at the United Nations: Reparative Progress in the Declaration on the Rights of Indigenous Peoples," *Journal of the History of International Law* 23 (2021): 206–28; Andrea Muehlebach, ' "Making place" at the United Nations: Indigenous cultural politics at the UN Working Group on Indigenous Populations', *Cultural Anthropology* 16, no. 3 (2001): 415–48.

clear objective of safeguarding and further promoting our genuine, long-term collective interests."[138]

As the World Summit on Sustainable Development in Johannesburg approached in August 2002, some observers and governments anticipated that the tenth anniversary of the Rio Summit and the climate convention would be marked by the ratification of the Kyoto Protocol. Finalizing the Protocol would allow for the climate negotiations to proceed at last to the convention's longer-term objectives, particularly with regard to sustainable development and attending to the interests of the developing countries. From the perspective of observers in the South, Adil Najam, Saleelmul Huq, and Youba Sokona, "there is a danger that Kyoto has now become so much of a mechanism for managing the global carbon trade that the issue of emission cuts for atmospheric carbon stabilization could be neglected, or at least delayed."[139] Vulnerable low-lying nations could least afford the climate regime to fail to achieve actual emissions cuts.

---

[138]Bagher Asadi, "Statement by Ambassador Bagher Asadi, Chairman of the Group of 77 (Islamic Republic of Iran), before the Second Committee of the General Assembly on Agenda item 98 (f)," November 28, 2001, Group of 77 at the United Nations, https://v.gd/dex1LK.
[139]Adil Najam, Saleemul Huq, and Youba Sokona, "Climate Negotiations beyond Kyoto: Developing Countries Concerns and Interests," *Climate Policy* 3 (2003): 226.

7

Global Security

2007: 383.56 ppm

In September 2005, world leaders gathered at the United Nations (UN) headquarters in New York City for the World Summit meeting. There, they planned to discuss the progress of the Millennium Development Goals and to consider ways of reforming the organization in its sixtieth year. Among them was Chinese president Hu Jintao, who took the opportunity to outline the four pillars of his vision for a "harmonious world of lasting peace and common prosperity." Since acceding to the World Trade Organization (WTO) in 2001, China had pursued an export-led growth strategy that had resulted in annual GDP growth rates of roughly 10 percent per year, and it would soon become the world's fourth largest economy after the United States, Japan, and Germany.[1]

Accompanying this rapid growth were the nation's rising greenhouse gas emissions: China alone constituted over half of the global increase in carbon dioxide emissions between 2001 and 2006, surpassing the United States as the world's largest emitter by 2007. Foremost, President Hu declared in New York, was the need to ensure peace, for "without peace, we can neither go for [a] new development agenda nor prevent the destruction of the achievements of our previous development." "We must abandon the Cold War mentality," he continued, "cultivate a new security concept featuring mutual trust, mutual benefit, equality and cooperation, and build a fair and effective security mechanism aimed at jointly preventing war and conflict and safeguarding world peace and security."[2]

In early February 2005, the Kyoto Protocol had come into force after Russia's ratification ninety days earlier. Without the United States, Russia, with its 17 percent of global greenhouse gas emissions, had become vital: the protocol required the ratification of fifty-five countries, accounting for

---

[1]Chunlai Chen, "China's Economy after WTO Accession: An Overview," in Chunlai Chen (ed.), *China's Integration with the Global Economy: WTO Accession, Foreign Direct Investment and Economic Impacts* (Cheltenham: Edward Elgar, 2009), 1–16.
[2]"Statement by H.E. Hu Jintao, President of the People's Republic of China," New York, September 15, 2005, (English translation) United Nations, https://v.gd/YPIIuS.

55 percent of the emissions of the Annex B members in 1990. By late 2004, over 120 countries had ratified the protocol but they together constituted less than 45 percent of emissions.[3] Keen to bring the protocol into effect, the European Union (EU) had dropped its objections to Russia joining the WTO, which paved the way for Moscow to sign the climate agreement.[4] The deal came in the wake of a record heat wave across Europe, which claimed the lives of at least twenty thousand people. For the first time, the emerging science of climate attribution could show that the likelihood of such a heat wave had doubled due to human influence.[5]

In the decade after the attacks of September 11, 2001, climate change became widely associated with questions of global security. As the United States and its allies invaded Afghanistan and then Iraq in 2003, Sir John Houghton, the former chair of the Intergovernmental Panel on Climate Change's (IPCC's) Working Group I, likened climate change to "a weapon of mass destruction."[6] Researchers associated with the Pentagon, meanwhile, analyzed a scenario of abrupt climate change that "although not the most likely, is plausible, and would challenge United States national security in ways that should be addressed immediately."[7] Amplifying the impact of their findings was the 2004 Hollywood blockbuster, *The Day After Tomorrow*, with its vivid depiction of New York City in the grip of a new ice age, which dramatized the conditions discussed in the BBC documentary *The Big Chill* a year earlier.

Exponents of the security implications of unmitigated climate change were alive to the weaknesses that the Kyoto Protocol entailed. Without the United States, and with mitigation efforts contained mostly to the Organisation for Economic Co-operation and Development countries, they feared that the prevailing regime would not meet the objective of the Framework Convention: to avert dangerous climate change. From the planetary perspective of some scientists, meeting this objective could necessitate further interventions to secure the earth system in its Holocene state, which prompted both regulatory and ethical concerns about the possibility of unilateral action.

[3]Alexander Gusev, "Evolution of Russian Climate Policy: From the Kyoto Protocol to the Paris Agreement," *L'Europe en Formation* 380, no. 2 (2016): 39–52.

[4]Nick Paton Walsh, "Putin Throws Lifeline to Kyoto as EU Backs Russia Joining WTO," *Guardian*, May 22, 2004, https://v.gd/7iaXtO.

[5]Peter A. Stott, D. A. Stone, and M. R. Allen, "Human Contribution to the European Heatwave of 2003," *Nature* 432 (2004): 610–14. On the study of the attribution of extreme weather, see Friederike E. L. Otto, "Attribution of Weather and Climate Events," *Annual Review of Environment and Resources* 42 (2017): 627–46.

[6]John Houghton, "Global Warming Is Now a Weapon of Mass Destruction," *Guardian*, July 28, 2003, https://v.gd/1RX5rG.

[7]Peter Schwartz and Doug Randall, "An Abrupt Climate Change Scenario and Its Implications for United States National Security," Emergency Management Institute, Federal Emergency Management Agency, October 2003, https://v.gd/ewHdzv.

Seeking greater and more ambitious multilateral cooperation, by contrast, European governments stressed the security implications of climate change for the "integrity" of the national borders of the developed countries. These arguments also resonated with the governments of small island nations, for whom the impacts of climate change physically threatened their entire territories and peoples. Although their deployment of the security framing sought to emphasize the "common" nature of climate change, it met with the considerable resistance of the developing countries and the United States.

Planetary Security: Intervening in the Earth System

The association of climate change with global security stretched back to the 1988 Toronto meeting, where delegates had declared, "These changes represent a major threat to international security and are already having harmful consequences over many parts of the globe" (Chapter 4).[8] Before the Kyoto Protocol had come into force in 2005, former executive secretary of the UN Framework Convention Michael Zammit Cutajar wondered if a "possible hook" for accelerating efforts to ameliorate climate change might be global security. Gesturing to the Pentagon's study, he pointed out, "Long-term climate change will be a factor of global inequity, dumping the adverse fall-out of global economic growth on poor people least able to cope. It will thus add to global instability, if only because of its impacts on food, water, and migration."[9] It followed that those for whom such scenarios were unconvincing would avoid gestures to security: a strategist, for instance, advised the Bush administration in 2002 to substitute the "more benign" term "climate change" for the "more frightening" "global warming" with its "catastrophic connotations."[10] Likewise, readers of Michael Crichton's 2004 novel *State of Fear* could feel reassured that climate change was merely an environmentalist plot and not worthy of concern or action.[11] Accordingly, the

---

[8]For an overview of the association of climate change and variability with violent conflict in the 1990s, see Angela Oels, "Climate Security as Governmentality: From Precaution to Preparedness," in Johannes Stripple and Harriet Bulkeley (eds.), *Governing the Climate* (Cambridge: Cambridge University Press, 2014), 197–218.

[9]Michael Zammit Cutajar, "Reflections on the Kyoto Protocol – Looking Back to See Ahead," *International Review for Environmental Strategies* 5, no. 1 (2004): 68.

[10]Luntz Research Companies, "The Environment: A Cleaner, Safer, Healthier America," c. 2002, archived by *Mother Jones*, https://v.gd/bXxGQY.

[11]Michael J. Janofsky, "Michael Crichton, Novelist, Becomes Senate Witness," *New York Times*, September 29, 2005, https://v.gd/51OAoW . See also Naomi Oreskes and Erik Conway, "Challenging Knowledge: How Climate Science Became a Victim of the Cold War," in Robert Proctor and Linda

Bush administration removed the phrase "environmental security" from its National Security Strategy in 2005 and moved to exempt military activity from environmental legislation.[12]

Across the Atlantic, however, Blair's government had set climate change as one of the top priorities for the UK's presidency of the Group of Eight (G8) and the EU in 2005. To signal this approach, the government sponsored in February an International Symposium on Stabilisation of Greenhouse Gas Concentrations, focused on "avoiding dangerous climate change"—the goal of the Framework Convention.[13] Although the conference participants did not quantify a specific threshold, they did warn of abrupt or rapid changes that might cause the collapse of the West Antarctic Ice Sheet, the melting of the Greenland ice sheet, or the ocean's thermohaline circulation: "We do not fully understand the thresholds that might lead to such a dramatic effect, nor the time frame over which this might happen."[14]

For UK climate scientist Mike Hulme, this conference signaled what he deemed an unwelcome "step-change in the ways in which the risks associated with climate change were conceived, presented and debated in the public sphere."[15] The conference's language of thresholds, "tipping points," and "nonlinear responses" that had helped to foster this step change drew from the conceptual vocabulary of its organizer, German physicist Hans-Joachim "John" Schellnhuber and the field of earth systems science. In the late 1990s, he had introduced the notion that global change might not unfold in a linear or smooth fashion; instead, human activities could trigger rapid and irreversible shifts of the earth system disastrous for human well-being.[16] Hoping that "global research programmes will soon enable us to identify and respect 'guardrails' for responsible planetary management," he subsequently helped steer the International Geosphere–Biosphere Program toward the integration of human activities into the study of the earth system.[17] An intellectual descendant of Vladimir Vernadsky's biosphere concept, cybernetics, ecosystem ecology, and

---

Schiebinger (eds.), *Agnotology: The Making and Unmaking of Ignorance* (Palo Alto, CA: Stanford University Press, 2008), 55–89.

[12]Angela Oels, "From 'Securitization' of Climate Change to 'Climatization' of the Security Field: Comparing Three Theoretical Perspectives," in Jürgen Scheffran et al. (eds.), *Climate Change, Human Security and Violent Conflict* (Berlin: Springer, 2012), 189.

[13]Hans-Joachim Schellnhuber et al. (eds.), *Avoiding Dangerous Climate Change* (Cambridge: Cambridge University Press, 2006).

[14]"Key Vulnerabilities of the Climate System and Critical Thresholds," in Schellnhuber et al. (eds.), *Avoiding Dangerous Climate Change*, 2.

[15]Mike Hulme, *Why We Disagree about Climate Change: Understanding Controversy, Inaction and Opportunity* (Cambridge: Cambridge University Press, 2009), 381.

[16]Will Steffen et al., "The Emergence and Evolution of Earth System Science," *Nature Reviews Earth & Environment* 1 (2020): 54–63.

[17]Hans-Joachim Schellnhuber, "'Earth System' Analysis and the Second Copernican Revolution," *Nature* (1999): C19–C23.

Gaia theory (Chapters 1, 2, and 3), the "Earth System" became understood by Schellnhuber and his collaborators as the suite of interlinked physical, chemical, biological, and human processes that transport and transform materials and energy in complex ways within the system.[18]

From Schellnhuber's characterization of humanity as Prometheus, his associate Nobel Laureate Paul Crutzen derived another term, the Anthropocene. Although the Anthropocene concept has a longer genealogy, it was Crutzen's use of the term in 2000 that became widely associated with the vast planetary scale of humanity's impacts. Over the past couple of centuries, humanity had acquired the means to transform the earth such that it had become a geological force itself, pushing the planet out of the Holocene and into the Anthropocene. In an early exploration of the concept, Crutzen himself had suggested the future possibility of "internationally accepted, large-scale geoengineering projects ... to 'optimize' climate."[19] In August 2006, Crutzen revived this idea: in light of what he described as the "grossly disappointing international political response to the required greenhouse gas emissions," research on the "feasibility and environmental consequences of climate engineering ... should not be tabooed." Reworking his nuclear winter hypothesis (Chapter 4), Crutzen advocated the deployment of sulfates into the stratosphere, which would quickly cool the climate by reflecting sunlight.[20]

In putting forward such technologies, these scientists "opened a can of worms," as historian James Fleming puts it. They had emboldened "the control fantasies of the climate engineers" who already had the ear of US and Russian government officials.[21] That sentiment was shared among environmental organizations and other critics who considered that, at best, geoengineering was a distraction from reducing greenhouse gas emissions, while, at worst, highly risky and uncertain—not unlike the "nonlinear" process of climate change itself.[22] Aside from the scientific and ethical questions arising from such proposals, the regulation or governance of geoengineering occupied legal minds. What was there to stop a country or an individual from taking unilateral action? As political scientist David Victor speculated in 2008, "one large nation might justify and fund an effort on its own. A lone

---

[18]Steffen et al., "The Emergence and Evolution," 54–63.

[19]Paul J. Crutzen, "Geology of Mankind," *Nature* 415 (2002): 23.

[20]Paul Crutzen, "Albedo Enhancement by Stratospheric Sulfur Injections: A Contribution to Resolve a Policy Dilemma?," *Climatic Change* 77 (2006): 211–20. See also Tom Wigley, "A Combined Mitigation/Geoengineering Approach to Climate Stabilization," *Science* 314 (2006): 52–4.

[21]James Fleming, *Fixing the Sky: The Checkered History of Weather and Climate Control* (New York: Columbia University Press, 2010), 2; James Fleming, "The Climate Engineers," *Wilson Quarterly* 31, no. 2 (2007): 58.

[22]Melinda Cooper, "Turbulent Worlds: Financial Markets and Environmental Crisis," *Theory, Culture & Society* 27 (2010): 167–90.

Greenfinger, self-appointed protector of the planet and working with a small fraction of the [Microsoft cofounder Bill] Gates bank account, could force a lot of geoengineering on his own."[23]

The challenge was territorial: nation-states could regulate land-based mitigation activities, such as reforestation or biochar, but the fluid conditions of the atmosphere and the ocean could not be so easily contained or governed—a challenge not dissimilar to the international governance of climate change more generally.[24] Existing international regulatory instruments were unsuited to the prevention of climate engineering: the UN Environmental Modification Convention applied to only hostile uses (Chapter 3), while the UN Environment Programme's 1980 guidelines for weather modification had been designed for regional efforts, not planetary interventions.[25] For their part, the architects of the Framework Convention had neither considered solar geoengineering of Crutzen's variety in their negotiations nor provided for its regulation.[26]

The convention's objective to stabilize atmospheric concentrations of greenhouse gases was relevant to carbon dioxide removal, however. The exclusion of ocean sinks from the Kyoto Protocol, which recognized only land-based sink activities such as forests, had not deterred those who saw opportunity in the world's oceans. By the time of Crutzen's writing in 2006, researchers had undertaken at least twelve iron-enrichment studies since 1993 to test the so-called Iron Hypothesis, while entrepreneurs lobbied for access to the lucrative carbon market.[27] Drawing on observations from the 1925–27 *Discovery* expedition, oceanographer John Martin had proposed in the late 1980s that a lack of iron in some ocean areas was responsible for limiting the growth of phytoplankton. Adding iron, he reasoned, would not only stimulate their growth but also draw down carbon dioxide from the atmosphere into the ocean in the process.[28] The implications for addressing climate change were promising: "Give me half a tanker of iron and I'll give you an ice age," he had joked at a 1988 meeting at the Woods Hole Oceanographic Institution.[29]

---

[23]David G. Victor, "On the Regulation of Geoengineering," *Oxford Review of Economic Policy* 24, no. 2 (2008): 324.

[24]Clive Hamilton, *Earthmasters: The Dawn of the Age of Climate Engineering* (New Haven: Yale University Press, 2013), 145–6.

[25]Daniel Bodansky, "May We Engineer the Climate?," *Climatic Change* 33 (1996): 309–21.

[26]Daniel Bodansky, "Solar Geoengineering and International Law," in Robert N. Stavins and Robert C. Stowe (eds.), *Governance of the Deployment of Solar Geoengineering* (Cambridge, MA: Harvard Project on Climate Agreements, 2019), 119–23.

[27] Kemi Fuentes-George, "Consensus, Certainty and Catastrophe: Discourse, G, and Ocean Iron Fertilization," *Global Environmental Politics* 17, no. 2 (2017): 125–43.

[28]John H. Martin, R. Michael Gordon, and Steve E. Fitzwater, "Iron in Antarctic Waters," *Nature* 345 (1990): 156–8; John H. Martin, "Glacial-Interglacial CO_2 Change: The Iron Hypothesis," *Paleoceanography and Paleoclimatology* 5, no. 1 (1990): 1–13.

[29]John Martin, cited in Leslie Roberts, "Report Nixes 'Geritol' Fix for Global Warming," *Science* 253 (1991): 1490.

In the months after Crutzen's 2006 intervention, a Greenfinger made his intentions known. Russ George, a Californian entrepreneur, was no newcomer to ocean iron fertilization, but the unprecedented scale of his 2007 plan rang alarm bells. The largest scientific experiments to date had fertilized or seeded patches of 225 km²; his proposal entailed seeding some 10,000 km² of the equatorial Pacific near the Galapagos Islands.[30] Citing a plethora of unanswered legal and scientific questions, Greenpeace and the International Union for the Conservation of Nature called on the London Convention and Protocol's Scientific Groups to subject such projects to closer scrutiny. The US Environmental Protection Agency, meanwhile, made it clear at the June meeting of the London Convention and Protocol that it would not permit George to proceed with a US-flagged vessel; Ecuador, likewise, raised its concerns about the proposal's proximity to the Galapagos, a UN Educational, Scientific and Cultural Organization (UNESCO) World Heritage Site. In the first step toward the international regulation of ocean iron fertilization, the full Conference of the Parties to the London Convention and Protocol issued a statement of concern about the risks of projects like George's to the marine environment and human health.[31] With its vessel stuck at Madeira and now short of investor confidence, George's company folded in early 2008.[32]

Meeting in Ecuador in May, the parties to the London Convention and Protocol reiterated their concerns and specifically asserted that ocean iron fertilization projects fell under their jurisdiction. Following these decisions closely, the members of the UN Convention on Biological Diversity effectively imposed a moratorium on such activities, stating that governments should prevent their taking place "until there is an adequate scientific basis on which to justify such activities."[33] The growing appeal of geoengineering warranted its governance, nevertheless: a poll of eighty "leading scientists" conducted by a UK newspaper found that over half saw geoengineering as a viable plan should efforts to curb greenhouse gas emissions fall short.[34] The 2007 report of the IPCC had concluded that "warming of the climate system is unequivocal" and projected temperature rises of up to 6°C by the end of the

---

[30] Aaron L. Strong, John J. Cullen, and Sallie W. Chisholm, "Ocean Fertilization: Science, Policy and Commerce," *Oceanography* 22, no. 3 (2009): 236–61.

[31] Strong, Cullen, and Chisholm, "Ocean Fertilization," 253.

[32] Russ George renewed his efforts, nevertheless. See Bruce Falconer, "Can Anyone Stop the Man Who Will Try Just about Anything to Put an End to Climate Change?," *Pacific Standard*, January 16, 2018, https://v.gd/7Dk8gM; Kelsey Piper, "The Climate Renegade," *Vox*, June 4, 2019, https://v.gd/tvZHUD .

[33] Sherry P. Broder, "International Governance of Ocean Fertilization," in Carlos Espósito et al. (eds.), *Ocean Law and Policy: Twenty Years of Development under the UNCLOS Regime* (Leiden: Brill, 2016), 322–3.

[34] Steve Connor, "Climate Scientists: It's Time for 'Plan B'," *Independent*, January 2, 2009, https://v.gd/9eA4rZ.

century.[35] Such a temperature increase would breach what Schellnhuber and his collaborators would describe in 2009 as a "planetary boundary"—one of nine such boundaries that define "the safe operating space for humanity."[36]

Arguing that humanity "should be the heart and mind of the Earth not its malady," UK scientist James Lovelock also speculated on the value of geoengineering. In his 2009 book *The Vanishing Face of Gaia*, Lovelock presented a scenario of 2030 where sea level and global temperature rise would be such that "orderly survival … may require, as in war, the suspension of democratic government for the duration of the survival emergency."[37] He hoped that the application of geoengineering techniques and "the most practical and available geoengineering procedure of all"—nuclear energy—would "buy us a little time" to avoid catastrophic climate change.[38] Meanwhile, British entrepreneur Sir Richard Branson announced a generous prize for a commercially viable means of carbon removal, pointing out that "with a geoengineering answer to this problem … [w]e could carry on flying our planes and driving our cars."[39]

Later in 2009, the Royal Society published a report on geoengineering that concluded, "There are serious and complex governance issues which need to be resolved if geoengineering is ever to become an acceptable method for moderating climate change."[40] Welcoming the report's findings, the UK's Brown Labour government supported the House of Commons Science and Technology Committee to explore further the Royal Society's recommendation to review existing international and regional mechanisms that might regulate geoengineering.[41] Having heard evidence for need to engage with both "the emerging powers and the poorer countries" in questions of geoengineering, its report recommended that the UK government "press" proposals for the regulation of geoengineering through the UN.[42]

Following its 2009 report on geoengineering and with philanthropic support from both Branson and Gates, the Royal Society joined with the Environmental Defense Fund and the Academy of Sciences for the Developing World to explore ways to govern the kind of geoengineering that Crutzen had advocated.[43] Their

[35]IPCC, *Climate Change 2007: Synthesis Report* (Geneva: IPCC, 2007), 2.

[36]Johan Rockström et al., "A Safe Operating Space for Humanity," *Nature* 461 (2009): 472–5.

[37]James Lovelock, *The Vanishing Face of Gaia: A Final Warning* (London: Penguin, 2009), 61.

[38]James Lovelock, "A Geophysiologist's Thoughts on Geoengineering," *Philosophical Transactions of the Royal Society A: Mathematical, Physical and Engineering Sciences* 366 (2008): 3883–90.

[39]Richard Branson, cited in Andrew Revkin, "Branson on the Power of Biofuels and Elders," *New York Times Blog: Dot Earth*, October 15, 2009, https://v.gd/yMZ4Gi.

[40]Royal Society, *Geoengineering the Climate: Science, Governance and Uncertainty* (London: Royal Society, 2009), ix.

[41]Royal Society, *Geoengineering the Climate*, 53.

[42]Science and Technology Committee, *Fifth Report: The Regulation of Geoengineering*, House of Commons, 2010, https://v.gd/y0jBsP .

[43]Royal Society, *Solar Radiation Management: The Governance of Research* (London: Royal Society, 2011); "Interview – Richard Branson," *Alliance*, June 1, 2012, https://v.gd/tuSSYg.

report, published in 2011, pointed to the recent decision of the UN Convention on Biological Diversity, which strengthened its earlier stance to now ensure "no climate-related geo-engineering activities that may affect biodiversity take place," aside from small-scale scientific research studies in controlled settings.[44] It concluded that research into this form of geoengineering would likely continue apace—even "generate its own momentum"—and its effective governance demanded close consideration and international conversations both in and outside the UN.[45] Six months later at the Rio+20 summit in mid-2012, the delegates agreed to a final communique that called on "relevant intergovernmental bodies ... to continue addressing with utmost caution ocean fertilization."[46] The threat of climate change did not justify the risks that unilateral action entailed, and intergovernmental regulation was necessary to protect the earth system from the interventions of geoengineering.

State Security: Protecting Territorial Borders

The growing share of greenhouse gas emissions among the developing nations was an issue that UK prime minister Tony Blair had sought to address as host of the G8 in early July 2005. Owing to the contrasting positions of the G8 members, climate change had been off the summit agenda since the 1997 Denver meeting. Hoping to broker talks with the developing world, the prime minister had invited Chinese president Hu as well as the leaders of Brazil, India, Mexico, and South Africa to the G8 summit in Gleneagles, Scotland. Climate change had become a foreign policy priority for the Blair government, which increasingly framed the issue as a threat to global security in order to reinforce the need for industrialized countries to meet their Kyoto commitments.[47] The Labour government's chief science advisor, for instance, had in 2004 declared in *Science* that "climate change is the most severe problem that we are facing today—more serious even than the threat of terrorism."[48] Although bombings in London overshadowed the talks at Gleneagles, the leaders gathered there

---

[44]Royal Society, *Solar Radiation Management*, 25. Kyle Powys Whyte observes that no Indigenous peoples were consulted for the Royal Society's 2011 report, Kyle Whyte, "Indigenous Peoples, Solar Radiation Management, and Consent," in Christopher J. Preston (ed.), *Engineering the Climate: The Ethics of Solar Radiation Management* (Lanham: Lexington Books, 2012), 65–76.
[45]Royal Society, *Solar Radiation Management*, 10.
[46]United Nations, *The Future We Want: Outcome Document of the UN Conference on Sustainable Development* (New York: United Nations, 2012), 43.
[47]Prime Minister's Strategy Unit, *Investing in Prevention: An International Strategy to Manage Risks of Instability and Improve Crisis Response* (London: Strategy Unit, 2005).
[48]David King, "Climate Change Science: Adapt, Mitigate, or Ignore," *Science* 303 (2004): 176–7. King's association of climate change with terrorism encouraged the UK media to report on climate change in such terms; see Hulme, *Why We Disagree about Climate Change*, 422.

agreed to what Prime Minister Blair called "a pathway to a new dialogue when Kyoto expires in 2012."[49]

In line with the Blair government's approach at Gleneagles, the UK used its presidency of the UN Security Council in April 2007 to amplify its association of climate change with security concerns. In contrast to other organs of the UN, the Security Council has the unparalleled power to make decisions binding on its members. Introducing the issue to the Security Council, UK foreign secretary Margaret Beckett argued that climate change constituted a key issue in "our collective security in a fragile and increasingly interdependent world."[50] This day-long debate was the first time that climate change had been explicitly discussed in the council, where some small island states and African nations shared their own concerns about human dislocation due to drought and rising sea levels.[51] Echoing President Maumoon Abdul Gayoom's 1987 address to the UN General Assembly (Chapter 4), the foreign minister of the Maldives, Abdulla Shahid, observed, "A mean sea-level rise of 2 meters would suffice virtually to submerge the entire country ... that would be the death of a nation," reminding delegates of the enormous human toll of the recent Indian Ocean tsunami (Figure 7.1).[52]

In Germany, where "Klimakatastrophe" was the 2007 word of the year, the Merkel government pursued a similar strategy through both the Group of Seven (G7) and the EU during its presidency.[53] Already, the government had commissioned a report on the security implications of climate change from its German Advisory Council on Global Change, which warned of "ever-deeper lines of division and conflict in international relations."[54] Following Germany's term, the EU presented a report on *Climate Change and International Security*, which described climate change as a "threat multiplier which exacerbates existing trends, tensions and instability." Citing the risks of mass migration and conflict "at the European Union's borders," the report called on governments

---

[49]Tony Blair, "British Prime Minister Tony Blair Reflects on 'Significant Progress' of G8 Summit," July 8, 2005, Gleneagles Official Documents, G7 Research Group, University of Toronto, https://v. gd/xrX7Kd.

[50]Cited in Uttam Kumar Sinha, "Climate Change and Foreign Policy: The UK Case," *Strategic Analysis* 34, no. 3 (2010): 404.

[51]Antto Vihma, Yacob Mulugetta, and Sylvia Karlsson-Vinkhuyzen, "Negotiating Solidarity? The G77 through the Prism of Climate Change Negotiations," *Global Change, Peace and Security* 23, no. 3 (2011): 315–34; Oli Brown, Anne Hammill, and Robert McLeman, "Climate Change as the 'New' Security Threat," *International Affairs* 83, no. 6 (2007): 1141–54.

[52]Cited in "UN Security Council, 5663rd Meeting, Tuesday 17 April 2007," New York, S/PV.5663, https://v.gd/G3BQRJ.

[53]Matthias Dörries, "Climate Catastrophes and Fear," *WIREs Climate Change* 1 (2010): 885; Thomas Fues and Julia Leininger, "Germany and the Heiligendamm Process," in Andrew F. Cooper and Agata Antkiewicz (eds.), *Emerging Powers in Global Governance: Lessons from the Heiligendamm Process* (Waterloo: Wilfrid Laurier University Press, 2008), 235–63.

[54]Renate Schubert et al., *Climate Change as a Security Risk* (London: Earthscan, 2007).

FIGURE 7.1 *Abdulla Shahid, minister of state for foreign affairs of the Maldives, addresses the UN Security Council, April 17, 2007, New York. UN Photo by Mark Garten.*

to foster a post-Kyoto treaty that would be internationally binding.[55] Nobel Laureate and former archbishop of Cape Town Desmond Tutu similarly warned in 2007 against complacency in the North: "No country—however rich or powerful—will be immune to the consequences [of climate change]. In the long-run, the problems of the poor will arrive at the doorstep of the wealthy, as the climate crisis gives way to despair, anger and collective security threats."[56]

Although these deployments of security aimed to mobilize the more reluctant industrialized states into action, many developing countries rejected this framing of climate change. In response to the UK's efforts, India and other developing countries expressed strong objections to discussing climate change in the UN Security Council, as did Russia and the United States, where members of its own military establishment had recently voiced concerns

[55]Oels, "From 'Securitization'," 189; "Climate Change and International Security: Paper from the High Representative and the European Commission to the European Council," European Council, March 14, 2008, S113/08, https://v.gd/0IAVTx.

[56]Desmond Tutu, "We Do Not Need Climate Change Apartheid in Adaptation," in *UN Human Development Report 2007/2008* (New York: UN Development Programme, 2007), 166.

about the threat of climate change to national security.[57] The Group of 77 (G77) argued that climate change was an issue of sustainable development only for consideration under the Framework Convention or the UN General Assembly, where developing countries have a majority voice. In elevating climate change into the Security Council, Egypt argued that the UK was attempting to promote the notion of shared responsibility, contrary to the notion of "common but differentiated responsibilities" outlined in the Framework Convention.[58]

Among the areas for concern identified in the EU's 2007 report was the rapid changes underway in the Arctic and their implications for "European security interests."[59] Since the end of the Cold War, science and environmental protection had become means to "desecuritize" the Arctic, as the International Geophysical Year had proved for the Antarctic (Chapter 1). A new International Polar Year loomed, which would involve studies of Arctic sea ice, satellite programs, and engagement with the knowledge of Arctic Indigenous peoples.[60] Its coincidence with a record sea ice minimum, worse than what climate models had projected, brought new attention to the region, as researcher Henry Huntington observed, mainly from "people outside the Arctic, concerned about conservation of bears, looking for new business opportunities, or worrying about the fate of the planet."[61]

For the Arctic's Indigenous peoples, the record ice conditions were no surprise. Already in 2005 the Arctic Climate Impact Assessment had reported, "For Inuit, warming is likely to disrupt or even destroy their hunting and food sharing culture as reduced sea ice causes the animals on which they depend to decline, become less accessible, and possibly become extinct."[62] Soon afterward, the chair of the Inuit Circumpolar Council, Sheila Watt-Cloutier, and over sixty other Inuit people from the United States and Canada submitted a petition to the Inter-American Commission on Human Rights. Demanding "the right to be cold," they requested "relief from human rights violations resulting from the impacts of global warming and climate change caused by acts and omissions of the United States."[63] Although the commission did

---

[57]CNA Corporation, *National Security and the Threat of Climate Change* (Alexandria: CNA Corporation, 2007).

[58]Vihma, Mulugetta, and Karlsson-Vinkhuyzen, "Negotiating Solidarity?," 315–34.

[59]"Climate Change and International Security."

[60]Nina Wormbs et al., "Bellwether, Exceptionalism and Other Tropes: Political Coproduction of Arctic Climate Modeling," in Matthias Heymann, Gabriele Gramelsberger, and Martin Mahony (eds.), *Cultures of Prediction in Atmospheric and Climate Science* (London: Routledge, 2017), 159–77.

[61]Henry P. Huntington, "A Question of Scale: Local versus Pan-Arctic Impacts from Sea-Ice Change," in Miyase Christensen, Annika E. Nilsson, and Nina Wormbs (eds.), *Media and the Politics of Arctic Climate Change* (London: Palgrave Macmillan, 2013), 114–27.

[62]Arctic Climate Impact Assessment, *Arctic Climate Impact Assessment* (Cambridge: Cambridge University Press, 2005), 1000.

[63]Cited in Paul Crowley, "Interpreting 'Dangerous' in the UNFCCC and the Human Rights of Inuit," *Regional Environmental Change* 11 (2010): 265–74.

FIGURE 7.2 *UN secretary-general Ban Ki-moon (left) presents the 2007 UNDP Mahbub ul Haq Award for Excellence in Human Development to Sheila Watt-Cloutier (right), UN Headquarters, New York, June 20, 2007. UN Photo by Mark Garten.*

not accept to hear the petition, it invited Watt-Cloutier to attend a hearing to provide additional information that would strengthen the association of climate change with human rights (Figure 7.2).

As Marybeth Long Martello observes, "The authority, credibility, and visibility of the [Inuit Circumpolar Council] in global environmental institutions derives largely from an at-risk expert status attained by way of science and scientific recognition, validation and reliance on Indigenous knowledge". "With this status," Martello continues, "the [Council] is attempting to redefine the terms by which scientists, policymakers, and others imagine, debate and manage the global environment."[64] Prior to submitting the petition, Watt-Cloutier had attended the Milan climate conference in 2003 and shared the concerns of many Inuit people. There, she framed the significance

---

[64]Marybeth Long Martello, "Arctic Indigenous Peoples and Climate Change as Representations and Representatives of Climate Change," *Social Studies of Science* 38, no. 3 (2008): 370–1.

of their cause in global terms: "The Arctic is the barometer of the globe's environmental health. You can take the pulse of the world in the Arctic. Inuit, the people who live further north than anyone else, are the canary in the global coal mine."[65] As chair of the Inuit Circumpolar Council, she had been tasked to "bring Arctic/Inuit perspectives" on climate change to the attention of the parties to the climate convention, "with the aim of positioning Inuit to influence international discussions and decisions."[66] In the climate regime, however, the experiences or knowledge of Indigenous Peoples were yet to be formally acknowledged. Prior to the 2005 conference in Montreal, there had been no mention of Indigenous Peoples in any decision texts produced by the UNFCCC.[67]

Taking a lead from the Inuit petition, representatives from twenty-six of the small island states met in the Maldives capital of Malé in November 2007 to negotiate a declaration identifying climate change as a threat to their human rights. A month later, the UN deputy high commissioner for human rights, Kyung-wha Kang, addressed the Bali climate conference. Citing the Malé Declaration, she noted, "Emerging evidence suggests that the livelihoods and cultural identities of [I]ndigenous peoples of North America, Europe, Latin America, Africa, Asia, and the Pacific are already being threatened by the impact of climate change."[68] Over the following eighteen months, the UN Human Rights Council passed a resolution recognizing the "threat" climate change poses to "people and communities around the world" and determined that states have obligations under human rights law to protect those rights from the impacts of climate change. The council further affirmed the need to ensure the "full, effective and sustained implementation" of the Framework Convention on Climate Change.[69]

Undeterred by the resistance the UK's efforts had met in the Security Council, small island nations revived the issue in the UN General Assembly in June 2009 as the Copenhagen meeting approached. Acting on behalf of the Alliance of Small Island Nations (AOSIS), Nauru's ambassador Marlene Moses introduced the first resolution to be led by a Pacific island nation, "Climate change and its possible security implications." Moses deployed

[65]Cited in Heather A. Smith, "Disrupting the Global Discourse of Climate Change: The Case of Indigenous Voices," in Mary E. Pettenger (ed.), *The Social Construction of Climate Change: Power, Knowledge, Norms, Discourses* (London: Routledge, 2007), 210.

[66]Cited in Noor Johnson and David Rojas, "Contrasting Values of Forests and Ice in the Making of a Global Climate Agreement," in Ronald Niezen and Maria Sapignoli (eds.), *Palaces of Hope: The Anthropology of Global Organizations* (Cambridge: Cambridge University Press, 2017), 228.

[67]Meghan Shea and Thomas Thornton, "Tracing Country Commitment to Indigenous Peoples in the UN Framework Convention on Climate Change," *Global Environmental Change* 58 (2019): 101973.

[68]Kyung-wha Kang, "Climate Change and Human Rights," Office of the High Commissioner for Human Rights, December 14, 2007, https://v.gd/RECVDX .

[69]Crowley, "Interpreting 'Dangerous' in the UNFCCC and the Human Rights of Inuit," 265–74.

the language of international law to frame the issue: "Clearly the survival of States, their sovereignty and territorial integrity, and the impact on their neighbors are matters of international peace and security." The resolution enjoyed wide support in the General Assembly, with the representative for the EU describing climate change as a "universal threat that will create new security dynamics and risks in all regions."[70] In deference to Beijing's concerns that the resolution would implicitly undermine the notion of "differentiated responsibilities," the adopted text duly noted the responsibility of the Security Council for such matters of international peace and security.[71] On that basis, however, Germany returned the issue to the agenda of the Security Council when it assumed the presidency in July 2011, observing that "the effects of climate change go beyond the mandate of the United Nations Framework Convention."[72]

Washington Agenda: Beyond the Framework Convention

In 2007, the Norwegian Nobel Committee awarded the Nobel Peace Prize to the IPCC and former US vice president Al Gore Jr. for their joint efforts to foster greater knowledge about anthropogenic climate change. In his Nobel Lecture on behalf of the panel, its chair Rajendra K. Pachauri observed, "Peace can be defined as security and the secure access to resources that are essential to living … How climate change will affect peace is for others to determine, but we have provided scientific assessment of what could become the basis for conflict." He concluded with a message for those gathered in Bali for the latest climate conference, "The world's attention is riveted on that meeting and hopes are alive that unlike the sterile outcome of previous sessions in recent years, this one will provide some positive results."[73]

Pachauri had recently presided over the IPCC's recent Fourth Assessment Report, his first since replacing chemist Robert Watson as chair in 2002.

---

[70]Cited in *General Assembly Official Records, 63rd Session: 85th Plenary Meeting, Wednesday 3 June 2009, New York*, United Nations, A/63/PV.85, https://v.gd/xohQmX.

[71]*63/281 Climate Change and Its Possible Security Implications: A Resolution*, Resolution adopted by the General Assembly on June 3, 2009, United Nations A/RES/63/281, https://v.gd/AobbBI.

[72]UN Security Council, "Security Council, in Statement, Says 'Contextual Information' on Possible Security Implications of Climate Change Important when Climate Impacts Drive Conflict," *United Nations*, July 20, 2011, https://v.gd/SWQEgl; Stuart Beck and Elizabeth Burleson, "Inside the System, Outside the Box: Palau's Pursuit of Climate Justice and Security at the United Nations," *Transnational Environmental Law* 3, no. 1 (2014): 17–29.

[73]Rajendra K. Pachauri, "Intergovernmental Panel on Climate Change: Nobel Lecture," *The Nobel Prize*, December 10, 2007, https://v.gd/I9103R.

According to some of his colleagues, the former chief scientist of NASA's Mission to Planet Earth (Chapter 4), who had served as chair since 1997 and headed the World Bank's environment department, was no friend of the Bush administration, which chose not to support his reappointment on account of his "taking too much of an 'activist role' in exploring the causes of global climate change."[74] Their suspicions were not unfounded: already, on the Bush administration's request, the National Academy of Sciences had undertaken a month-long review of the panel's work in the lead up to the Bonn meeting.[75] Then, in April 2002, the Natural Resources Defense Council had obtained a memo from an ExxonMobil lobbyist addressed to the White House Council on Environmental Quality. Written just a month after Bush's election, the memo asked, "Can Watson be replaced now at the request of the U.S.?" "Hand picked by Al Gore," as the memo alleged, Watson's days in the role looked to be numbered under an administration with especially close connections to the fossil fuel industry. The memo's author recommended that "none of the Clinton/Gore proponents [should be] involved in any decisional activities" of panel meetings in the future.[76]

Backing IPCC vice-chair Pachauri, the head of the New Delhi-based Tata Energy Research Institute, might have appeared an unusual decision for the United States. He too had been critical of the Bush administration. Perhaps this criticism, scientists such as Stephen Schneider wondered, had worked in his favor—an IPCC chair that supported the position of developing countries to limit emissions reductions to the developed world would enable the United States to maintain its rejection of the Kyoto Protocol.[77] At the very least, journalists speculated, Pachauri might "embroil the IPCC in a paralyzing spat between Western and non-Western nations" and thus derail further international action on climate change.[78] After delegates cast their vote in Geneva in April 2002, Pachauri won 76 to 49. In addition to the United States, he had also attracted support from Asian and African nations, which had long sought better representation of the developing world in the panel.[79]

Having declared the Kyoto Protocol "an unfair and ineffective means of addressing global climate change concerns," the Bush administration

[74]Kevin Trenberth, cited in Don C. Smith, "Science Considers Global Climate Change," *Refocus* 3, no. 6 (2002): 51.

[75]Mark Schrope, "Consensus Science, or Consensus Politics?," *Nature* 412 (2001): 112–14.

[76]"Exxon Lobbyist's Memo to the White House (2001)," *Inside Climate News*, October 22, 2015, https://v.gd/yv445t.

[77]Debora Mackenzie, "Too Hot for Head of Climate Panel," *New Scientist*, April 20, 2002, https://v.gd/i0rc5H.

[78]Timothy Noah, "Dix Exxon Mobil Get Bush to Oust the Global Warming Chief?," *Slate*, April 22, 2002, https://v.gd/kxvL8q.

[79]Andrew Lawler, "Pachauri Defends Watson in New Chapter for Global Panel," *Science* 296 (2002): 632.

embarked on its own program of climate diplomacy.[80] Just months after the protocol entered into force, the United States joined with Australia—the only other Annex I country yet to sign the protocol—to create the Asia-Pacific Partnership for Clean Development and Climate. This nonbinding agreement also involved China, India, Japan, and South Korea, covering almost half the world's population and almost half of global greenhouse gas emissions. The focus of the partnership was on the development of clean energy and energy-efficient technologies, which Australia's foreign minister described as a "complementing" the Kyoto Protocol.[81] After the 2006 Canadian federal election, the Harper government also joined what critics had come to see as "the anti-Kyoto club," intended to weaken the UN process.[82] Seeking a larger circle of allies, the Bush administration formed with sixteen nations in 2007 the Major Economies Meeting which sought to develop rules for a more flexible approach to reducing emissions.[83] This grouping effectively became the more constructive Major Economies Forum on Energy and Climate under the Obama administration.

Carbon Economy: The European Union Emissions Trading Scheme

The discussions underway in the UN Security Council and the UN General Assembly echoed the findings of the UK's Stern Review, published in 2006. Commissioned by the chancellor of the exchequer as part of the G8 Gleneagles Dialogue and supervised by Sir Nicholas Stern, former chief economist of the World Bank, the report concluded that the costs of preventing dangerous climate change would be greatly outweighed by the costs of unmitigated climate change.[84] To stabilize the climate at about 500 ppm, the costs of mitigation to economic growth, Stern's report estimated, would be about

---

[80]George W. Bush, "Text of a Letter from the President to Senators Hagel, Helms, Craig, and Roberts," *The White House*, March 13, 2001, https://v.gd/m0rfe2.

[81]Cited in Jeffrey Mcgee and Ros Taplin, "The Asia-Pacific Partnership on Clean Development and Climate: A Complement or Competitor to the Kyoto Protocol?," *Global Change, Peace & Security* 18 (2006): 173–92.

[82]Allan Woods, "Canada in 'Anti-Kyoto' club," *Toronto Star*, September 25, 2007, https://v.gd/ViF xhm; Steinar Andresen, "Exclusive Approaches to Climate Governance: More Effective Than the UNFCCC?," in Todd L. Cherry, Jon Hovi and David M. McEvoy (eds.), *Toward a New Climate Agreement: Conflict, Resolution and Governance* (London: Routledge, 2014), 155–66.

[83]Robert O. Keohane and David G. Victor, *The Regime Complex for Climate Change*, Discussion Paper 10-33 (Cambridge, MA: Harvard Project on International Climate Agreements, 2010).

[84]Nicholas Stern, *The Economics of Climate Change: The Stern Review* (Cambridge: Cambridge University Press, 2007).

1 percent—a figure that gained wide currency in European policy circles.[85] By contrast, the likely damage from climate change, Stern suggested, would be "larger than the two world wars of the last century. The problem is global and the response must be a collaboration on a global scale." An important means of such collaboration, he argued, should be emissions trading "because it can provide the international incentives for participation, and promote efficiency and equity, while controlling quantities of emissions."[86]

By the time Stern's review was published, the EU's Emissions Trading Scheme had been operational for over a year. Having resisted the idea of emissions trading during the negotiations of the Kyoto Protocol (Chapter 6), Brussels had now crafted its own scheme to help the EU reach its reduction target of 8 percent compared to 1990 levels.[87] The architects of this scheme drew inspiration from the UK, where the industry-led Emissions Trading Group had been established at BP headquarters in 1999 to counter the prospect of tax-based carbon legislation. Leading the way there were BP and Shell, which had launched their own internal emissions trading schemes after the Kyoto meeting as part of their efforts to rally business and government support for emissions trading in Europe. With a carbon tax ruled out, thanks to the strong resistance from industry in the early 1990s (Chapter 6), emissions trading had become the most politically expedient means for the EU to sustain the legitimacy of the Kyoto framework in the absence of the United States.[88]

A key feature of the EU's Emissions Trading Scheme was the ability to trade credits internationally, which enables European producers to buy cheaper credits in the South via the Clean Development Mechanism (CDM). The market rationale behind this strategy was that these credits would ensure that carbon prices remained lower, allowing polluting companies time to alter their operations and encouraging sustainable development in the developing world.[89] The implementation of cleaner technologies in developing countries

---

[85]Dieter Helm, "Climate-Change Policy: Why Has so Little Been Achieved?," in Dieter Helm and Cameron Hepburn (eds.), *The Economics and Politics of Climate Change* (Oxford: Oxford University Press, 2010), 9–35.

[86]Cited in Alison Benjamin, "Stern: Climate Change a 'Market failure'," *Guardian*, November 29, 2007, https://v.gd/Jvacip.

[87]Karin Backstränd and Eva Lövbrand, "Climate Governance Beyond 2012: Competing Discourses of Green Governmentality, Ecological Modernization and Civic Environmentalism," in Mary E. Pettenger (ed.), *The Social Construction of Climate Change: Power, Knowledge, Norms, Discourses* (Aldershot: Ashgate, 2007), 123–48.

[88]Jonas Meckling, *Carbon Coalitions: Business, Climate Politics, and the Rise of Emissions Trading* (Cambridge, MA: MIT Press, 2011), 103–17; Jon Birger Skjærseth and Jørgen Wettestad, "Making the EU Emissions Trading System: The European Commission as an Entrepreneurial Epistemic Leader," *Global Environmental Change* 20 (2010): 314–21.

[89]Samuel Randalls, "Assembling Climate Expertise: Carbon Markets, Neoliberalism and Science," in Vaughan Higgins and Wendy Larner (eds.), *Assembling Neoliberalism: Expertise, Practices, Subjects* (New York: Palgrave Macmillan, 2017), 67–85.

would then diminish overall carbon emissions, its architects hoped. As former Framework Convention executive secretary Michael Zammit Cutajar put it, "The essence of the Kyoto Protocol—its genius—is that it encourages recourse to the market to achieve environmental objectives at the least economic cost."[90] By 2008, the World Bank estimated that the scheme was responsible for about 90 percent of the overall demand for Certified Emissions Reduction credits in the South, particularly in China and India.[91]

Allowing European polluters to offset their emissions, however, meant allowing them to continue to operate under the terms of "business as usual," critics argued. Projects hosted by the South, meanwhile, were responsible for delivering emissions reductions, but even the World Bank reported in 2010 that the "treatment of sustainable development in (CDM) documents is sketchy and uneven."[92] Renewable energy projects accounted for less than a third of the emissions reductions from CDM projects, which mostly funded schemes to reduce hydrochlorfluorocarbons.[93] For the growing climate justice movement, emissions trading represented a false solution to climate change, and the CDM represented "sustainable development window-dressing."[94] This critique had undergone consolidation at the Durban Climate Justice Summit, hosted by the local organization Carbon Trade Watch in October 2004, where representatives from around the world produced the "Durban Declaration on Carbon Trading."[95] Its signatories argued, "Through [the] process of creating a new commodity—carbon—the Earth's ability and capacity to support a climate conducive to life and human societies is now passing into the same corporate hands that are destroying the climate."[96]

This intervention reflected a wider fragmentation among the environmental organizations that had been engaged in the climate regime since the early 1990s. Assembled under the umbrella network of the Climate Action Network, the rapidly expanding number of organizations introduced a growing array of perspectives and priorities to the coalition. Those groups oriented against globalization and liberal environmentalism became increasingly disillusioned with the more established environmental organizations, such

---

[90]Cutajar, "Reflections on the Kyoto Protocol," 61.
[91]Cited in Peter Newell and Matthew Paterson, *Climate Capitalism: Global Warming and the Transformation of the Global Economy* (Cambridge: Cambridge University Press, 2010), 104.
[92]World Bank, *Development and Climate Change: World Development Report 2010* (Washington, DC: World Bank: 2010), 265; Kathleen McAfee, "The Contradictory Logic of Global Ecosystem Services Markets," *Development and Change* 43, no. 1 (2012): 105–31.
[93]Meckling, *Carbon Coalitions*, 198.
[94]"The Durban Declaration on Carbon Trading," Durban Group for Climate Justice, October 10, 2004, https://v.gd/RyR77m.
[95]James Goodman, "From Global Justice to Climate Justice? Justice Ecologism in an Era of Global Warming," *New Political Science* 31, no. 4 (2009): 499–514.
[96]"The Durban Declaration on Carbon Trading."

as the Environmental Defense Fund, that had supported carbon trading as a politically expedient approach to achieving emissions reductions.

The anti-globalization movement more broadly, meanwhile, became engaged in the issue of climate change in response to the G8 agenda that the UK government promoted in 2005, which Germany then reprised in 2007.[97] At the 2007 climate conference in Bali, climate justice groups finally parted ways with the Climate Action Network, forming Climate Justice Now! and Climate Justice Action. Aligning with a wider network of environmental justice organizations, including Oilwatch and the Indigenous Environmental Network, these groups were focused on the intersections of climate change with issues of social, ecological, and gender justice, and the protection of livelihoods and the environment.[98]

Adaptation Ascendant: Preparing for Climate Change

Touted as the "Africa COP"—the first to be held in southern Africa, the challenges facing developing countries were at the forefront of the discussions in Nairobi in 2006.[99] There, UN secretary-general Kofi Annan reprised the association of climate change with security concerns, declaring climate change as "not just as environmental issue, but [as] an 'all-encompassing threat'," citing "growing threats" to human health, global food supply, low-lying settlements, and "to peace and security."[100] Although many commentators concluded that the conference had been largely ineffectual, delegations from the least developed countries celebrated its success.[101] The parties had agreed that the Marrakech Adaptation Fund would be filled by a 2 percent levy on the carbon emissions

---

[97]Philip Bedall and Christoph Görg, "Antagonistic Standpoints: The Climate Justice Coalition Viewed in Light of a Theory of Societal Relationships with Nature," in Matthias Dietz and Heiko Garrelts (eds.), *Routledge Handbook of the Climate Change Movement* (London: Routledge, 2013), 44–65.

[98]David Schlosberg and Lisette B. Collins, "From Environmental to Climate Justice: Climate Change and the Discourse of Environmental Justice," *WIREs Climate Change* 5 (2014): 359–74; Jennifer Hadden, *Networks in Contention: The Divided Politics of Climate Change* (New York: Cambridge University Press, 2015), 114–41; Fernando Tormos-Aponte and Gustavo A. García-López, "Polycentric Struggles: The Experience of the Global Climate Justice Movement," *Environmental Policy and Governance* 28 (2018): 284–94.

[99]Chukwumerije Okereke et al., *Assessment of Key Negotiating Issues at Nairobi Climate COP/MOP and What It Means for the Future of the Climate Regime*, Working Paper 106 (Norwich: Tyndall Centre for Climate Change Research, 2007). .

[100]Cited in "Climate Change Sceptics 'Out of Step, Out of Arguments and Out of Time', Annan Tells UN Meeting," *UN News*, November 15, 2006, https://v.gd/Q72jRB.

[101]Benito Müller, "The Nairobi Climate Change Conference: A Breakthrough for Adaptation Funding," Oxford Energy & Environment Comment, January 2007, https://v.gd/84j5ma.

permits in the CDM. The likes of Brazil, China, and India, which hosted the lion's share of these Mechanism projects, saw this levy as an "innovative solidarity fund" from the larger developing countries to the least developed countries. The view from some parts of Africa was different, however: Desmond Tutu argued in 2007, "Adaptation is becoming a euphemism for social injustice on a global scale. ... Leaving the world's poor to sink or swim with their own meagre resources in the face of the threat posed by climate change is morally wrong. ... We are drifting into a world of 'adaptation apartheid'." [102]

Adaptation had been among the recommendations of the Stern Review. Stern argued, "Adaptation to climate change—that is, taking steps to build resilience and minimise costs—is essential. ... Adaptation efforts, particularly in developing countries, should be accelerated." Yet in Nairobi, just six of the forty-four proposals for a post-Kyoto regime dealt with adaptation as a policy issue. [103] Adaptation had long been the poor relation to mitigation efforts in climate diplomacy: the Framework Convention has no article solely on the concept, and it is only mentioned five times in the negotiation text. In terms of the convention's "ultimate objective" of Article 2, adaptation is framed rather vaguely—the stabilization level "should be achieved within a time frame sufficient to allow ecosystems to adapt naturally to climate change, to ensure that food production is not threatened and to enable economic development to proceed in a sustainable manner." [104] Likewise, the early reports of the IPCC paid little attention to questions of adaptation, vulnerability, or equity. Its first report had focused exclusively on mitigation, while a review of its Second Assessment Report had found, "Of the 728 pages of substantive text, about two thirds are devoted to impacts, one third to mitigation, and only 32 pages to Adaptation." [105]

With the focus of climate diplomacy on negotiating the Kyoto Protocol and the commitments of industrialized nations to curb their emissions, adaptation had been widely regarded as a distraction at best, or defeatist at worst. Al Gore Jr., for instance, in his 1992 *Earth in the Balance* had asked, "Do we have so much faith in our own adaptability that we will risk destroying the integrity of the entire global ecological system? Believing that we can adapt to just about anything is ultimately a kind of laziness, an arrogant faith in our ability to react in time to save our skin." [106] Accordingly for the industrialized nations,

---

[102]Tutu, "We Do Not Need Climate Change Apartheid in Adaptation," 166.

[103]David Ciplet, J. Timmons Roberts, and Mizan Khan, *Power in a Warming World: The New Global Politics of Climate Change and the Remaking of Environmental Inequality* (Cambridge, MA: MIT Press, 2015), 103.

[104]Cited in E. Lisa F. Schipper, "Conceptual History of Adaptation in the UNFCCC Process," *Review of European Community & International Environmental Law* 15, no. 1 (2006): 89.

[105]Robert W. Kates, "Review of 'Climate Change 1995: Impacts, Adaptations, and Mitigation'," *Environment* 39, no. 9 (1997): 31.

[106]Al Gore Jr., *Earth in the Balance: Ecology and the Human Spirit* (New York: Penguin, 1992), 240.

raising adaptation in the context of the difficult negotiations of the 1990s might have implied a soft approach to emissions reduction, or perhaps acceptance of responsibility for contributing to climate change, thus associating the concept with matters of liability and compensation.[107] The convention's narrow focus on anthropogenic climate change alone, furthermore, discouraged some industrialized nations from considering funding in developing countries adaptation projects that might be more associated with climate variability. The local benefits of adaptation, relative to the global (or Northern) benefits of mitigation, had likewise discouraged the developed world's interest during the 1990s (Chapter 6).

Adaptation's cause had been further hindered by the oil-producing nations. During the negotiations of the Kyoto Protocol in the 1990s, they mounted a case for compensation on the grounds of their interpretation of the convention's provision of the "impact of the implementation of response measures" on those economies "highly dependent" on fossil fuel production. They claimed compensation to assist what they deemed to be adaptation.[108] This interpretation, which had little support within the EU or among other developing countries, also pushed adaptation to the sidelines of the negotiations.[109] Only AOSIS, for whom adaptation was especially urgent, had tabled the issue during the convention negotiations in the early 1990s. Their vulnerability to sea level rise had been granted recognition in Agenda 21 at the 1992 Earth Summit, which noted that they were "facing the increasing threat of the loss of their entire national territories." AOSIS advocated the need to adopt strategies to minimize damage and cope with climate changes and rising sea levels, and called for an insurance fund as a means to support adaptation measures.[110]

In the wake of the IPCC's Third Assessment Report, however, adaptation entered the negotiations at the Marrakech climate conference in late 2001. The report defined adaptation as "adjustment in natural or human systems in response to actual or expected climatic stimuli or their effects, which moderates harm or exploits beneficial opportunities."[111] That report had also declared that there was now stronger evidence that most of the warming observed over the past fifty years was attributable to human activities. This conclusion, together with the Bush administration's withdrawal from the Kyoto Protocol, encouraged the least developed countries to pursue adaptation support since

[107]Schipper, "Conceptual History of Adaptation," 84–5.
[108]Ciplet, Roberts, and Khan, *Power in a Warming World*, 58.
[109]Ciplet, Roberts, and Khan, *Power in a Warming World*, 103–5.
[110]Schipper, "Conceptual History of Adaptation," 88.
[111]"Annex B: Glossary of Terms," in James J. McCarthy et al. (eds.), *Climate Change 2001: Impacts, Adaptation, and Vulnerability* (Cambridge: Cambridge University Press, 2001), 982.

it appeared increasingly uncertain that the industrialized countries would meet their emissions reductions commitments.

During the New Delhi conference in 2002, the specter of developing country commitments returned when the EU attempted to raise the issue of the Kyoto Protocol's future after its first commitment period ended in 2012. Deepening the divide in the Green Alliance that had persisted since the first climate conference in Berlin, the United States sided with the largest of the developing countries, with its delegation head Harlan Watson noting, "We must also recognize that it would be unfair—indeed, counterproductive—to condemn the developing nations to slow growth or no growth by insisting that they take on impractical and unrealistic greenhouse gas targets."[112] For both the United States and much of the G77, adaptation helpfully shifted the focus of the negotiations away from the difficult subject of mitigation commitments. The Delhi Declaration affirmed the position of the developing countries and emphasized the principle of common but differentiated responsibilities: "Parties have a right to, and should, promote sustainable development. Policies and measures to protect the climate system against human-induced change should be appropriate for the specific conditions of each Party and should be integrated with national development programmes, taking into account that economic development is essential for adopting measures to address climate change."[113]

With adaptation now closely tied to the development aspirations of the South, it assumed equal billing with mitigation efforts in the lead up to the Nairobi climate meeting in 2006. Within the Framework Convention secretariat's Subsidiary Body for Scientific and Technological Advice, work got underway on the scientific, technical, and socioeconomic aspects of vulnerability and adaptation, while developing clear policies for implementation measures and technology transfer (the Nairobi Work Programme).[114] From the perspective of AOSIS and the least developed countries, however, "a strategic response to adaptation is clearly lacking."[115] On their behalf, Tuvalu submitted at the 2007 conference in Bali an "International blueprint for adaptation," which set the agenda on adaptation politics for the next five years.[116] Their case for funding to support their adaptation to climate change marked a renewal of

[112]Cited in Stavros Afionis, *The European Union in International Climate Change Negotiations* (London: Routledge, 2017), 100.

[113]"The Delhi Ministerial Declaration on Climate Change and Sustainable Development," UNFCCC, October 28, 2002, https://v.gd/MnmeMQ.

[114]Mizan R. Khan and J. Timmons Roberts, "Adaptation and International Climate Policy," *WIREs Climate Change* 4, no. 3 (2013): 171–89.

[115]Tuvalu, "International Blueprint on Adaptation," United Nations Framework Convention on Climate Change, Conference of the Parties, Thirteenth Session, December 8, 2007, https://v.gd/JBWZRY.

[116]Ciplet, Roberts, and Khan, *Power in a Warming World*, 107.

the South's efforts to ensure that the climate regime upheld the principle of additionality they had established at Stockholm in 1972 (Chapter 2).

Forest Protection: Climate Change, Deforestation, and Indigenous Peoples

The UK's Stern Review had envisioned a role for carbon markets in providing incentives to curb deforestation. The subsequent Eliasch Review concluded that "as long as forest carbon or other ecosystem services are not reflected in the price of commodities produced from converted forest land, forests will—in financial terms—generally be worth more to landowners cut rather than standing."[117] Having earlier resisted questions of avoided deforestation on sovereignty grounds (Chapter 5), Brazil proposed at the 2006 Nairobi climate conference a forest fund that would compensate countries for slowing the destruction of their forests.[118] The Lula government meanwhile amended the national forest law to clarify ownership of carbon rights on public concessions, following a raft of similar changes in Indonesia.[119]

Although developed countries were permitted under the Kyoto Protocol to include forest activities in their own accounting, concerns that the forests of the South might become a loophole to avoid emissions reductions for the North had meant forests had been set aside for the first Kyoto commitment period. Once the Kyoto Protocol came into force, however, the forest question came into view. Already, Brazilian scientists had estimated that "annual rates of tropical deforestation from Brazil and Indonesia alone would equal four-fifths of the emissions reductions gained by implementing the Kyoto Protocol in its first commitment period."[120] Then at the Montreal climate conference in 2005, after a preliminary proposal presented by Amazonian scientists and non-governmental organizations, Papua New Guinea and Costa Rica headed a Coalition of Rainforest Nations in a proposal for "Reducing Emissions from Deforestation in Developing Countries." Stressing that their interest lay in

---

[117]In 2007, UK prime minister Gordon Brown commissioned British businessman and conservationist Johan Eliasch to undertake a review on the role of international finance mechanisms in the protection of forests for climate change mitigation. See Johan Eliasch, *Climate Change: Financing Global Forests – the Eliasch Review* (London: Earthscan, 2008), 35.

[118]For further alternatives raised by contemporaries, see Ian Fry, "Reducing Emissions from Deforestation and Forest Degradation: Opportunities and Pitfalls in Developing a New Legal Regime," *Review of European Community and International Environmental Law* 17, no. 2 (2008): 166–82.

[119]Lars Friberg, "Varieties of Carbon Governance: The Clean Development Mechanism in Brazil – a Success Story Challenged," *Journal of Environment and Development* 18, no. 4 (2009): 395–424.

[120]Márcio Santilli et al., "Tropical Deforestation and the Kyoto Protocol," *Climatic Change* 71 (2005): 267–76.

standing forests (not sinks), their proposal declared, "Lasting climatic stability will depend upon the equitable expansion of the market systems initiated following the Kyoto Protocol that actively facilitate and integrate developing nation participation."[121]

Among the factors that accounted for this significant tree change was the looming need to negotiate Kyoto's successor, as deforestation might afford the parties the means to finally engage developing countries to make meaningful emissions reductions. Significantly, the scientific and technical means to measure, monitor, and verify the reduced emissions from deforestation and forest degradation had markedly improved since the negotiation of the Kyoto agreement during the 1990s.[122] Remote-sensing techniques had undergone refinement and become more widely adopted in developing countries, which allowed for the assessment of changes in forest cover as well as their quantification in terms of carbon stocks. Highlighting the extent that the planetary view had become instantiated in climate diplomacy, ecologist Ruth DeFries argued, "The synoptic view from remote sensing has transformed the perceived role of terrestrial vegetation in the Earth system. Rather than a site-specific characteristic studied at the plot level, the global role of vegetation in carbon, water, and energy exchange is now the norm in Earth system models."[123] This scientific transformation provided a vehicle to reframe forests as a "common concern," that is, in global, rather than national, terms, which would help to usher them into the international climate change regime.

With voluntary efforts to curb emissions from deforestation encouraged at the Bali climate meeting in 2007, the World Bank launched its Forest Carbon Partnership Facility in 2008. By supporting developing countries to prepare legally and technically for trading their forest carbon, the Facility's goal was to "jump-start a forest carbon market" consistent with the Framework Convention.[124] A UN-REDD (Reducing Emissions from Deforestation and Forest Degradation) Programme soon followed, bringing together the expertise of the UN Development Programme, the UN Environment Programme, and the Food and Agriculture Organization, as well as the Forest Investment Programme as part of the World Bank's Climate Investment Fund.[125]

---

[121] Papua New Guinea and Costa Rica, "Reducing Emissions from Deforestation in Developing Countries: Approaches to Stimulate Action," United Nations Framework Convention on Climate Change, Eleventh Session, November 11, 2005, https://v.gd/R2PXD4. See Johnson and Rojas, "Contrasting Values of Forests and Ice in the Making of a Global Climate Agreement," 219–45.

[122] William Boyd, "Ways of Seeing in Environmental Law: How Deforestation Became an Object of Climate Governance," *Ecology Law Quarterly* 37 (2010): 843–916.

[123] Ruth DeFries et al., "Earth Observations for Estimating Greenhouse Gas Emissions from Deforestation in Developing Countries," *Environmental Science and Policy* 10, no. 4 (2007): 385–94.

[124] Cited in Julia Dehm, *Reconsidering REDD+: Authority, Power and Law in the Green Economy* (New York: Cambridge University Press, 2021), 81.

[125] Dehm, *Reconsidering REDD+*, 86–7.

Despite this enthusiasm among the UN, the World Bank, and participating nations, concerns about the social impact of these programs were growing among Indigenous peoples and climate justice advocates. Shortly after the signing of the UN Declaration on the Rights of Indigenous Peoples in 2007, the chair of the UN Permanent Forum on Indigenous Issues, Victoria Tauli-Corpuz, issued a statement, pointing out that "we (Indigenous Peoples) remain in very vulnerable situations because most States still do not recognize our rights to these forests and resources found therein."[126] Likewise, a Forest Peoples' Program report argued that an effective policy "on forests and climate change mitigation must be based on the recognition of rights, respect for the principle of free, prior and informed consent and requirements for progressive forests sector tenure and governance reforms."[127] The Friends of the Earth International also noted, "Indigenous Peoples and other forest-dependent communities have no guarantee that they will receive any form of REDD 'incentive' or reward for their extensive forest conservation efforts."[128] These concerns led to the inclusion in the negotiating texts of language in support of the UN Declaration on the Rights of Indigenous Peoples at the 2008 Poznań climate conference. There, the International Indigenous Peoples' Forum on Climate Change (IIPFCC) was formally established as the caucus for Indigenous peoples to participate in the climate convention meetings. But these references to Indigenous rights in the negotiating texts were removed at the insistence of the United States, Canada, Australia, and Aotearoa New Zealand, all of which had refused to sign the Declaration on the Rights of Indigenous Peoples.[129]

Conclusion

In April 2009, over four hundred Indigenous representatives from eighty countries across the Arctic, North America, Asia, Latin America, Africa, Russia, and the Caribbean gathered on the lands of the Ahtna and Dena'ina

---

[126]Victoria Tauli-Corpuz, "Statement on the announcement of the World Bank Forest Carbon Partnership Facility," United Nations Permanent Forum on Indigenous Issues, December 11, 2007, https://v.gd/TNMuA7.

[127]Tom Griffiths, *Seeing 'REDD'?: Forests, Climate Change Mitigation and the Rights of Indigenous Peoples and Local Communities* (Moreton-in-Marsh: Forest Peoples Programme, 2008), 1.

[128]Ronnie Hall, *REDD Myths: A Critical Review of Proposed Mechanisms to Reduce Emissions from Deforestation and Degradation in Developing Countries* (Amsterdam: Friends of the Earth, 2008), 6.

[129]Ciplet, Roberts, and Khan, *Power in a Warming World*, 192; "Indigenous Peoples, NGOs Protest at UN Climate Change Talks," UN Framework Convention on Climate Change, December 9, 2008, https://v.gd/7G1PDR. The United States, Canada, Australia and Aotearoa New Zealand have since endorsed the Declaration. The United States, Canada, Australia and Aotearoa New Zealand have since endorsed the Declaration.

Athabascan Peoples in Anchorage, Alaska, for the Indigenous Peoples Global Summit on Climate Change. They had assembled to exchange their knowledge and experience in adapting to the impacts of climate change, and to develop recommendations for the upcoming climate conference in Copenhagen at the end of the year. In their Anchorage Declaration, they expressed their alarm at "the accelerating climate devastation brought about by unsustainable development" and stressed, "We are experiencing profound and disproportionate adverse impacts on our cultures, human and environmental health, human rights, well-being, traditional livelihoods, food systems and food sovereignty, local infrastructure, economic viability, and our very survival as Indigenous Peoples." Finally, the Declaration argued for the recognition of their rights with regard to REDD programs, asserting their "fundamental rights as intergenerational guardians of [Indigenous] Knowledge [must be] full recognized and respected."[130]

Resonating with the security framing that had persisted over the course of the decade, the message of the Anchorage Declaration brought into stark relief the kinds of conditions that different actors sought to secure through engaging in climate diplomacy. By invoking security, the UK, Germany, and the EU had sought to uphold the climate regime and ensure that other developed countries honored their commitments. Carbon markets offered a means for governments and international organizations to minimize the economic risks that those commitments could entail. Planetary managers, meanwhile, entertained interventions in the earth system to counter anthropogenic climate change and to secure Holocene conditions. For Indigenous peoples and small island states, however, their acute vulnerability to the effects of climate change meant that only actual emissions reductions would afford them the chance of a secure future.

---

[130]Indigenous Peoples' Global Summit on Climate Change, "The Anchorage Declaration," April 24, 2009, https://unfccc.int/resource/docs/2009/smsn/ngo/168.pdf.

8

Climate Crisis

2015: 400.86 ppm

Speaking before United Nations (UN) secretary-general Ban Ki-moon and the Group of 77 (G77) at the UN in May 2010, Bolivian president Evo Morales reflected on the First World People's Conference on Climate Change and the Rights of Mother Earth that his country had hosted the previous month. Four months after the disappointment of Copenhagen, where Bolivia had been one of just seven nations to object to the meeting's Accord and scutter its adoption, Morales's government had invited activists, scientists, and governments to the Bolivian city of Cochachamba. Now, in New York, he declared, "What we are seeing is not just a climate crisis, an energy crisis, a food crisis, a financial crisis ... but also the systemic crisis of capitalism itself, which is bringing about the destruction of humanity and nature. If the cause is systemic, then the solution must be systemic as well."[1]

In the secretary-general at least, Morales might have found a sympathetic audience. He too had observed a troubling international scene: "We are living through an age of multiple crises. Fuel, flu and food, and most seriously financial. Each is something not seen for years, even for generations. But now they are hitting us all at once." The extent of these crises demanded a "spirit of renewed multilateralism," he had told the UN General Assembly in 2009.[2] The future looked much brighter in late 2015, however, when the 196 parties adopted the Paris Agreement. Its overarching goal is to hold "the increase in the global average temperature to well below 2°C above pre-industrial levels" and to pursue efforts "to limit the temperature increase to 1.5°C above pre-industrial levels." For the first time, *all* nations are expected to reduce

---

[1]Cited in Evo Morales, "Evo Morales' Speech to the G77 + China at the UN: Climate Change and the Rights of Mother Earth," *Voltaire Network*, May 7, 2010, https://v.gd/kLmJtG; Radoslav S. Dimitrov, "Inside Copenhagen: The State of Climate Governance," *Global Environmental Politics* 10, no. 2 (2010): 18–24.
[2]Cited in "With Crises in Food, Energy, Recession Hitting All at Once, 'the World Looks to Us for Answers', Secretary-General Says, Opening General Debate," UN Department of Public Information, September 23, 2009, https://v.gd/5TMb6h.

their greenhouse gas emissions through their own nationally determined contributions, an approach often called "pledge and review."

This final chapter considers the aftermath of the 2009 Copenhagen conference, where a pervading sense of crisis engulfed governments, environmental organizations, and the UN Framework Convention. In light of the failure to devise Kyoto's successor, they feared that the future of the climate regime was in the balance and, with it, the prospect of avoiding dangerous climate change. Yet this sense of crisis, this chapter shows, belied the changing dynamics of climate diplomacy as the interests of some of the largest emitters from the South and North began to align more closely. Despite the unorthodox negotiations at Copenhagen, what they had revealed was a shared investment in the established procedures of climate diplomacy among governments as well as among the facilitators and beneficiaries of the growing carbon market.

Copenhagen Disappointment: Challenging the Climate Regime

The 2009 climate conference in Copenhagen had represented the culmination of the Bali Action Plan—to determine the nature of the climate regime that would follow the Kyoto Protocol. It was now "time to seal a deal," as UN secretary-general Ban Ki-moon declared at the launch of the "Hopenhagen" promotional campaign.[3] Even "Climate-gate"—after hackers posted online more than a thousand emails from the Climatic Research Unit at the University of East Anglia—barely cast a pall over the negotiations.[4] As the final day of the conference approached, however, the prospect that the parties would agree to a legally binding treaty looked increasingly remote as procedural difficulties and questions of transparency derailed the negotiations. A final declaration looked in the balance, only for a five-page text to be brokered at the eleventh hour between the leaders of just five nations: Brazil, China, India, South Africa, and the United States (Figure 8.1). The following day, delegates approved a motion to only "take note" of the Copenhagen Accord—without all countries signing on, it would not be a legally binding agreement.[5]

---

[3]Mark Sweney, "Copenhangen Climate Change Treaty Backed by 'Hopenhagen' Campaign," *Guardian*, June 24, 2009, https://v.gd/0ekih6.

[4]On "Climate-gate," see Reiner Grundmann, "The Legacy of Climategate: Revitalizing or Undermining Climate Science and Policy?," *WIREs Climate Change* 3 (2012): 281–8.

[5]John Drexhage and Deborah Murphy, *Copenhagen: A Memorable Time for All the Wrong Reasons?* International Institute for Sustainable Development, 2009, https://v.gd/us4bwh.

FIGURE 8.1 *The heads of state of Brazil, China, India, and South Africa during informal consultations at COP 15 in Copenhagen, December 2009. From left to right: President Jacob Zuma of South Africa (back to camera), Premier Wen Jiabao of China, Prime Minister Manmohan Singh of India, and President Lula da Silva of Brazil. Photo by* IISD/Earth Negotiations Bulletin, *Leila Mead.*

Having emerged bruised from the Copenhagen meeting, the UN Framework Convention on Climate Change rallied over the following months, and in December 2010, the parties agreed by consensus to adopt the Cancún Agreements. Since Copenhagen, officials from both the United States and the European Union (EU) had strongly encouraged all countries to commit their support to the agreement. Diplomatic cables revealed by Wikileaks in late 2010 showed that in the wake of the Copenhagen conference, US deputy special envoy for climate change Jonathan Pershing and EU climate action commissioner Connie Hedegaard had "agreed that the US-EU cooperation remains important" and "on the need to operationalize the Copenhagen Accord and ensure it is incorporated into the UNFCCC process."[6]

This process of "operationalization," the leaked cables indicated, involved finance. The Obama administration had already rescinded funding promised to Bolivia and Ecuador on account of their opposition to the Copenhagen Accord, while Ethiopia's prime minister Meles Zenawi and Maldives ambassador Abdul Ghafoor Mohamed sought what the latter described to

[6] "US Embassy Cables: EU raises 'Creative Accounting' with US over Climate Aid," *Guardian*, December 4, 2010, https://v.gd/65cntD.

US officials as "tangible assistance from the larger economies" for their support.[7] The EU's Hedegaard also suggested to Pershing that the "countries of the Alliance of Small Island States 'could be our best allies',' given their need for financing.[8] Upholding the UN Framework Convention on Climate Change (UNFCCC) process was also important to ensuring the support of developing countries: Saudi Arabia's lead negotiator Mohammad Al-Sabban told US officials that "climate change negotiations should remain under the UNFCCC and not be pursued under alternative frameworks."[9] Resembling much of what had been contained in the Copenhagen Accord, the Cancún Agreements marked a new chapter for climate diplomacy. Avoiding any specific commitments, the agreements blurred the distinction between industrialized and developing countries for the first time and affirmed the voluntary pledge and review approach introduced a year earlier.[10] The parties also now agreed to formally accept the goal to keep temperature increases below a global average of 2°C.

Although the EU had been excluded from the last-minute Copenhagen deal, its commitment to a temperature target of 2°C had entered the final 2009 text and then became formally adopted in the Cancún Agreements. With its origins in the general circulation model experiments undertaken by Syukuro Manabe and Richard Wetherald in the late 1960s (see Chapter 2), this target emerged in the nascent climate diplomacy of the late 1980s at the Villach and Bellagio meetings of the Advisory Group on Greenhouse Gases (see Chapter 4). Participants Pier Vellinga and Rob Swart had subsequently published their recommendation in 1991 for a "global strategy" in which they framed this figure as a threshold to guide global climate policy.[11] As the First Conference of the Parties approached in Berlin in 1995 (Chapter 5), the

---

[7]Suzanne Goldenberg,"US Denies Climate Aid to Countries Opposing Copenhagen Accord," *Guardian*, April 10, 2010, https://v.gd/RVzXMj; "US Embassy Cables: US Urges Ethiopia to Back Copenhagen Climate Accord," *Guardian*, December 4, 2010, https://v.gd/xF49TV; "US Embassy Cables: Maldives Tout $50m Climate Projects to US," *Guardian*, December 4, 2010, https://v.gd/D0o649.

[8]"US Embassy Cables: EU Raises 'Creative Accounting' with US over Climate Aid," *Guardian*, February 17, 2010, https://v.gd/65cntD.

[9]"US Embassy Cables: Saudi Arabia Fears Missed Trick on Copenhagen Climate Accord," *Guardian*, December 4, 2010, https://v.gd/s5lXK8.

[10]David Ciplet, J. Timmons Roberts, and Mizan R. Khan, *Power in a Warming World: The New Global Politics of Climate Change and the Remaking of Environmental Inequality* (Cambridge, MA: MIT Press, 2015), 90.

[11]Piero Morseletto, Frank Biermann, and Philipp Pattberg, "Governing by Targets: *Reductio ad unum* and Evolution of the Two-Degree Climate Target," *International Environment Agreements* 17 (2016): 655–76; Pier Vellinga and Robert Swart, "The Greenhouse Marathon: Proposal for a Global Strategy," in Jill Jäger and Howard Ferguson (eds.), *Climate Change: Science, Impacts and Policy – Proceedings of the Second World Climate Conference* (New York: Cambridge University Press, 1991), 129–34.

German Advisory Council on Global Change set out a "tolerance window" that recommended limiting a global mean temperature rise to less than 2°C.[12] The council's chair, Hans-Joachim Schellnhuber, would convince the German minister for environment Angela Merkel of this target's policy significance, which the European Commission and the International Climate Change Taskforce (a consortium of thinktanks) affirmed a decade later.[13] Proliferating beyond scientific circles and Europe, the target now became "transformed" at Copenhagen as "the only quantitative element which proved consensual, as long as no measures or prescriptions were attached to it."[14]

The 2°C temperature target was nevertheless a cause for concern among vulnerable developing nations and climate justice advocates. In a letter to heads of state and Christian leaders in the conference aftermath, Nobel laureate and former Cape Town archbishop Desmond Tutu cited the Intergovernmental Panel on Climate Change's (IPCC's) Fourth Assessment Report and pointed to Africa's particular vulnerability to climate change, where the median temperature rise would be greater than the global mean response. Given this sensitivity, he argued, "A global goal of about 2 degrees C [sic] is to condemn Africa to incineration and no modern development."[15] Drawing the archbishop's attention to these figures had been Sudan's Lumumba Di-Aping, chair of the G77 in Copenhagen. During the troubled proceedings, he had criticized the financial aid on offer from the industrialized countries as too little "to buy the developing countries' citizens enough coffins" for the millions that would die as a result of 2°C of global warming.[16] Di-Aping had also appealed to the Framework Convention's "common but differentiated responsibilities" in his rejection of a draft proposal, declaring, "The text robs developing countries of their just and equitable and fair share of the atmospheric space. It tries to treat rich and poor countries as equal."[17] The climate finance proposed at Copenhagen, with a goal of mobilizing $100USD billion each year from 2020 onwards, was reasserted by the parties at Cancún, where the Green Climate Fund was established to administer the funding.

---

[12]German Advisory Council on Global Change (WBGU), *Scenario for the Derivation of Global CO2 Reduction Targets and Implementation Strategies* (Bremerhaven: WBGU, 1995).

[13]Carlo Jaeger and Julia Jaeger, "Three Views of Two Degrees," *Climate Change Economics* 1, no. 3 (2010): 145–66.

[14]Morseletto, Biermann, and Pattberg, "Governing by Targets," 655–76.

[15]Tom Athanasiou, "Desmond Tutu (with a Little Help from the Scientific Community) Explains Africa's Position," *Grist*, December 16, 2009, https://v.gd/f9td2E. On the uneven experience of geophysical thresholds and their implications, see Petra Tschakert, "1.5°C or 2°C: A Conduit's View from the Science-Policy Interface at COP20 in Lima, Peru," *Climate Change Responses* 2 (2015): 10.1186/s40665-015-0010-z.

[16]Cited in John M. Broder, "Poor and Emerging States Stall Climate Negotiations," *New York Times*, December 16, 2009, https://v.gd/p02zDa.

[17]Cited in John Vidal and Dan Milmo, "Copenhagen: Leaked Draft Deal Widens Rift between Rich and Poor Nations," *Guardian*, December 9, 2009, https://v.gd/ratucp.

Climate Crossroads: Questioning the Climate Regime

"Copenhagen has shown us the limits of what can be achieved on climate change through centralizing and hyperbolic multilateralism," argued the authors of *The Hartwell Paper*. An interdisciplinary group of researchers from the UK, North America, Europe, and Japan had gathered in 2009 to devise a new direction for the climate regime. In addition to their criticism of the Kyoto model, they advocated for the reorientation of the IPCC, suggesting instead "a more indirect yet encompassing approach via the attainment of different objectives which bring contingent benefits is, indeed, the only one that is likely to be materially (in contrast to rhetorically) successful."[18] Researchers who had been involved in its assessment reports to varying degrees called variously for its disbandment and reform, citing the need for a more rapid, transparent, and responsive assessment process that would "better serve the world, and its peoples," as Mike Hulme, founder of the UK's Tyndall Centre for Climate Change Research, argued.[19] Ken Caldeira echoed these sentiments when he stepped down as a lead author of the IPCC's Fifth Assessment Report in late 2011, noting, however, "It may turn out to be a far more efficient and effective vehicle for scientific communication than I now anticipate."[20]

The shortcomings of the prevailing climate order that *The Hartwell Paper* highlighted were well known. As early as the 2006 climate conference in Nairobi, Japan's delegation had raised concerns about the limitations of the Kyoto approach (Chapter 6). "Are we really right to say that we've established in Kyoto all the right answers to this challenge?," asked Ambassador Mutsuyoshi Nishimura. He continued,

Shouldn't we be a little more humble to the awesome might of nature and human action and start exploring many more tools and strategies on top of the Kyoto tool box? Simply, we feel we cannot accept that we can win this colossal planetary battle by numericals, timeline, flexible mechanisms and punishment. I simply cannot accept that strategies we invented in '97

---

[18]Steve Rayner et al., *The Hartwell Paper: A New Direction for Climate Policy after the Crash of 2009*, Institute for Science, Innovation and Society, May 2010, 6, 10, https://v.gd/pFrjHv. See also Mike Hulme, *Why We Disagree about Climate Change: Understanding Controversy, Inaction and Opportunity* (Cambridge: Cambridge University Press, 2009).
[19]Mike Hulme, Eduardo Zorita, Thomas F. Stocker, Jeff Price, and John R. Christy, "IPCC: Cherish It, Tweak It or Scrap It?', *Nature* 463 (2010): 730–2.
[20]Ken Caldeira, cited in Andrew C. Revkin, "New Directions for the Intergovernmental Climate Panel," *Dot Earth: New York Times Blog*, December 21, 2011, https://v.gd/wKk9N4.

are all valid and effective in years and decades to come and nothing should be added.[21]

Elsewhere, a former staffer of the UNFCCC secretariat observed in 2006 that the climate change regime had become "ossified" around "longstanding political fissures" between the North and the South, and between the EU and the rest of the industrialized world. In her view, "to get out of the rut in which it finds itself, the climate change regime needs positive input from outside the rather stifling confines of the negotiation process."[22] As though to add insult to injury, the Montreal Protocol, which had been a model for climate diplomacy, had been found in 2007 to have been more effective than the reduction target of the Kyoto Protocol at curbing greenhouse gas emissions.[23]

Nevertheless, the recommendations of *The Hartwell Paper* left many veterans of climate diplomacy unconvinced. Bill Hare, the Australian founder of the nonprofit organization Climate Analytics, told the BBC, "The Kyoto Protocol is one of the few things that have worked, in that it's given momentum to low-carbon energy development—we wouldn't have had the explosion in wind power without it."[24] Between 2004 and 2011, 190 gigawatts of wind-generating capacity had been installed, which was three quarters of the additional hydroelectric capacity (250 gigawatts), and almost three times as much as the amount of solar photovoltaic-generating capacity (66 gigawatts) installed during the same period.[25] For all its shortcomings, the Kyoto Protocol had also provided a "normative influence" on the voluntary initiatives at the city and state levels, particularly in the United States.[26] Elsewhere, Hare also pointed to the weakness of the national pledges arising from Copenhagen's so-called bottom-up approach, which would fall well short of the Framework Convention's objective: preventing dangerous climate change.[27] A 2010 study by the UN Environment Programme (UNEP), for example, indicated that these

---

[21]Japan, "Work by Annex I Parties on the Scientific Basis for Determining Their Future Commitments," UN Framework Convention on Climate Change, 12th Conference of the Parties, Nairobi, 2006, https://v.gd/DV7QfN.

[22]Joanna Depledge, "Opposite of Learning: Ossification in the Climate Change Regime," *Global Environmental Politics* 6, no. 1 (2006): 1–22.

[23]Guus J. M. Velders et al., "The Importance of the Montreal Protocol in Protecting Climate," *PNAS* 104, no. 12 (2007): 4814–19.

[24]Bill Hare, cited in Richard Black, "Academics Urge Radical New Approach to Climate Change," *BBC News*, May 11, 2010, https://v.gd/LJWRle.

[25]Govinda R. Timilsina, G. Cornelis van Kooten, and Patrick A. Narbel, "Global Wind Power Development: Economics and Policies," *Energy Policy* 61 (2013): 642–52.

[26]Raymond Clémençon, "The Bali Road Map: A First Step on the Difficult Journey to a Post-Kyoto Protocol Agreement," *Journal of Environment and Development* 17, no. 1 (2008): 3–96.

[27]William Hare et al., "The Architecture of the Global Climate Regime: A Top-Down Perspective," *Climate Policy* 10 (2010): 600–14.

national pledges were not sufficiently ambitious to limit warming to the agreed Copenhagen target of 2°C.[28]

For others, the challenges of the global regime would not be resolved by limiting negotiations to a smaller group of the largest emitters or clubs with shared ambitions.[29] Veterans of the climate negotiations since the 1990s, Joanna Depledge and Farhana Yamin, gestured to the proceedings in Copenhagen, observing that "small emitters *are not the countries that are currently slowing down the negotiations*." The Alliance of Small Island States (AOSIS), with whom Yamin had served as a legal expert, had led ambitious initiatives for insurance and adaptation since the convention's inception in the early 1990s. "The main victims of climate change cannot be left out of the process simply because of their low emissions, limited leverage, and weak power," they argued.[30] Former UNFCCC executive secretary Yvo de Boer, who had stepped down after the Copenhagen meeting, took a wider view in the lead up to the Cancún conference in late 2010:

> If coming to grips with climate change were only about reducing emissions, then it would make a lot of sense to just bring the 20 or so major economies of the world together in a room and get them to focus on emissions reduction, but the climate change agenda isn't only about reducing emissions. It's about adapting to the impacts of climate change as well, and then you're talking about the 100 or so developing countries who did absolutely nothing to contribute to climate change but will be confronted with the bulk of the impacts. So, you need a larger group at the table.[31]

For all the frustrations with the prevailing system of climate negotiations, its defenders believed its strengths were those very aspects that its detractors had observed: in its longevity lay an established order of international environmental governance, and in its inclusion of all states, the means to negotiate the means of both mitigation and adaptation.

---

[28]United Nations Environment Programme, *The Emissions Gap Report: Are the Copenhagen Accord Pledges Sufficient to Limit Global Warming to 2°C or 1.5°C? A Preliminary Assessment* (Nairobi: United Nations Environment Programme, 2010). Note, Hare was a lead author of this report.

[29]See, for example, David G. Victor, *Global Warming Gridlock: Creating More Effective Strategies for Protecting the Planet* (Cambridge: Cambridge University Press, 2011); Robin Eckersley, "Moving Forward in the Climate Negotiations: Multilateralism or Minilateralism?," *Global Environmental Politics* 12, no. 2 (2012): 24–42.

[30]Joanna Depledge and Farhana Yamin, "The Global Climate-Change Regime: A Defence," in Dieter Helm and Cameron Hepburn (eds.), *The Economics and Politics of Climate Change* (Oxford: Oxford University Press, 2009), 451–52. Emphasis in text.

[31]Yvo de Boer with Rebecca Lutzy, "Yvo de Boer on the Future of the UNFCCC," *Change Oracle*, November 29, 2010, https://v.gd/lxn7W1.

Carbon Capital: Marketing Climate Mitigation

There had been two Copenhagens underway in December 2009. One at the Bella Conference Center for the official negotiations of the conference of the parties. The other was unfolding across the city—such as in the "Hopenhagen" exhibition area and the KlimaForum, where the International Emissions Trading Association was hosting its own event.[32] Consensus may have been wanting in the Bella Center, but in the KlimaForum, discussions suggested clear agreement that separate cap and trade systems might become linked so as to enable permit trading across systems.[33] This sentiment carried substantial weight in climate diplomacy: the Association's lobbying budget was roughly two to four times the size of the largest environmental organization, the Climate Action Network.[34] The Reducing Emissions from Deforestation and Forest Degradation (REDD) program, meanwhile, had been a "bright spot" in the darkness of Copenhagen, according to *Time* magazine.[35] Parties agreed to establish a new mechanism that would provide "positive incentives" to developing countries for national-level reductions in emissions from deforestation. To finance the measures necessary to render Southern forests legible to carbon accounting and markets, the United States, France, Japan, and Norway pledged substantial financial and technical assistance.[36]

As the discord that erupted at the Bella Center lingered and a successor to Kyoto remained unclear, the carbon market entered freefall. Recession, over-allocation in the EU's Emissions Trading System, and scandals associated with the Clean Development Mechanism (CDM) projects were taking their toll.[37] By 2013, the value of the global carbon market had fallen to less than half its value in 2011, while investments in new CDM projects had nearly dried up

---

[32]Matthew J. Hoffmann, *Climate Governance at the Crossroads: Experimenting with a Global Response after Kyoto* (New York: Oxford University Press, 2011), 3–26.

[33]Michele Betsill and Matthew J. Hoffman, "The Contours of 'Cap and Trade,'" *Review of Policy Research* 28, no. 1 (2011): 83–106; Steven Bernstein et al., "A Tale of Two Copenhagens: Carbon Markets and Climate Governance," *Millennium* 39, no. 1 (2010): 161–73.

[34]Ciplet, Roberts, Khan, *Power in a Warming World*, 137; Emma Lund, "Environmental Diplomacy: Comparing the Influence of Business and Environmental NGOs in Negotiations on Reform of the Clean Development Mechanism," *Environmental Politics* 22, no. 5 (2013): 739–59.

[35]Bryan Walsh, "In Copenhagen's Dark Mood, a Ray of Light for Forests," *Time*, December 17, 2009, https://v.gd/7m0MKp.

[36]William Boyd, "Climate Change, Fragmentation, and the Challenges of Global Environmental Law: Elements of a Post-Copenhagen Assemblage," *University of Pennsylvania Journal of International Law* 32, no. 2 (2010): 457–550; William Boyd, "Ways of Seeing in Environmental Law: How Deforestation Became an Object of Climate Governance," *Ecology Law Quarterly* 37 (2010): 843–916.

[37]Peter Newell, "Dialogue of the Deaf? The CDM's Legitimation Crisis" (2014).

altogether.[38] As a market analyst observed in early 2014, "without ambitious climate targets there is no need for deep emission reductions and carbon prices will remain at low levels."[39]

Elsewhere, carbon markets remained attractive. Their future looked bright even after the collapse of carbon prices in 2012, with national and regional schemes moving ahead in China, Kazakhstan, the United States (California), and Canada (Quebec), as well as in Chile, Turkey, and some Brazilian states.[40] The EU, meanwhile, retained broad political and industry support as it battled to keep its flagging carbon-trading scheme afloat.[41] "Our flagship policy is the EU's Emissions Trading System," the European commissioner for climate action Connie Hedegaard had declared on the eve of the 2011 Durban climate meeting. "One of our primary concerns is to develop climate policies that are environmentally efficient, in a way that will not hamper economic growth in Europe, but which leaves companies maximum flexibility to cut emissions at least cost."[42]

Spearheading the proliferation of carbon markets was the World Bank. At the Cancún climate conference in 2010, it announced its Partnership for Market Readiness program to encourage middle-income countries to devise domestic emissions trading systems. The International Emissions Trading Association followed with its complementary Business Partnership for Market Readiness scheme in 2012, which reportedly represented for its president Henry Derwent "a sectoral overcoming of common but differentiated responsibilities."[43] China was among the first eight countries to receive funding from the World Bank's Partnership program, which the deputy director of its National Development and Reform Commission, Wang Shu, welcomed to support the development of the nation's own carbon market. In 2013, he explained, "China is determined to pursue low carbon development. Addressing climate change is an important

---

[38]Kathleen McAfee, "The Post- and Future politics of Green Economy and REDD+," in Benjamin Stephan and Richard Lane (eds.), *The Politics of Carbon Markets* (London: Routledge, 2014), 237–60.

[39]Anders Nordeng, cited in "Value of Global CO2 Markets Drop 38 pct in 2013," *Reuters*, January 3, 2014, https://v.gd/kOvgJw.

[40]Richard Lane and Benjamin Stephan, "Zombie Markets or Zombie Analyses?: Revivifying the Politics of Carbon Markets," in Stephan and Lane (eds.), The Politics of Carbon Markets, 1–24; Matthew Paterson and Johannes Stripple, "Virtuous Carbon," *Environmental Politics* 21, no. 4 (2012): 563–82.

[41]Misato Sato et al., "Allocation, Allocation, Allocation! The Political Economy of the Development of the European Union Emissions Trading System," *WIREs Climate Change* 13, no. 5 (2022): e796.

[42]Connie Hedegaard, "Climate Protection Is not Deindustrialisation, but Reindustrialisation: Doing Things Smarter and More Efficiently!," European Commission, November 21, 2011, https://v.gd/cFGRAT.

[43]Cited in Oscar Reyes, "World Bank Partnership for Market Readiness: A Critical Introduction," *Carbon Trade Watch*, January 20, 2011, https://v.gd/6rJKMW; Jørgen Wettestad, Lars H. Gulbrandsen, and Steinar Andresen, "Calling In the Heavyweights: Why the World Bank Established the Carbon Pricing Leadership Coalition, and What It Might Achieve," *International Studies Perspectives* 22, no. 2 (2021): 201–17.

opportunity for transformation to a green and low carbon economy. We believe market-based instruments, such as carbon emissions trading, is an efficient approach to achieving this."[44] China had already engaged with carbon markets through its endorsement of the CDM. Just prior to the Kyoto Protocol's entry into force in 2005, China had enacted draft legislation to manage projects associated with the CDM, indicating that it was willing to engage in the process. Although India registered a greater number of CDM projects than China, the credits that China produced accounted for about 60 percent of the value of the world's carbon credits in 2007.[45] Such was the scale of the Chinese market that the state's establishment of a price floor effectively became a global price floor for Certified Emission Reductions.[46]

At the Durban climate conference in 2011, Beijing had made it clear that the CDM would continue to feature in its approach to climate diplomacy. Speaking on behalf of the BASIC countries, China's delegation emphasized that "defining a second commitment period should be the main priority for Durban."[47] During a second commitment period of the Kyoto Protocol, China would have the potential to benefit even more from Mechanism projects, as its potential supply of Certified Emission Reduction credits were expected to nearly double the 2012 amount by 2020.[48] Renewable energy projects, particularly wind power, and technology transfers had dominated China's engagement with the Mechanism during the first commitment period.[49] For Premier Wen Jiabao, attuned to the mounting environmental and health toll of the nation's rapid growth, such projects offered the prospect of restructuring China's economy from one dominated by energy-intensive heavy industry to a more low-carbon model.[50] It was firmly in Beijing's interest, then, for the Kyoto arrangement to continue, as it would afford China further opportunities for investment and technology transfers from the North.

The World Bank anticipated that carbon-market revenues would finance REDD+ programs: governments would accrue some of the profits from the sale of REDD+ credits, which they would distribute to landholders, or they would

---

[44]Cited in "Breathing New Life into Carbon Markets," *World Bank*, March 13, 2013, https://v.gd/XvmhTU.

[45]Gang Chen, "China's Diplomacy on Climate Change," *Journal of East Asian Affairs* 22, no. 1 (2008): 145–74.

[46]Anita Engels, Tianbao Qin, and Eva Sternfeld, "Carbon Governance in China by the Creation of a Carbon Market," in Stephan and Lane (eds.), *The Politics of Carbon Markets*, 150–70.

[47]Soledad Aguilar et al., "Summary Report, 28 November – 11 December 2011," *Earth Negotiations Bulletin*, https://v.gd/Bx3hGu.

[48]Fuzuo Wu, "Shaping China's Climate Diplomacy: Wealth, Status, and Asymmetric Interdependence," *Journal of Chinese Political Science* 21 (2016): 199–215.

[49]Joanna I. Lewis, "The Evolving Role of Carbon Finance in promoting Renewable Energy Development in China," *Energy Policy* 38, no. 6 (2010): 2875–86.

[50]Chen, "China's Diplomacy on Climate Change," 145–74.

use those funds directly to strengthen forest protection and other conservation programs.[51] Despite Greenpeace revelations that the highly-renowned Noel Kempff forest offset project in northeast Bolivia (a collaboration between The Nature Conservancy and BP) had overestimated the quantity of emissions saved and had simply led logging in adjacent forests, REDD+'s star continued to rise.[52] In 2011, UNFCCC executive secretary Christiana Figueres described the REDD initiative as "the spiritual core [of a] global business plan for the planet," while the UNEP saw it as "at the heart of its climate strategy."[53] At Cancún in 2010, REDD had expanded to become REDD+, which added the conservation of forest carbon stocks, sustainable management of forests, and enhancement of forest stocks. Through its Forest Carbon Partnership Facility, with Germany and Norway its largest donors, the World Bank was "setting the post-Cancún agenda in terms of how forests are integrated into a global climate regime, how REDD will be implemented and how finance will be sourced," as the environmental justice groups Fern and the Forest Peoples Programme observed warily in 2011.[54]

In Durban in December 2011, the parties agreed to negotiate a "treaty, another legal instrument or an agreed outcome with legal force," with 2015 the deadline for their negotiations. Although the EU agreed to the South's insistence that the Kyoto Protocol should be extended for another five to eight years, few industrialized countries committed to a second term. The exit of Canada, Japan, and Russia led Bolivia's former negotiator Pablo Solon to describe the protocol as a "zombie" that "will keep on walking until 2020 so that carbon markets don't disappear."[55] Negotiations in the meantime would embark on a new chapter—a climate regime in which there would be no difference in the obligations of developed and developing countries.[56]

Building Bridges: Finding Common Ground

President Obama's ascent to the White House in 2009 had helped to bring a fresh approach to international climate diplomacy. Since the mid-1990s at least,

[51]McAfee, "The Post- and Future Politics of Green Economy and REDD+," 237–60.

[52]Ariana Densham, Roman Czebiniak, Daniel Kessler and Rolf Skar, *Carbon Scam: Noel Kempff Climate Action Project and the Push for Sub-national Forest Offsets* (Amsterdam: Greenpeace International, 2009).

[53]Cited in Julia Dehm, *Reconsidering REDD+: Authority, Power and Law in the Green Economy* (New York: Cambridge University Press, 2021), 5.

[54]Kate Dooley et al., *Smoke and Mirrors: A Critical Assessment of the Forest Carbon Partnership Facility* (Moreton in Marsh: FERN and Forest Peoples Programme, 2011), 32.

[55]Cited in Jeff Conant, "African Soil in the Blood and Dust of COP17," *Earth Island Journal*, December 10, 2011, https://v.gd/h5EbJU/.

[56]Lavanya Rajamani, "The Durban Platform for Enhanced Action and the Future of the Climate Regime," *International & Comparative Law Quarterly* 61, no. 2 (2012): 501–18.

the options available to US negotiators had been severely limited by domestic politics (Chapter 6), which made Senate ratification of any international agreement unlikely. To sidestep the Senate entirely, the Obama administration sought an agreement that did not contain any new legal obligations for the United States. This position favored the negotiation of a pledge-and-review system based on both nationally determined and voluntary (rather than legally binding) pledges. This system closely resembled a proposal by Japan's negotiators, with support from the UK and France, during negotiations of the climate convention in 1991, whereby "each country (or regional group) makes public a pledge, consisting of past performance strategies to limit greenhouse gas emissions and targets or estimates for such emissions as the result of the strategies."[57] In Tokyo's view, this would have offered a compromise between the European countries, which sought emission reductions targets, and the United States, which was reluctant to do so. Critics, however, dismissed the plan as more like a strategy for "hedge and review"—a means to avoid emissions reductions.[58] Although the notion of "pledge and review" had floundered then and again at Copenhagen, the Obama administration persevered, finding an unlikely ally in China, which had been the world's largest emitter of greenhouse gas emissions since at least 2007.[59]

China's delegation had left Copenhagen stung by international criticism of its role in the unusual proceedings. The British climate minister, for instance, had accused Beijing and its allies of "hijacking" the talks as the delegation held fast to its position to refuse binding targets to reduce emissions, or for developing countries to share the burden of mitigation.[60] Defending Beijing's stance, climate envoy Yu Qingtai argued, "We cannot blindly accept that protecting the climate is humanity's common interest—national interests should come first."[61] Besides, as his colleague Li Gao observed, gesturing to a renewed discussion about consumption emissions, "we produce products and these products are consumed by other countries, especially the developing countries. This share of emissions should be taken by the consumers, but not

---

[57]Japan, "Pledge and Review Process as Possible Mechanism to Implement Commitments Defined on the Basis of the Convention," Preparation of a Framework Convention on Climate Change, Addendum 7, A/AC.237/Misc.1/Add.7, June 20, 1991, Intergovernmental Negotiating Committee for a Framework Convention on Climate Change, Second Session, Geneva, https://v.gd/sC0uL1.

[58]Yasuko Kawashima, "Japan and Climate Change: Responses and Explanations," *Energy and Environment* 12, no. 2–3 (2001): 167–79.

[59]Manjana Milkoreit, "The Paris Agreement on Climate Change – Made in USA?," *Perspectives on Politics* 17, no. 4 (2019): 1019–37.

[60]John Vidal, "Ed Miliband: China Tried to Hijack Copenhagen Climate Deal," *Guardian*, December 21, 2009, https://v.gd/tgDO7N .

[61]Yu Qingtai, "China's Interests Must Come First," *China Dialogue*, August 27, 2010, https://v.gd/xXMTxv.

the producers."[62] Beijing's negotiators could also point to their constructive role in Copenhagen, having signed a Memorandum of Understanding with New Delhi in 2009 to cooperate on the issue of climate change. Accordingly, both countries set emissions reductions targets at Copenhagen.[63]

During the Copenhagen negotiations, China's delegation had also sought an agreement that was neither legally binding nor obligated economically costly mitigation measures. Growing political unrest over urban air pollution, as well as its role in developing renewable energy markets, emissions trading, and support for international climate finance had increasingly inured Beijing to the Obama administration's vision of a post-Kyoto regime. Bilateral efforts before and after Copenhagen eventually bore fruit, as the rise of Xi Jinping fostered a shift away from a "developing country" identity to one of international leadership.[64] President Obama and President Xi reached an agreement in mid-2013 to phase out the consumption and production of hydrofluorocarbons—a potent greenhouse gas that had become a substitute for ozone-depleting substances in the 1980s—that then became an amendment to the Montreal Protocol.[65] A year later, with the Paris conference approaching, the two presidents made a remarkable announcement of cooperation on climate change in Beijing's Great Hall of the People.[66] Their nations responsible for 42 percent of global greenhouse gas emissions in 2012, the leaders announced their plans to reduce emissions and, in a first for China, a cap on its output of emissions.[67]

In the meantime, the Obama administration had been working on overcoming its credibility deficit at home. Despite Obama's 2008 Nobel prize for his commitment to international diplomacy, Copenhagen had arrived too early in his term for his administration to overturn over a decade of distrust toward the United States in climate negotiations. Todd Stern, the US lead negotiator, had already noted, "Unlike what happened in Kyoto we need to have our international and domestic postures in sync, so there is domestic and political support for whatever it is we do. We don't want the cart before

[62]Cited in "China Seeks Export Carbon Relief," *BBC News*, March 17, 2009, https://v.gd/ZtrYqw.

[63]Robert Mizo, "India, China and Climate Cooperation," *India Quarterly* 72, no. 4 (2016): 375–94; Björn Conrad, "China in Copenhagen: Reconciling the 'Beijing Climate Revolution' and the 'Copenhagen Climate Obstinacy'," *China Quarterly* 210 (2012): 435–55.

[64]Jilong Yang, "Understanding China's Changing Engagement in Global Climate Governance: A Struggle for Identity," *Asia Europe Journal* 20 (2022): 357–76.

[65]Steven Mufson, "President Obama and Chinese President Xi Jinping Agree to Wind Down Production and Use of Hydrofluorocarbons, or HFCs," *Washington Post*, June 8, 2013, https://v.gd/WkE5Rr.

[66]Suzanne Goldenberg, "Secret Talks and a Personal Letter: How the US–China Climate Deal Was Done," *Guardian*, November 13, 2014, https://v.gd/tHRPJi.

[67]Milkoreit, "The Paris Agreement on Climate Change – Made in USA?," 1019–37.

the horse this time."[68] Although President Obama's cap-and-trade bill failed in Congress in 2010, his reelection in 2012 brought improvements in energy and fuel efficiency standards, investments in renewable energy, limits on methane emissions, and the effort to reduce carbon dioxide emissions from existing coal power plants with the planned Clean Power Act in 2014.[69] Buoying these efforts were changes in the national energy market, as the rapid expansion of natural gas production made domestic coal production increasingly uncompetitive. As natural gas surpassed coal, US greenhouse gas emissions declined during the three years leading up to the Paris negotiations. With Paris approaching, the Obama administration could now point to its domestic record on reducing emissions, transforming the nation's energy system, and demonstrating that its climate diplomacy was not tethered to Congressional support.[70]

Yet the climate agreement "could not be a G2 agreement extended to all," as pointed out by Laurent Fabius, whose deft handling of the presidency of the Paris meeting would earn him wide acclaim.[71] The other parties also needed to be courted to ensure that there would be no repeat of Copenhagen, which had demonstrated the significance of multilateralism to the climate regime's parties. In addition to meeting with President Xi, France's president François Hollande also met with India's prime minister Narendra Modi to convince him of the need to assent to transparency mechanisms, and promised to include references to climate justice in return. The pair also launched India's International Solar Alliance on the eve of the Paris conference, which the Indian leader envisioned as a means to unite the "Suryaputra" or "sons of the sun".[72]

As Copenhagen had earlier indicated, New Delhi was demonstrating a willingness to engage differently in the climate regime, reflecting the nation's global ambitions. India in 2009 had signed the Declaration of the Major Economies Forum on Energy and Climate, held alongside the G8 summit in Italy, which required all parties to identify a global goal for substantially reducing emissions by 2050. Aligning with New Delhi's recent launch of a National Action Plan on Climate Change, the declaration also reflected India's aspirations to permanent membership in the UN Security Council and greater

---

[68]Cited in John M. Broder and James Kanter, "Europeans Say U.S. Lacks Will on Climate," *New York Times*, September 20, 2009, https://v.gd/xW51V9.

[69]A sustained campaign against this bill, led by the US Chamber of Commerce and the National Association of Manufacturers, forced the Obama administration to delay this plan in 2015, and in February 2016, the Supreme Court stopped its implementation. See Christian Downie, "Ad Hoc Coalitions in the United States Energy Sector," *Business and Politics* 20, no. 4 (2018): 643–68.

[70]Milkoreit, "The Paris Agreement on Climate Change – Made in USA?," 1019–37.

[71]Laurent Fabius, cited in Jen Iris Allan, Charles B. Roger, Thomas N. Hale, Steven Bernstein et al., "Making the Paris Agreement: Historical Processes and the Drivers of Institutional Design," *Political Studies* (2021): 13.

[72]Aashish Chandorkar, "The International Solar Alliance: India's Own 'OPEC' of Future," *Swarajya*, March 11, 2018, https://v.gd/iFhd9B.

voting rights in international financial institutions, such as the International Monetary Fund and the World Bank. As Copenhagen approached, India's delegation had navigated the thorny question of transparency that China and the United States were yet to resolve. Its introduction of the concept of "international consultations and analysis" would allow both India and China to remove sensitive references to "verification," while ensuring that transparency would enter the Copenhagen Accord.[73] Since then, Beijing had welcomed Prime Minister Modi in early 2015, where he and President Xi issued a joint statement urging "the developed countries to raise their pre-2020 emissions reductions targets and honor their commitment to provide US$100 billion per year by 2020 to developing countries."[74]

By this time, the high-level Advisory Group on Climate Change Financing had identified several potential sources of such climate finance. Formed by the UN secretary-general after the Copenhagen conference and comprising the likes of Nicholas Stern and Mutsuyoshi Nishimura, the group had found in 2010 that the removal or redirection of fossil fuel subsidies could be a particularly effective way to raise revenue.[75] Having been considered by the World Bank in the early 1980s in the spirit of liberalization, the role of such subsidies were later considered in terms of the ascendant notion of sustainable development in a 1991 report for the Trilateral Commission coauthored by the secretary-general of the World Commission on Environment and Development, Jim MacNeill (Chapter 4). "They often rig the market not only against the economy, but also against the environment and, ultimately, against development itself," wrote MacNeill with Pieter Winsemius and Taizo Yakushiji in *Beyond Interdependence*.[76] As the negotiations for the climate convention proceeded, both Vanuatu (on behalf of AOSIS) and Sweden proposed provisions to curb fossil fuel subsidies, with the latter calling for them to be abolished at the latest by the year 2000.[77] Although the World Bank also returned to the subject in the context of carbon emissions in 1992, the removal of fossil fuel subsidies would be entirely absent from the

[73]Namrata Patodia Rastogi, "Winds of Change: India's Emerging Climate Strategy," *International Spectator* 46, no. 2 (2011): 127–41; Rajnish Saryal, "Climate Change Policy of India: Modifying the Environment," *South Asia Research* 38, no. 1 (2018): 1–19.
[74]Mizo, "India, China and Climate Cooperation," 375–94.
[75]Advisory Group on Climate Change Financing, *Report of the Secretary-General's High-Level Advisory Group on Climate Change Financing* (New York: United Nations, 2010).
[76]For example, World Bank, *The Joint UNDP/World Bank Energy Sector Assessment Programme and Energy Sector Management Programme* (Washington, DC: World Bank, 1982); Jim MacNeill, Pieter Winsemius, and Taizo Yakushiji, *Beyond Interdependence: The Meshing of the World's Economy and the Earth's Ecology* (New York: Oxford University Press, 1991), 23.
[77]Harro van Asselt and Kati Kulovesi, "Seizing the Opportunity: Tackling Fossil Fuel Subsidies Under the UNFCCC," *International Environmental Agreements* 17 (2017): 357–70.

UNFCCC. The Kyoto Protocol for its part only referred to the removal of such subsidies as one of the measures that parties could implement to meet their commitments.[78]

Since then, phasing out these subsidies had only recently returned to the international climate change agenda in the wake of the global financial crisis. At the Pittsburgh summit of the Group of 20 (G20) in September 2009, the assembled governments pledged "to turn the page on an era of irresponsibility" and agreed "to phase out and rationalize over the medium term inefficient fossil fuel subsidies while providing targeted support for the poorest." Citing the OECD and the International Energy Agency, the governments noted that "eliminating" the subsidies by 2020 would reduce global greenhouse gas emissions by 10 percent in 2050.[79] This alignment of financial interests of the G20 with addressing climate change was spearheaded by the Obama administration's Larry Summers, head of the US National Economic Council and former chief economist of the World Bank in the early 1990s.[80] The governments of the Asia-Pacific Economic Cooperation followed suit shortly afterward at their November meeting in Singapore, representing, with the G20, the majority of the world's coal, oil, and gas consumption, as well as the majority of the world's coal and gas production.[81]

As the Paris meeting approached, however, these initiatives and others struggled to make such commitments to reduce fossil fuel subsidies materialize, let alone feature explicitly in the negotiating texts.[82] Both the World Bank and the International Energy Agency already anticipated that the funds necessary to achieve the 2°C target would far exceed the annual $100 billion commitment of the Cancún Agreements.[83] Other sources of climate finance, namely, private sources of funding for mitigation and adaptation measures in developing countries, would be "essential for the transition to a low-carbon, climate-resilient future," as the Advisory Group on Climate

---

[78]Bjorn Larsen and Anwar Shah, *World Fossil Fuel Subsidies and Global Carbon Emissions* (Washington, DC: World Bank, 1992). See Thijs van de Graaf and Mathieu Blondeel, "Fossil Fuel Subsidy Reform: An International Norm Perspective," in Jakob Skovgaard and Harro van Asselt (eds.), *The Politics of Fossil Fuel Subsidies and their Reform* (Cambridge: Cambridge University Press, 2018), 83–99.

[79]"G20 Leaders Statement: The Pittsburgh Summit, September 24–25, 2009, Pittsburgh," *G20 Information Centre*, University of Toronto, https://v.gd/iFBmny. Emphasis in original.

[80]Van de Graaf and Blondeel, "Fossil Fuel Subsidy Reform," 91.

[81]Joseph E. Aldy, "Policy Surveillance in the G-20 Fossil Fuel Subsidies Agreement: Lessons for Climate Policy," *Climatic Change* 144 (2017): 97–110.

[82]On the delaying role of the largest emerging economies, as well as oil-producing nations, see Jakob Skovgaard, *The Economization of Climate Change: How the G20, the OECD and the IMF Address Fossil Fuel Subsidies* (Cambridge: Cambridge University Press, 2021), 87–104.

[83]See, for example, World Bank, *World Development Report 2010: Development and Climate Change* (Washington, DC: World Bank, 2010); International Energy Agency, *World Energy Outlook 2009* (Paris: IEA, 2009).

Change Financing put it.[84] Flows from such private sources had been dwarfing Official Development Assistance since the Rio Summit, and the delegations of Aotearoa New Zealand, Australia, the European Union, and the United States flagged in 2011 their role in meeting their countries' climate finance commitments.[85]

By now, it had become clear to Christiana Figueres and the UNFCCC secretariat that the time had come to embrace and promote the subnational activities that had proliferated since the Copenhagen meeting. Cities, corporations, and other organizations had developed their own initiatives in its wake, building on the handful of transnational programs that had been underway since the early 1990s, such as the Climate Alliance, a network between European cities and Indigenous peoples.[86] With funding from sources such as the Bill and Melinda Gates Foundation, the Rockefeller Foundation, and the World Economic Forum, Figueres and the secretariat launched the "Momentum for Change" program at the Durban meeting in 2011 to showcase the growing range of climate action initiatives.[87] The AOSIS was especially welcoming of these non-state actors in the second work stream of the newly established Ad-hoc Working Group on the Durban Platform for Enhanced Action, which focused on raising the mitigation ambitions of the parties before 2020. Just months after the UN secretary-general's 2014 Climate Summit (Introduction), where both state and non-state leaders pledged firm commitments on mitigation, adaptation, and finance, their role in the intergovernmental climate regime was formalized in the Lima-Paris Action Agenda and their pledges tracked in the online Nonstate Actor Zone for Climate Action portal.[88]

Although the participation of some corporations was largely cynical, others called on the parties to the climate convention to take stronger action as Paris

---

[84]Advisory Group, *Report of the Secretary-General's High-Level Advisory Group on Climate Change Financing*, 10.

[85]Bradford S. Gentry and Daniel C. Esty, "Private Capital Flows: New and Additional Resources for Sustainable Development," in Luis Gomez-Echeverri (ed.), *Bridges to Sustainability: Business and Government Working Together for a Better Environment* (New Haven, CT: Yale University Press, 1997), 18–45. See Submissions from Parties, "Views on the Items Relating to a Work Programme for the Development of Modalities and Guidelines Listed in Decision 1/CP.16, Paragraph 46," Ad Hoc Working Group on Long-Term Cooperative Action under the Convention, March 2011, UNFCCC, https://v.gd/qE4mp8.

[86]Karin Bäckstrand, Jonathan W. Kuyper, Björn-Ola Linnér, and Eva Lövbrand, "Non-State Actors in Global Climate Governance: From Copenhagen to Paris and Beyond," *Environmental Politics* 26, no. 4 (2017): 561–79; Harriet Bulkeley and Peter Newell, *Governing Climate Change* (London: Routledge, 2010), 54–70.

[87]Thomas Hickmann, Oscar Widerberg, Markus Lederer, and Philipp Pattberg, "The United Nations Framework Convention on Climate Change Secretariat as an Orchestrator of Global Climate Policymaking," *International Review of Administrative Sciences* 87, no. 1 (2019): 21–38.

[88]Thomas Hale, "'All Hands on Deck': The Paris Agreement and Nonstate Climate Action," *Global Environmental Politics* 16, no. 3 (2016): 12–22.

neared.[89] Among them was the We Mean Business coalition, which formed in New York in 2014 involving representatives of multinational companies such as IKEA, Mars, Unilever, and Swiss Re, as well as organizations such as the World Business Council for Sustainable Development. Their first report, published on the eve of the Climate Summit and featuring a foreword by Figueres, declared, "It's clear that the business climate has changed."[90] And so it had. In the lead up to the Paris meeting, US secretary of state John Kerry told business leaders, "We [governments] are going to set a stage … but in the end it is business and the choices that you make, the buildings you build and the investments you make [that will make the difference]."[91] The We Mean Business group went even further as part of its wider "goal to be constructive partners in this process," preparing negotiating texts for Paris that called for "aggressive emissions reductions" from the parties.[92] Its emphasis on progressively strengthening the ambition of their nationally determined commitments reflected business practice—of "setting long-term plans but then revising and accelerating them as circumstances allow," as one of its leaders later noted.[93] Such efforts to ensure that Paris would be a success reflected a wider emphasis within the Secretariat and among government leaders since Copenhagen to craft a more positive and hopeful narrative of the climate negotiating process that they believed would help to accelerate the reduction of carbon emissions.[94] "If 150 nations are taking it seriously and setting targets, even if they don't make them, that will generate massive investment and a huge amount of private-sector activity," John Kerry explained after the Paris meeting. "And then you have to hope that somebody comes up with clean-energy technology, which makes it competitive with fossil fuel, and then, boom, you get your low-carbon economy."[95]

---

[89]Sarah Benabou, Nils Moussu, and Birgit Müller, "The Business Voice at COP21: The Quandaries of a Global Political Ambition," in Stefan C. Aykut, Jean Foyer, and Edouard Morena (eds), *Globalising the Climate: COP21 and the Climatization of Global Debates* (London: Routledge, 2017), 57–74.

[90]We Mean Business, *The Climate Has Changed: Why Bold, Low Carbon Action Makes Good Business Sense* (New York: CDP, 2014), xii.

[91]Cited in Michael Stothard and Kiran Stacey, "COP21: Big Polluters See No Short-term Change," *Financial Times*, December 14, 2015, https://v.gd/TtTJCM.

[92]We Mean Business, *The Business Brief: Shaping a Catalytic Paris Agreement* (New York: CDP, 2015), 2–15.

[93]Stephen Howard with Tim Smedly, "Business: Creating the Context," in Henrik Jepsen, Magnus Lundgren, Kai Monheim and Hayley Walker (eds), *Negotiating the Paris Agreement: The Insider Stories* (New York: Cambridge University Press, 2021), 278.

[94]Stefan C. Aykut, Edouard Morena, and Jean Foyer, "'Incantatory' Governance: Global Climate Politics' Performative Turn and its Wider Significance for Global Politics," *International Politics* 58 (2021): 519–40.

[95]Cited in Jeff Goodell, "Will the Paris Climate Deal Save the World?," *Rolling Stone*, January 13, 2016, https://v.gd/PsZvfx.

Climate Justice: Negotiations of the Vulnerable

Just days after Typhoon Haiyan devastated the Philippines in November 2013, the leader of the nation's delegation wept during the opening ceremony of the Warsaw climate conference. Yeb Saño pointed to the convention's record of climate diplomacy: "20 years hence we continue to fail in fulfilling the ultimate objective of the Convention. ... [I]f we have failed to meet the objective of the Convention, we have to confront the issue of loss and damage."[96] The parties were meeting at the end of a two-year study by the secretariat's Subsidiary Body for Scientific and Technological Advice (SBSTA) on the issue of "loss and damage." According to the Climate Action Network, "Tackling loss and damage is about climate justice. It is about protecting people, their livelihoods, and most importantly, their human rights and dignity. It is time for those who are mainly responsible for climate change to act."[97] The lead negotiator for the Alliance of Small Island States, Palau's Olai Uludong, backed Saño's call, as "it has become clear that there are now impacts from climate change that can no longer be avoided."[98]

Earlier that year at the Pacific Islands Forum summit, delegates gathered in the Marshall Islands' capital had announced the Majuro Declaration for Climate Leadership. The declaration opened with the following: "On 9 May 2013, atmospheric concentrations of carbon dioxide measured near the summit of Mauna Loa in Hawai'i exceeded 400 parts per million for the first time since measurements began. In crossing this historic threshold, the world entered a new danger zone."[99] That milestone had followed within days of the Marshall Islands government's declaration of a "state of disaster," as drought gripped its northern atolls.[100] The Majuro Declaration's signatories committed themselves to be "Climate Leaders"—to do more to accelerate their own mitigation and adaptation efforts because of their "unique vulnerability to climate change."[101] Shortly after Marshall Islands president Christopher Loeak presented the Majuro Declaration to the UN secretary-general, the IPCC's Fifth Assessment Report reported that sea levels would continue to rise,

---

[96]Yeb Saño, "'It's Time to Stop This Madness' – Philippines Plea at UN Climate Talks," *Climate Home News*, November 11, 2013, https://v.gd/7CqOjS.
[97]"The Loss and Damage Mechanism: Don't Leave Warsaw Without It!," *ECO*, November 19, 2013, https://v.gd/XDLCum.
[98]"Statement," *Alliance of Small Island States*, 2013, https://v.gd/tY7Pd1.
[99]Pacific Islands Forum Secretariat, *Forum Communique of the Forty-Fourth Pacific Islands Forum, Majuro, Republic of the Marshall Islands, 3–5 September 2013*, https://v.gd/1iuknb.
[100]Tony de Brum, "Marshall Islands Call for 'New Wave of Climate Leadership' at Upcoming Pacific Islands Forum," *Climate and Development Knowledge Network*, July 2, 2013, https://v.gd/RESDRB.
[101]Pacific Islands Forum Secretariat, *Forum Communique of the Forty-Fourth Pacific Islands Forum*.

warning, "Most aspects of climate change will persist for many centuries even if emissions of CO_2 are stopped."[102] A month later, a Kiribati man argued before a New Zealand court that he was a refugee from climate change.[103] In Fiji, meanwhile, preparations were underway for the relocation of the village of Vunidogoloa, where rising sea levels, soil erosion, and flooding had led villagers to seek higher ground.[104]

The support of AOSIS for elevating the question of loss and damage on Warsaw's agenda dated back to the early 1990s. Soon after its formation in the wake of the Second World Climate Conference (Chapter 5), the Alliance had tabled a proposal in 1991 for an "insurance mechanism" to "compensate the most vulnerable small island and low-lying coastal developing countries for *loss and damage* resulting from sea level rise."[105] Recourse to international law for compensation was also an option that Fiji, Kiribati, Nauru, and Papua New Guinea had emphasized in their declarations upon approving the Framework Convention in the early 1990s.[106] These endeavors had found common cause in the notion of the North's "ecological debt," an idea that Chile's Instituto de Ecologia Politica put forward in the lead up to the 1992 Earth Summit.[107] In echoes of the South's position at the UN Stockholm Conference twenty years earlier (Chapter 2), the grounds for this debt were twofold: developed countries owed an "emissions debt" to compensate for using most of the atmospheric space for their carbon emissions, and an "adaptation debt" to compensate for the impacts of climate change on developing countries that were not of their own making.[108]

---

[102]IPCC, "Summary for Policymakers," in T. F. Stocker et al. (eds.), *Climate Change 2013: The Physical Science Basis – Contribution of Working Group I to the Fifth Assessment Report of the Intergovernmental Panel on Climate Change* (Cambridge: Cambridge University Press, 2013), 27.

[103]" 'Climate Change Refugee' Fights to Stay in New Zealand," *Guardian*, October 1, 2013, https://v.gd/tSv62D.

[104]"Fiji: A Village Forced to Move by Rising Seas," International Labour Organization, September 13, 2013, https://v.gd/Lryjf8.

[105]Vanuatu, "Draft Annex Relating to Article 23 (Insurance) for Inclusion in the Revised Single Text on Elements Relating to Mechanisms (A/AC.237/WG.II/Misc.13) Submitted by the co-Chairmen of Working Group II," Intergovernmental Negotiating Committee for a Framework Convention on Climate Change, Working Group II, Geneva, December 17, 1991, https://v.gd/RaYxkY. Emphasis added. This 1991 proposal, in which states would contribute to an insurance fund relative to their historical emissions and share of global gross national product, was modeled on the 1963 Brussels Supplementary Convention on Third Party Liability in the Field of Nuclear Energy. See Krishnee Appadoo, "A Short History of the Loss and Damage Principle," *Revue juridique de l'Océan Indien* 31 (2021): 315–23.

[106]Lavanya Rajamani, "Addressing Loss and Damage from Climate Change Impacts," *Economic and Political Weekly* 50, no. 30 (2015): 17–21.

[107]Andrew Ross, "Climate Debt Denial," *Dissent* 60, no. 3 (2013): 33–7.

[108]Joan Martinez-Alier, "The Ecological Debt," *Kurswechsel* 4 (2002): 5–16; Mizan Khan et al., "Twenty-Five Years of Adaptation Finance through a Climate Justice Lens," *Climatic Change* 161 (2020): 251–69.

As the negotiation of the Kyoto Protocol dominated the following decade, it was not until after it came into force that the question of the North's financial liability resurfaced. In Montreal in 2005, for instance, Bangladesh had called for compensation for climate change damages on behalf of the least developed countries.[109] As AOSIS continued to press for insurance and compensation, the World Wildlife Fund called in 2008 for an additional compensation mechanism under the Framework Convention "to address the issue of what happens when damage becomes too severe for adaptation to become possible."[110] With the formalization of the phrase "loss and damage" at the Bali conference in late 2007, developing countries were increasingly of the view that compensation for harm was not only separate from discussions of adaptation and mitigation but also a concern that extended beyond the small island states to the least developed nations.[111]

In the wake of Copenhagen, meanwhile, the notion of climate justice was gaining greater currency among environmental organizations and small island states.[112] Palau turned in 2011 to the International Court of Justice to produce an advisory opinion on climate change, building on its experience of the security framing in the UN General Assembly (Chapter 7). Tuvalu had earlier threatened to sue Australia and the United States over the impacts of climate change but did not proceed with the litigation. Summoning principles from the Stockholm Conference, Palau's goal was to determine the applicability of existing norms of international law prohibiting transboundary harm to greenhouse gas emissions. Strong opposition from China and the United States, however, led Palau to shelve the proposal.[113]

These interests aligned with the ascendant idea of a global "carbon budget," which offered a new representation of urgency that also accommodated the distributive notions of equity and fairness that developing countries had advocated for decades.[114] Based on historical emissions, the

---

[109]Lisa Vanhala and Cecilie Hestabaek, "Framing Climate Change Loss and Damage in UNFCCC Negotiations," *Global Environmental Politics* 16, no. 4 (2016): 111–29.

[110]Roda Verheyen and Peter Roderick, *Beyond Adaptation: The Legal Duty to Pay Compensation for Climate Change Damage* (Surrey: WWF-UK, 2008).

[111]Vanhala and Hestabaek, "Framing Climate Change Loss and Damage in UNFCCC Negotiations," 111–29.

[112]Jen Iris Allan and Jennifer Hadden, "Exploring the Framing Power of NGOs in Global Climate Politics," *Environmental Politics* 26, no. 4 (2017): 600–20.

[113]Stuart Beck and Elizabeth Burleson, "Inside the System, Outside the Box: Palau's Pursuit of Climate Justice and Security at the United Nations," *Transnational Environmental Law* 3, no. 1 (2014): 17–29.

[114]Bård Lahn, "A History of the Global Carbon Budget," *WIREs Climate Change* 11, no. 3 (2020): e636; Martin Khor, *The Equitable Sharing of Atmospheric and Development Space: Some Critical Aspects* (Geneva: South Centre, 2010); Jiahua Pan and Ying Chen, "Carbon Budget Proposal: A Framework for an Equitable and Sustainable International Climate Regime," in Wang Weiguang,

carbon budget allows for the calculation of remaining atmospheric space for future emissions, according to temperature targets or thresholds. Taking shape during the 1990s, scientists saw the concept as a useful tool to assist decision-making. Schellnhuber's German Advisory Council on Global Change, for instance, had drawn on "per capita" proposals that India and China had long backed to calculate a global emissions budget up to the year 2050, which would be compatible with the 2°C "guardrail."[115] On the eve of the Warsaw climate meeting in 2013, the carbon budget also featured in the IPCC's Fifth Assessment Report, which warned that the world's warming could exceed 2°C within two to three decades. Wary of the budget concept's potential to revive old distributional issues in climate negotiations, the UNFCCC executive secretary Christiana Figueres was quick to rule them out of the Warsaw talks: "Politically it would be very difficult," she told journalists.[116]

The United States, meanwhile, had joined the High-Ambition Coalition, which had emerged from the Cartagena Dialogue and the Climate Vulnerable Forum with the EU's backing. Having shouldered much of the second commitment period of the Kyoto Protocol without the United States, Australia, and other industrial countries, the EU's negotiators now sought a truly global regime that would include all the major emitters. Led by Tony Brum, the head of the Marshall Islands' delegation, the Coalition brought together some 150 countries that broadly supported an ambitious, legally binding agreement with strong transparency rules and the provision of assistance for developing countries. Despite reservations within the EU, engaging the United States was necessary to pursue the Coalition's agenda—as Yamin observed, "in the run-up to Paris, it was clear it was the United States wooing and herding the remaining industrialized-economies cats."[117]

Among the objectives of the High-Ambition Coalition was to garner support for a more ambitious long-term goal of 1.5°C. Although the 2°C target had become a familiar figure in international climate diplomacy, this new goal had emerged in the wake of the UN General Assembly's adoption of AOSIS's resolution on climate change and security (Chapter 7).[118] Calling for "1.5 to stay

Guoguang Zheng, and Jiahua Pan (eds.), *China's Climate Change Policies* (New York: Routledge, 2012), 133–59.

[115]German Advisory Council on Global Change (WBGU), *Solving the Climate Dilemma: The Budget Approach* (Berlin: WBGU, 2009).

[116]Cited in Fiona Harvey, "IPCC's 'Carbon Budget' Will Not Drive Warsaw Talks, Says Christiana Figueres," *Guardian*, October 24, 2013, https://v.gd/zy1foP.

[117]Farhana Yamin, "The High Ambition Coalition," in Henrik Jepsen, Magnus Lundgren, Kai Monheim and Hayley Walker (eds.), *Negotiating the Paris Agreement: The Insider Stories* (New York: Cambridge University Press, 2021), 230.

[118]"AOSIS Summit on Climate Change," United Nations, 2009, https://v.gd/mtDt3r; Lisa Benjamin and Adelle Thomas, "1.5 to Stay Alive? AOSIS and the Long Term Temperature Goal in the Paris Agreement," *SSRN* (2016), https://v.gd/sHLZnM.

alive," AOSIS nations emphasized the acute vulnerability of member countries to the impacts of climate change and, in Copenhagen, forced a suspension of the proceedings to make their case heard. A representative from Tuvalu's delegation, Ian Fry, wept as he told the conference, "The fate of my country rests in your hands."[119] As a result of the Alliance's campaign, the Copenhagen Accord included text to ensure further negotiations included the "consideration of strengthening the long-term goal referencing various matters presented by the science, including in relation to temperature rises of 1.5 degrees Celsius."[120]

Having introduced this temperature limit at Copenhagen, the Alliance advocated for a study of its cost and benefits in the face of strong opposition from Saudi Arabia and Kuwait.[121] Condemning the stance of the members of the Organisation of the Petroleum Exporting Countries, a representative of the small island states from Barbados declared, "It is ironic that developing country brothers have blocked us. Is this the solidarity that they so eloquently talk about? … This is not a game. The lives of people, the future of nations is at stake."[122]

The scientific foundation for this more ambitious temperature goal, however, remained uncertain. Preparations for the IPCC's Fifth Assessment Report had become too far advanced to satisfactorily address this question. Moreover, owing to its only recent appearance on the international scene, there was scant scientific research to inform either the climate negotiations or the IPCC's assessment process. Some scientists attributed the paucity of research on the impacts of 1.5°C warming to the view that this "target" was

[119]Carol Farbotko and Helen V. McGregor, "Copenhagen, Climate Science and the Emotional Geographies of Climate Change," *Australian Geographer* 41, no. 2 (2010): 159–66. AOSIS members had also called for the return of atmospheric carbon dioxide concentrations to return to 350 ppm, which had been surpassed in the late 1980s. The President of the Maldives, Mohamed Nasheed, for instance, told a crowd at the KlimaForum that his country was "part of a growing bloc of nations, all committed to keeping Three - Five - Oh as the central guiding goal of our global survival plan". Endorsed by Pachauri and the authors of the 2009 'planetary boundaries' paper, this figure had been calculated by US climate scientist James Hansen after environmentalist Bill McKibben asked him in 2007 what might be a "safe" level of atmospheric carbon dioxide. This figure represented a departure from the widely held target of 450ppm as the level to avoid, as per the objective of the UN Framework Convention on Climate Change. See "President Warns Against Undermining the Importance of 350 In Climate Debate," Republic of Maldives, Press Office, December 14, 2009, https://v.gd/CFvCuY; Richard Monastersky, "Climate Crunch: A Burden Beyond Bearing," *Nature* 458 (2009): 1091–4.

[120]"Copenhagen Accord," UN Framework Convention on Climate Change, December 18, 2009, https://v.gd/sWZdHc.

[121]At the Copenhagen Conference, Bolivia's delegation had ventured a further target of just 1°C. See Tomilola Akanle et al., "Summary Report, 7–19 December 2009: Copenhagen Climate Change Conference," *Earth Negotiations Bulletin*, December 2009, https://v.gd/VbEuL4 ; Tomilola Akanle et al., "Summary Report, 31 May-11 June 2010: Bonn Climate Change Talks," *Earth Negotiations Bulletin*, June 2010, https://v.gd/AZarlY.

[122]Cited in Antto Vihma, Yacob Mulugetta, and Sylvia Karlsson-Vinkhuyzen, "Negotiating Solidarity? The G77 through the Prism of Climate Change Negotiations," *Global Change, Peace and Security* 23, no. 3 (2011): 315–34.

not realistic, given the sluggish international efforts to reduce greenhouse gas emissions. Others pointed instead to the lack of resources and infrastructure available to researchers from small island states and developing countries to elevate their concern for the 1.5°C threshold through traditional scholarly channels, which could then undergo IPCC assessment. For these researchers and their governments, drawing the attention of scientists associated with the IPCC to 1.5°C was a strategic means to advance research on this area and to underpin calls for earlier and higher loss and damage compensation. The EU's successful advancement of its own 2°C target, as discussed earlier, threw their predicament into sharp relief.[123]

In response to the advocacy of AOSIS, the Framework Convention commenced in 2013 a Structured Expert Dialogue review of the scientific knowledge on the long-term targets of 1.5°C and 2°C. Despite the paucity of scholarship on the differences between these targets, a consensus had emerged by the end of this two-year process. As participating scientist Petra Tschakert observed on the eve of the Paris meeting, it was clear that "a 2°C danger level seemed utterly inadequate given the already observed impacts on ecosystems, food, livelihoods, and sustainable development, and the progressively higher risks and lower adaptation level with rising temperatures, combined with disproportionate vulnerability."[124] AOSIS advocated for this review's findings to be shared at the Paris conference and successfully built a coalition of support for the 1.5°C threshold among fellow members of the High-Ambition Coalition and nongovernmental organizations.[125]

As the Structured Expert Dialogue got underway in 2013, the Framework Convention's SBSTA had solicited guidance on the use of Indigenous Knowledge in climate change assessment and adaptation.[126] The Nairobi Work Programme adopted in 2006 had earlier encouraged the consideration of "local and Indigenous Knowledge" in adaptation measures.[127] The Cancún Adaptation Framework adopted at the 2010 meeting had also noted the "need

[123]Samuel Randalls, "History of the 2°C Target," *WIREs Climate Change* 1 (2010): 598–605; Jasmine E. Livingstone and Markku Rummukainen, "Taking Science by Surprise: The Knowledge Politics of the IPCC Special Report on 1.5 Degrees," *Environmental Science and Policy* 112 (2020): 10–16. See also Reto Knutti et al., "A Scientific Critique of the Two-Degree Climate Change Target," *Nature Geoscience* 9 (2016): 13–18.

[124]Petra Tschakert, "1.5°C or 2°C: A Conduit's View from the Science-Policy Interface at COP20 in Lima, Peru," *Climatic Change Responses* 2 (2015), https://doi.org/10.1186/s40665-015-0010-z.

[125]Béatrice Cointe and Hélène Guillemot, "A History of the 1.5C Target," *WIREs Climate Change* 14, no. 3 (2023): e824.

[126]"Nairobi Work Programme on Impacts, Vulnerability and Adaptation to Climate Change," Subsidiary Body for Scientific and Technical Advice, UN Framework Convention on Climate Change, November 16, 2013, https://v.gd/CDXWCX.

[127]Ben Powless, "An Indigenous Movement to Confront Climate Change," *Globalizations* 9, no. 3 (2012): 411–23.

for adaptation to be based on and guided by the best available science and, as appropriate, traditional and Indigenous knowledge." The IPCC's formal requirements of scientific documentation and peer-reviewed publication had, to date, largely marginalized the observations and assessments of Indigenous peoples and local communities. By contrast, the Arctic Council's 2005 *Arctic Climate Impact Assessment* had involved the participation of Indigenous Knowledge holders throughout its five-year study. Likewise, the nascent "IPCC for biodiversity"—the Intergovernmental Platform on Biodiversity and Ecosystem Services (IPBES)—would, by according greater value to regional and local scales, engage more closely with holders of Indigenous knowledge and local knowledge upon its establishment in 2012.[128] UNESCO's Local and Indigenous Knowledge Systems Programme, led by Canadian Douglas Nakashima, had joined with the UN University's Traditional Knowledge Initiative to convene an international conference in Mexico City in 2011 to inform the preparation of the IPCC's Fifth Assessment Report.

The IPCC's previous assessment report in 2007 had noted that "Indigenous Knowledge is an invaluable basis for developing adaptation and natural resource managements strategies in response to environmental and other forms of change."[129] By gathering knowledge holders from among Indigenous peoples, Indigenous knowledge experts, and developing country scientists in Mexico City in preparation for the upcoming Fifth Assessment Report, Nakashima's program sought to provide the IPCC authors with a review of more than three hundred publications from the scientific and "grey" literature, and to foster "the co-production of new knowledge sets".[130] Such efforts contributed to significantly greater attention to Indigenous knowledge, experiences, and livelihoods in the assessment of Working Group II on climate change impacts, adaptation, and vulnerability. Although observers welcomed this development, they urged the IPCC to situate climate change in its "socio-political-historical-cultural context"

---

[128]James D. Ford, "Including Indigenous knowledge and Experience in IPCC Assessment Reports," *Nature Climate Change* 6 (2016): 349–53; Silke Beck et al., "Towards a Reflexive Turn in the Governance of Global Environmental Expertise," *GAIA: Ecological Perspectives for Science and Society* 23, no. 2 (2014): 80–7.

[129]Oleg A. Anisimov et al., "Polar Regions (Arctic and Antarctic)," in Martin L. Parry et al. (eds.), *Climate Change 2007: Impacts, Adaptation and Vulnerability* (Cambridge: Cambridge University Press, 2007), 673–4.

[130]Douglas Nakashima, Jennifer T. Rubis, and Igor Krupnik, "Indigenous Knowledge for Climate Change Assessment and Adaptation: Introduction," in Douglas Nakashima, Igor Krupnik, and Jennifer T. Rubis (eds.), *Indigenous Knowledge for Climate Change Assessment and Adaptation* (New York: Cambridge University Press, 2018), 1–20; Douglas Nakashima et al., *Weathering Uncertainty: Traditional Knowledge for Climate Change Assessment and Adaptation* (Paris: UNESCO, 2012). On UNESCO's approach to the co-production of climate knowledge, see Andrés López Rivera, "Re-encountering Climate Change: Indigenous Peoples and the Quest for Epistemic Diversity in Global Climate Change Governance," PhD dissertation (International Max Planck Research School on the Social and Political Constitution of the Economy, 2022), 85–91.

to shine a light on the "root causes of vulnerability" and to show "the potential for linking adaptation to broader policy goals or decolonizing processes."[131]

On the eve of the 2015 conference, Nakashima's program organized a large conference at UNESCO's Paris headquarters, "Resilience in a Time of Uncertainty: Indigenous Peoples and Climate Change." The program then provided a forum to facilitate a common agenda among Indigenous organizations, while another meeting aimed to foster ties between Indigenous organizations and states as part of a wider mission to make the climate regime "more receptive to the demands of Indigenous peoples."[132] Already Bolivia had by now emerged as a state that was willing to use its influence in the negotiations to advance the cause of Indigenous peoples. Having shifted away from a state-centric foreign policy toward a "diplomacy of the peoples," the Morales government had emphasized the need for their "full and effective participation" in the climate negotiations at the Cochabamba conference in 2010.[133] As the Paris conference approached, the Bolivian government reprised the event and called for "the creation of an 'international permanent platform', where the struggles of the peoples of the world for Mother Earth would come together."[134]

During the negotiations in Paris, the Bolivian delegation hastily prepared the text that would establish such a platform. What became the Local Communities and Indigenous Peoples Platform was framed in the Paris Agreement as follows:

> 135. *Recognizes* the need to strengthen knowledge, technologies, practices and efforts of local communities and indigenous peoples related to addressing and responding to climate change and *establishes* a platform for the exchange of experiences and sharing of best practices on mitigation and adaptation in a holistic and integrated manner.[135]

This text entered the Paris Agreement without revision, as did an acknowledgment in the document's preamble of the "rights of indigenous peoples."

---

[131]Ford, "Including Indigenous Knowledge and Experience in IPCC Assessment Reports," 349–53.

[132]Jean Foyer and David Dumoulin Kervran, "Objectifying Traditional Knowledge, Re-enchanting the Struggle against Climate Change," in Stefan C. Aykut, Jean Foyer, and Edouard Morena (eds.), *Globalising the Climate: COP21 and the Climatization of Global Debates* (London: Routledge, 2017), 153–72.

[133]Andrés López-Rivera, "Diversifying Boundary Organizations: The Making of a Global Platform for Indigenous (and Local) Knowledge in the UNFCCC," *Global Environmental Politics* (2023): https://doi.org/10.1162/glep_a_00706.

[134]Cited in López-Rivera, "Diversifying Boundary Organizations," 9.

[135]UNFCCC, *Report of the Conference of the Parties on Its Twenty-First Session, Held in Paris from 30 November to 13 December 2015*, January 29, 2016, FCC/CP/2015/10/Add.1, 19, https://v.gd/hoQ Hut. Emphasis in original.

The International Indigenous Peoples' Forum on Climate Change had, meanwhile, called for the parties to establish an "Indigenous Peoples' Experts and 'knowledge-holders' Advisory body elected by indigenous organizations and 'indigenous territorial governments' with regional balance."[136] Unlike the Bolivian text, this proposal did not gain the support of the parties and did not enter the Paris Agreement. The Platform became, instead, "something that we can live with," as an Indigenous interlocutor later told researcher Andrés López-Rivera.[137] Engaging directly in the climate negotiations had thus afforded Indigenous peoples not only a means to uphold their rights and prevent negative impacts, but also, as anthropologists Noor Johnson and David Rojas argue, "to transform this arena into a platform to pursue Indigenous claims."[138]

Conclusion

In December 2015, UN secretary-general Ban Ki-moon declared the Paris Agreement "a monumental triumph for people and our planet."[139] Nearly two hundred countries had agreed: (1) to hold global warming below 2°C, (2) to a transparent system of nonbinding national pledges ("nationally determined contributions") in light of "different national circumstances", (3) to mobilize climate finance from both public and private sources, and (4) to extend the regime to encompass the majority of global emissions.[140] To ensure countries continue to make ambitious pledges, the Paris Agreement also established a "ratchet" mechanism through five-year cycles of pledge and review.[141]

What the formulation of the Paris Agreement had achieved, seasoned observer Daniel Bodansky argued, was to "tie a treaty ribbon around these key elements of the Copenhagen Accord" (Figure 8.2).[142] In doing so, the parties had come to a "Goldilocks solution to the climate change problem that is neither too strong (and hence unacceptable to key states) nor too weak

---

[136]"Our Proposals to Governments," *International Indigenous Peoples Forum on Climate Change*, November 27, 2015, https://v.gd/TOWDW0.

[137]Cited in López-Rivera, "Re-encountering Climate Change," 211.

[138]Noor Johnson and David Rojas, "Contrasting Values of Forests and Ice in the Making of a Global Climate Agreement," in Ronald Niezen and Maria Sapignoli (eds.), *Palaces of Hope: The Anthropology of Global Organizations* (Cambridge: Cambridge University Press, 2017), 227.

[139]Cited in "UN Chief Hails New Climate Change Agreement as 'Monumental Triumph'," *United Nations*, December 12, 2015, https://v.gd/ahwLjp.

[140]Daniel Bodansky, "Reflections on the Paris Conference," *OpinioJuris*, December 15, 2015, https://v.gd/KJrjo3.

[141]Allan et al., "Making the Paris Agreement."

[142]Bodansky, "Reflections on the Paris Conference."

FIGURE 8.2 *French climate change ambassador Laurence Tubiana, UNFCCC executive secretary Christiana Figueres, and French foreign minister and conference chair Laurent Fabius (left to right) celebrate the adoption of the Paris Agreement on December 12, 2015, in Paris. Photo by IISD/*Earth Negotiations Bulletin, *Kiara Worth.*

(and hence ineffective)."[143] This solution had been years in the making, the result of not only international negotiations within and beyond the Framework Convention but also the ongoing advocacy of Indigenous peoples and climate justice activists.

The Paris Agreement entered into force in November 2016, the hottest year yet on record, after at least fifty-five parties to the Framework Convention accounting in total for at least 55 percent of the total global greenhouse gas emissions had ratified the treaty. The distributional conflicts that had riven climate diplomacy since the negotiations of the Kyoto Protocol had been avoided by allowing each country to determine "in the light of its national circumstances" its own contribution to the collective mitigation effort to limit global temperature rise. Every five years, countries will hold a global stocktake of these contributions and their ambition, and to assess their effectiveness in reducing greenhouse gas emissions.[144] After the coronavirus pandemic

---

[143]Daniel Bodansky, "The Paris Climate Change Agreement: A New Hope?," *American Journal of International Law* 110, no. 2 (2016): 288–319.
[144]Jen Iris Allan, "Dangerous Incrementalism of the Paris Agreement," *Global Environmental Politics* 19, no. 1 (2019): 4–11.

disrupted negotiations, the final details of the Paris Agreement were only finalized at the Glasgow climate conference in November 2021, which heralded the return of the United States after its withdrawal during the term of the Trump presidency.[145] There, for the first time, parties agreed specifically to "phase-down" coal-fired power generation and to "phase-out inefficient fossil fuel subsidies", well over a decade after the pledges made at the Pittsburgh G20. Reaffirming their commitment to the Paris temperature goals, the parties "urged" developed countries to not only double their climate finance investments, but also provide support for addressing the loss and damage associated with the adverse effects of climate change. These commitments under the Glasgow Climate Pact will undergo close scrutiny as the first Global Stocktake approaches in late 2023.

---

[145]Joanna Depledge, Miguel Saldivia, and Cristina Peñasco, "Glass Half Full or Glass Half Empty? The 2021 Glasgow Climate Conference," *Climate Policy* 22, no. 2 (2022): 147–57.

Epilogue

The "science-society contract is irrevocably broken," declared three climate scientists in early 2022 as the Sixth Assessment of the Intergovernmental Panel on Climate Change (IPCC) neared completion. Working Group I had already reported that it is "unequivocal that human influence has warmed the atmosphere, ocean and land" and that "human-induced climate change is already affecting many weather and climate extremes in every region across the globe."[1] Having outlined two options for climate change science, the trio from Australia and Aotearoa New Zealand turned to a third:

> It would be wholly irresponsible for scientists to participate in a 7th IPCC assessment. We therefore call a halt to further IPCC assessments. We call for a moratorium on climate change research until governments are willing to fulfil their responsibilities in good faith and urgently mobilize coordinated action from the local to global levels. This third option is the only way to arrest the tragedy of climate change science … Other options are seductive but offer false hope.[2]

The scientists had deemed their other approaches "not tenable." These were, first, to "carry on" with scientific research as they had been, or, second, to intensify social science research and advocacy. Yet, they argued, "even with more social science research, scientific advocacy and significant support from civil society, there have been no signs of systemic change in government action. There is no evidence that more social science research and traditional forms of advocacy will lead to transformative action within the timeframes required to avert dire climate change consequences."[3] By publishing this provocation, as they described it, the trio hoped to mobilize other climate

---

[1]IPCC, "Summary for Policymakers," in Valérie Masson-Delmotte et al. (eds), *Climate Change 2021: The Physical Science Basis* (Cambridge: Cambridge University Press, 2021), 4.
[2]Bruce C. Glavovic, Timothy F. Smith, and Iain White, "The Tragedy of Climate Change Science," *Climate and Development* 14, no. 9 (2022): 832.
[3]Glavovic, Smith, and White, "The Tragedy of Climate Change Science," 832.

change scientists to take radical action—to break from their own "business as usual."[4]

Their call to strike, as it was reported in the *New York Times*, at once under- and overestimated the role of climate science and the IPCC in the ways that governments address climate change.[5] As this book has shown, the scientific study of both the atmosphere and climate was critical to the emergence of the global environment as a concept and to the understanding that humanity could affect planetary change. This planetary knowledge as it is employed for climate diplomacy is largely the result of a Cold War project to measure and monitor the earth's biological, chemical, and geophysical systems. The science and technology that this project produced continue to provide the reams of complex data that allow experts to take the pulse of the planet—to determine its temperature and its thresholds, for instance. This information renders the planet visible and meaningful to them, and it forms the scientific basis for understanding, documenting, and responding to climate change.

For decades, many people have relied on the translation of this evidence for their own sense of the large-scale environmental changes that lie far beyond individual human experience. This evidence also afforded new understandings of the relationships between humans and their wider environments, and the promise and perils of human activities. It was at this planetary scale that climate change, as a result of accumulating anthropogenic greenhouse gas emissions in the atmosphere, was first identified and its wider implications began to be understood.

Environmental problems at the planetary scale invited experts not only to understand them but also to govern and ameliorate them. As the span of postwar international institutions offered the closest proximation to the global extent of environmental challenges, their enrollment would afford national governments the channels and means to coordinate and mount collective responses. Those institutions, forged in the rubble of the Second World War, provided both the infrastructure of global science to study global change and the infrastructure of diplomacy and expertise to assess its implications for the community of nations.

The mistake lies in the assumption that the global environment has a global government with a global constituency that can respond directly to the insights of global science. The global climate that climate change science comprehends is instead subject to international treaties and the actions of states. Those treaties have been negotiated through the United Nations

---

[4]Nick Kilvert, "IPCC Reports Are a Climate Science Beacon: So Why Do These Scientists Say They Have to Stop?," *ABC News*, September 4, 2022, https://v.gd/cY9Uqm.
[5]Raymond Zhong, "These Climate Scientists Are Fed Up and Ready to Go on Strike," *New York Times*, March 1, 2022, https://v.gd/Qm6L2K.

(UN) and its agencies by nation-states with their own constituents. Those constituents reside in different parts of the world that are more or less vulnerable to the effects of climate change. Some have benefitted from the historical processes that have produced those effects, while many more have not. Since the end of the Second World War, international institutions like the UN have been forums where these structural inequalities might be addressed and overcome for the "economic and social advancement of all peoples," as its Charter reads. Each member of the UN General Assembly has one vote, no matter their population, size, or might.

The atmosphere, however, continues to be apportioned rather differently. This challenge of sharing the atmosphere between nations has existed long before humanity became aware of its own geological potential. By exploiting fossil fuel reserves within their territories and beyond, some societies laid early claim to the atmosphere. The greenhouse gas emissions accumulated from their industrialization and economic growth would be reproduced as other societies followed similar paths, assured that the atmosphere could accommodate all comers. Recognizing the limits of the atmosphere, as the scientists of the industrial world did from the 1960s, suggested that governments should cooperate to bring about its enclosure. Enclosing the atmosphere, however, meant limiting the use of its remaining space. Having not yet reached the kinds of prosperity of the societies that industrialized earlier, developing countries chafed against a regime that looked to them as an unjust means to limit the realization of their national aspirations. They were not alone, as some governments of the industrial countries similarly feared the economic and political effects that might arise from atmospheric enclosure.

Concerns as to the atmospheric and climatic consequences of human activity rose almost in parallel to the very processes that were laying claim to the atmosphere. Once pollution no longer represented wealth but waste, and worse, sickness, constituents to whom governments were accountable mobilized for their intervention. Pollutants that strayed beyond local or national boundaries, however, posed diplomatic challenges for governments to negotiate their amelioration, particularly when their emission profited one but polluted the other. Atmospheric science provided information for governments to newly account for and monitor these wayward emissions, to assess their impacts, and to legislate for their control within their own borders in order to fulfill their obligations under international agreements.

Although their accumulation in the atmosphere renders greenhouse gas emissions a global environmental problem, their origin, control, and impact are local and national concerns. Given their uneven distribution at both source and impact, coordinating the international means to curb their emission and diminish their repercussions has been no straightforward task, as this book has shown. As global atmospheric change was only legible to scientists, their challenge was to raise the alarm and convince both governments and their constituents that their activities would have ramifications in the near future. When those ramifications might become evident, and where and how they might manifest, were all important questions for governments, as they were relevant to determining the urgency and nature of their response. Here, the insights of climate science became translated into geopolitical triage, as governments weighed up the costs and benefits of further emissions to themselves and their allies.

Governments called on global climate science to determine how and when the atmosphere might respond to a menu of decisions, while they recruited further expertise to assess the implications of those decisions for their constituents and territories. These studies of the atmosphere and its impacts revealed that the comprehension of global phenomena was not a universally shared concern but rather a deeply social act of translation. Developing countries were particularly attuned to the ways that studies of global change tended to obscure the Northern origins of their findings and to gloss over their implications for the aspirations of the South. These insights have informed the composition of the IPCC, as its architects recognize the shortcomings of uneven national representation. The South's interpretation of the causes of climate change and the possibilities for its arrest encouraged those governments to advocate for their own national and territorial interests in the channels of climate diplomacy. Meanwhile, for governments whose nations atmospheric change would impact sooner and more severely than others, the insights of climate science provided critical leverage in their engagement in international negotiations. Their very existence depended on the ambitiousness of climate diplomacy to curb further atmospheric change and limit its local impacts.

The assembly of an international regime to measure and address global climate change at its nationally dispersed sources and effects has required climate diplomacy to navigate the geopolitical terrain that has developed since the end of the Cold War. Foremost a question of energy use, climate change has forced governments to reckon with each nation's reliance on particular forms of energy and to invest in the development of less-carbon-intensive alternatives. A reluctance to explore such alternatives has been encouraged in some nations by their own resource endowments and the dependence of their economies upon them. As governments have grappled with the shift to alternatives that will provide the means for their constituents to continue their preferred way of life, they have also sought out ways to counteract the atmospheric effects of their emissions. This search has also drawn on the insights of climate science and other areas of expertise to ascertain the function and location of natural "sinks" for their abatement. These sinks were found to reside in states that seek compensation for their maintenance and enhancement for global atmospheric benefit, which has further required scientists to render them legible to governments to assess their economic and atmospheric value. It took well over a decade after the adoption of the UNFCCC for the climate regime to begin to recognize that these sinks were in fact forests and other human environments endowed with cultural significance by Indigenous and local peoples.

Accelerating the shift to alternative or renewable energy sources in the South has become increasingly important as the major developing countries vie for access to the atmospheric space that remains. Governments in these countries have long advocated for the industrialized North to afford them the means to meet their aspirations in a less-carbon-intensive manner. They gesture to the benefits that the Northern nations have accrued from their early claims to the atmosphere to underscore the North's responsibility to not only curb its emissions first but also financially and technically support Southern nations to reduce their emissions. The longer nations have taken to negotiate the mitigation of atmospheric change, the sooner those changes have begun to manifest. Their uneven geographic distribution has further imbued them with geopolitical significance, as vulnerable countries and peoples seek financial and technical support to protect themselves from these effects and compensation for them. For evidence of culpability, these governments have turned to climate science to determine the extent that anthropogenic climate change is contributing to the changes that they are witnessing firsthand. Incorporating the needs of the most vulnerable into international measures to address climate change has thus raised both the ambition of mitigation efforts and the financial resources necessary to relieve their impact.

That climate science has suggested the accumulation of greenhouse gases in the atmosphere will now continue to affect some degree of climate change has mobilized concerns for their impact on future generations. Encouraging nations to avoid actions that would foreclose the possibilities available to those to come became framed in terms of sustainable development in the late 1980s. This outlook fostered an additional temporal dimension to how governments and their constituents have understood the problem of climate change, leading a growing number of people to ask, "If not now, then when?" Having resonated with the climate justice movement for twenty years, this question resonates particularly strongly now with younger people who have understood the insights of climate science to mean that the sooner actions to mitigate atmospheric change are taken today, the more limited the effects of climate change will be for them tomorrow. They are demanding governments to recognize that they are also accountable to their generation, the next, and those who follow.

As *Climate Change and International History* has shown, climate science has fundamentally shaped, and been shaped by, climate diplomacy, as governments and their constituents require its knowledge to negotiate the remaining atmospheric space and to moderate the impacts of atmospheric change in the present and future. The climate regime's belated recognition that holders of Indigenous knowledge and local knowledge also have expertise in understanding environmental change, while Indigenous peoples and their representatives have found in the international structures of climate diplomacy new means to assert their sovereignty and self-determination. The tragedy of climate science is for climate scientists to have the means to comprehend atmospheric change but not the means to manage the atmosphere directly. Under the prevailing climate regime, for better or worse, managing the atmosphere lies in the hands of governments for whom its space represents their nation's power and prosperity now and into the future. The negotiation of climate diplomacy requires climate knowledges across both global and local scales for governments to navigate the changing atmosphere in ways that will benefit them both today and tomorrow. Only time will tell whether these negotiations will turn the tide on a warming world.

Index